AF793241

Affix Ordering Across Languages and Frameworks

Affix Ordering Across Languages and Frameworks

Edited by Stela Manova

OXFORD
UNIVERSITY PRESS

OXFORD
UNIVERSITY PRESS

Oxford University Press is a department of the University of Oxford.
It furthers the University's objective of excellence in research, scholarship,
and education by publishing worldwide.

Oxford New York
Auckland Cape Town Dar es Salaam Hong Kong Karachi
Kuala Lumpur Madrid Melbourne Mexico City Nairobi
New Delhi Shanghai Taipei Toronto

With offices in
Argentina Austria Brazil Chile Czech Republic France Greece
Guatemala Hungary Italy Japan Poland Portugal Singapore
South Korea Switzerland Thailand Turkey Ukraine Vietnam

Oxford is a registered trade mark of Oxford University Press
in the UK and certain other countries.

Published in the United States of America by
Oxford University Press
198 Madison Avenue, New York, NY 10016

© Oxford University Press 2015

All rights reserved. No part of this publication may be reproduced,
stored in a retrieval system, or transmitted, in any form or by any means,
without the prior permission in writing of Oxford University Press,
or as expressly permitted by law, by license, or under terms agreed with the
appropriate reproduction rights organization. Inquiries concerning reproduction
outside the scope of the above should be sent to the Rights Department,
Oxford University Press, at the address above.

You must not circulate this work in any other form
and you must impose this same condition on any acquirer.

Cataloging-in-Publication data is on file with the Library of Congress

9780190210434

9 8 7 6 5 4 3 2 1

Printed in the United States of America on acid-free paper

CONTENTS

CONTRIBUTORS

Iwona Burkacka is assistant professor in the section of Lexicology, Theoretical Stylistics, and Culture of Polish Language of the Institute of Polish Language at the University of Warsaw (Poland). She obtained her PhD degree from the same university and also habilitated there with a thesis entitled *De-Nominal Suffix Combinatorics in Polish* in 2013. Her research interests include Polish morphology, specifically word formation and the theory of derivational nests; culture of Polish language; colloquial language; sociolinguistics; languages of professions; and new elements of vocabulary.

Wolfgang U. Dressler is professor emeritus in the Department of Linguistics at the University of Vienna (Austria), former head of the Department of Linguistics of the Austrian Academy of Sciences, and head of the Institute for Corpus Linguistics and Text Technology. He studied Linguistics and Classical Philology in Vienna, Rome, and Paris, taking his MA, PhD, and habilitation at the University of Vienna. His current research interests are theory of inflection and word formation (including morphopragmatics, morphonology, and morphonotactics), language acquisition, and corpus linguistics.

Katarzyna Dziubalska-Kołaczyk is professor and dean of the faculty of English at Adam Mickiewicz University, Poznań (Poland). She is also head of the Interdisciplinary Centre for Speech and Language Processing at AMU. She has published extensively on phonology, phonetics, and second language acquisition. In her works she has been pursuing and advocating the natural linguistic approach to language. Her books include *A Theory of Second Language Acquisition within the framework of Natural Phonology* and *Beats-and-Binding Phonology*. Her recent research focuses on phonotactics and morphonotactics. She is the editor of *Poznań Studies in Contemporary Linguistics*, published by de Gruyter Mouton, and organizer of *Poznań Linguistic Meetings* (PLM).

Angeliki Efthymiou is associate professor of linguistics at Democritus University of Thrace (Greece). She studied Greek at the University of Thessaloniki, Greece, and specialized in Linguistics (MA and PhD) at the University of Lille III, France. Her research and teaching interests fall within the areas of lexicology, lexical semantics, morphology, lexicography, educational linguistics, and language teaching. Recent publications include *Le suffixe -iá en grec moderne. La manifestation d'un degré maximal d'anthropocentricité* (2012, Saarbrücken: Éditions Universitaires Européenes), and "Productivity of Verb Forming Suffixes in Modern Greek:

A Corpus-Based Study" (with Georgia Fragaki and Angelos Markos), in *Morphology* 22 (2012).

Natalia Gagarina coordinates the research division on Multilingualism at the Center for General Linguistics in Berlin (ZAS) (Germany). She is the author and coeditor of a number of books on language acquisition in the *SOLA* (de Gruyter) and *Studies in Theoretical Psycholinguistics* (Springer) series; of the Russian language proficiency test for multilingual children; and of the Multilingual Assessment Instrument for Narratives (MAIN) that has been adapted in 26 languages. In 2011 she founded, with the support of the ZAS director, Prof. Manfred Krifka, the Berliner Interdisciplinary Association for Multilingualism. In her projects (Berliner Senate, BMBF, DFG, EU-funded) she investigates bilingual acquisition and develops tools for assessment and support of second and first (home) languages.

Songül Gündoğdu is a PhD student in the linguistics program at Boğaziçi University in Instanbul (Turkey). She is Turkish-Kurmanji bilingual and studies Kurdish linguistics, specifically the syntactic and morphological properties of Kurmanji Kurdish. She also works as a research assistant at the Department of Kurdish Language and Literature at Muş Alparslan University, Turkey. She received her MA from the linguistics program at Boğaziçi University with a thesis on the phrase structure and ergative constructions of two dialects of Kurmanji Kurdish, the Standard dialect and the Muş dialect.

Marianne Kilani-Schoch has been working for many years in natural phonology and morphology. Within this framework she has been doing research on aphasia, Arabic, and French. She has been involved in the Crosslinguistic Project on Pre- and Protomorphology in Language Acquisition since its beginning. She teaches linguistics at the Department of Language and Information Sciences and at the School of French as a Foreign Language of the University of Lausanne (Switzerland).

Yuni Kim is a lecturer in phonetics and phonology at the University of Manchester (UK), and holds a PhD in linguistics from the University of California, Berkeley. Her research focuses on the phonology and morphology of Mesoamerican languages, particularly Huave (isolate) and Amuzgo (Oto-Manguean). She is interested in prosody, phonological representations, and the phonetics–phonology and morphology–phonology interfaces.

Alexander Letuchiy (National Research University Higher School of Economics, Moscow, Russia) obtained his PhD from the Russian State University for Humanities with a dissertation entitled *Tipologija labil'nyx glagolov: semantičeskie i morfosintaksičeskie aspekty* ("A Typology of Labile Verbs: Semantic and Morphosyntactic Aspects"; advisor: Vladimir A. Plungian) in 2006. Since 2003 he has carried out research on Adyghe and taken part in the fieldtrips organized by the Russian State University for Humanities. He is also coeditor of a volume on

Adyghe grammar. His main research interests include transitivity, valency change, sentential arguments, zero copulas, and evidentiality in Caucasian languages and Russian from a typological perspective.

Ekaterina Lyutikova is associate professor at the Lomonosov Moscow State University (Russia). Her specialization is generative syntax and syntax–semantics interface. Current research areas include noun phase structure, syntax of relatives, and argument structure, and the Turkic, North-Caucasian, Iranian, and Uralic languages. She is also interested in formal approaches to Russian syntax. She received her MA and PhD degrees from the Moscow State University.

Stela Manova holds a PhD in general linguistics with a minor in Slavic linguistics from the University of Vienna (Austria). She is a senior researcher and has been involved in a number of projects at the University of Vienna and the Austrian Academy of Sciences in Vienna. She is the founder and organizer of the Vienna Workshops on Affix Order; the author of *Understanding Morphological Rules* (Springer 2011) and *Affixation* (2013, *Oxford Bibliographies in Linguistics*, ed. Mark Aronoff); as well as the editor of *Affix Combinations* (Part 1 and 2, Springer 2010, published as special issues of *Morphology*, with Mark Aronoff), and *Affixes and Bases* (Edinburgh University Press 2011, published as a special issue of *Word Structure*).

Luigi Talamo is a PhD candidate in linguistics at the University of Bergamo (Italy). He previously worked as a grant holder at Scuola Normale Superiore in Pisa (Italy), where he developed an annotated lexicon of Italian derivatives. His research interests include morphology and semantics of word formation, linguistic typology, and corpus linguistics.

Sergei Tatevosov is associate professor at Lomonosov Moscow State University (Russia). His specialization is morphology, formal semantics, and syntax–semantics interface. Current research areas include lexical and grammatical aspect, argument structure, and the Turkic, North-Caucasian, Iranian, and Uralic languages. He received his MA and PhD from the Moscow State University. Recent publications include "Complex Predicates, Eventivity, and Causative-Inchoative Alternation" (with Ekaterina Lyutikova), in *Lingua*; "Telicity, Measures, and Endpoints," in *Proceedings of Sinn und Bedeutung* 16; "Detelicization and Argument Suppression: Evidence from Godoberi," in *Linguistics*; and "Atelicity and Anticausativization" (with Ekaterina Lyutikova), in *Argument Structure and Syntactic Relations: A Cross-Linguistic Perspective* (John Benjamins).

Eva Zimmermann is a PhD student at Leipzig University (Germany). After receiving her MA from the same university in 2009, she was a research assistant in the project "Hierarchy Effects in Kiranti and Broader Algic" and spent time as a visiting student researcher at UC Berkeley. Since 2014, she holds a research assistant position in the project "Featural Affixes" at Leipzig University. Her research is in theoretical phonology, theoretical morphology, and its interfaces. Currently, she works mainly on nonconcatenative morphology (especially lengthening, subtraction, nonconcatenative allomorphy) and its analysis in phonological theory. An empirical focus in her work has been the nonconcatenative morphology in several Native American Indian languages and the verbal agreement morphology in Kiranti and Broader Algic.

Affix Ordering Across Languages and Frameworks

1

Introduction: Affix Ordering Across Languages and Frameworks

Stela Manova

Although affixation is the major morphological device for creating new words and word forms in the languages of the world there are heavy restrictions on the ordering of affixes, and of all phonologically possible and semantically conceivable affix combinations in a language, only a few exist.[1] The question of affix ordering is an important one in linguistic theory and there has been much research on the topic from various perspectives. Nevertheless, the exact nature of the factors involved in affix ordering and the ways those factors interact with one another are still a matter of debate. This volume contributes novel data from various typologically diverse languages, including languages whose affix order has not been examined so far, as well as original analyses to the research on affix ordering.

The present book is related to the Vienna Workshops on Affix Order and is the last of a sequence of collections of papers on affix ordering that have appeared since 2010. The other collections include Manova and Aronoff (2010a), *Affix Combinations*, and Aronoff and Manova (2010), *Affix Combinations, Part 2*, both published by Springer, as well as Manova (2011a), *Affixes and Bases*, published by Edinburgh University Press. The goals of this publication are similar to those of the previous ones: to provide better understanding of the mechanisms governing affix ordering through putting together research on affix ordering in typologically diverse languages that is carried out within different theoretical frameworks, and

[1]The author was funded by the Austrian Science Fund - FWF, grant V64-G03. The support is hereby gratefully acknowledged.

I would also like to thank the two anonymous reviewers of the volume for their thought-provoking comments on all papers, as well as the following colleagues who helped me with the selection of the papers: Werner Abraham (Munich/Vienna), Matthias Gerner (Hong Kong), Yuni Kim (Manchester), Natalia Korotkova (UCLA), Marijana Marelj (Utrecht), Angela Ralli (Patras), Andrea Sims (OSU, Columbus), Renata Szczepaniak (Hamburg), Bogdan Szymanek (Lublin), and Eva Zimmermann (Leipzig).

Many thanks for all their patience and support are due to the associate editor, Hallie Stebbins, the copy editor, Anne Sanow, and to Poovizhi Loganathan who oversaw the production.

to make the various analyses of affix ordering phenomena comparable and more easily accessible through unification of terminology and methodology.

As already mentioned, the present volume is related to the Vienna Workshops on Affix Order, a workshop series that was supported by the Austrian Science Fund and a few other institutions such as the Erste Bank Vienna, the Mayer of Vienna, and the University of Vienna. Specifically, this publication contains a selection of papers presented at the third workshop that had the topic of Advances in Affix Order Research and took place at the University of Vienna on January 15–16, 2011. All the papers but Kim's "Mobile Affixation Within a Modular Approach to the Morphology–Phonology Interface" were delivered at the workshop, which does not mean that Kim's contribution is not related to the Vienna workshops. Kim participated at the first workshop in 2008 and gave a talk on mobile affixation in Huave whose written version, "Phonological and Morphological Conditions on Affix Ordering in Huave," was published in Manova and Aronoff (2010a). However, Kim's Huave data and their OT analysis that ranks phonology over morphology (P >> M), as in McCarthy and Prince's (1993) paper, have raised some questions in the literature on affix ordering, specifically about the existence of phonologically driven affixation (see Paster 2009). Following a theoretical framework that assumes a modular architecture of grammar—that is, phonology and morphology are two different modules and the latter always precedes the former, which makes any interranking of the two impossible—Paster claims that there is no phonologically driven morphology. In other words, Paster's theoretical assumptions are incompatible with affix ordering based on phonological principles and, therefore, she offers alternative analyses of various examples of phonological affix ordering from the literature (Paster 2009). Kim's (2010) analysis of Huave is not invalidated, but a proposal is put forward how Kim's data could be reanalyzed so that they are no more considered instances of phonological affix ordering. Therefore, Kim was invited to contribute to the present volume and comment on possible alternative analyses of her Huave data in particular and on phonological affix ordering in general. In this way, Kim's contribution turns into a symbolic tie that goes from the first through the last publication in the series of publications on affix order Mark Aronoff and I have had the pleasure to edit over the years.

Virtually all papers included in this volume contribute novel data to the affix order debate, and the reader will find examples from Adyghe (Caucasian); Athpare and Camling (Kiranti); Bulgarian, Russian, and Polish (Slavic); English and German (Germanic); Greek; French and Italian (Romance); Huave (language isolate, Huavean); Karachay-Balkar (Turkic); Kurmanji (Kurdish); and Lithuanian (Baltic), among other languages.

As regards the analyses, they are couched within various theoretical frameworks and present the generative and nongenerative research paradigms. Additionally, the volume includes a contribution on the emergence of affix order in child language, the first of this type in the literature, as well as a paper that discusses affix ordering within the East European linguistic tradition. The latter, as the reader will see, significantly differs from what is known from the studies that follow West European and American linguistic traditions.

Unlike previous research on affix ordering, most of the contributions in this collection consider more than one language. Discussed are, among other things, cases of affix ordering that pose problems to linguistic theory such as affix repetition, variable ordering, and interaction of prefixes and suffixes in terms of parasynthesis and mobile affixation. Novel examples of affix repetition and variable ordering are given, and the volume provides evidence that these phenomena are neither rare nor typical only of lesser-studied languages with unstable grammatical organization, as has been claimed in the literature so far.

This introductory article has the following structure. In section 1, the approaches to affix ordering known from the literature are presented and the organization of the volume explained. In section 2, brief descriptions of the papers are provided. Section 3 accommodates a short discussion and conclusions.

1. Approaches to Affix Order

In linguistic theory, overviews of the existing approaches to affix order are rare. This is maybe due to the challenging character of the task, since there are not really standardized approaches and linguists working on affix ordering just try to explain the patterns they are confronted with, any argument being eligible. On the other hand, this situation results in confusions regarding comparison and interpretation of findings and calls for postulation of approaches.

The following articles represent those rare attempts toward putting some order in the research on affix order: Muysken (1986), Manova and Aronoff (2010b), and Rice (2011). They all provide helpful information on what could be seen as existing approaches to affix ordering. However, it should be mentioned that Muysken's work is also an argument against template morphology and draws on data from Quechua. Rice's work is on affix ordering that is not word-class changing, and the data come primarily from understudied languages. Therefore, we will follow the classification in Manova and Aronoff (2010b), which is the most general one and covers well-studied and lesser-studied languages. It takes the type of information involved in affix ordering as a criterion for definition of approaches. Thus, Manova and Aronoff (2010b) differentiate between grammatical and extragrammatical factors and define eight different approaches to affix order, that is, eight different ordering types: (1) phonological, (2) morphological, (3) syntactic, (4) semantic, (5) statistical, (6) psycholinguistic, (7) cognitive, and (8) templatic.

A phonological approach, or phonological ordering, relies on phonological information such as termination of the derivational base in a vowel/consonant, including the exact types of those vowels/consonants, and so on. Thus an example of a phonological-affix-ordering rule would be: 'If the base terminates in a palatalized consonant, attach the suffix X to it.' Purely phonologically ordered affixal systems are rare. Moreover, phonological affix ordering has been often reanalyzed as due to morphology and thus classified as morphological ordering, see the discussion in Kim, this volume.

A morphological approach uses rules of the type: 'If affix A, then affix B.' The approach is termed morphological since A and B are morphemes, be they classical morphemes that pair meaning and form or abstract morphemes in the sense of Distributed morphology (Halle and Marantz 1993). Abstract morphemes are usually presented in terms of morphosyntactic feature hierarchies or classes (see the analysis in Zimmermann, this volume).

The fact that in 58 languages with subject agreement the affixes that mark person and number usually attach in the order Person >> Number, i.e., person before number, (see the discussion in Trommer 2003) is a statistical observation, irrespective of how one accounts for it. Manova and Aronoff (2010b) term affix ordering based on statistical information statistical affix ordering. Constraints such as Person >> Number are also seen as hierarchical ordering in the literature as in an Optimality theory analysis Person >> Number reads "rank Person over Number." Hierarchies such as Person >> Number combined with ALIGNMENT constraints allow for formulation of generalizations even in cases when the affixes that mark the respective features are not on the same side of the root. For example, Trommer (2003) observes that in the majority of the languages he scrutinizes person markers are usually prefixes while number markers are usually suffixes. He maintains that the order is Person >> Number as fused person/number markers share the position of the person marker. A hierarchical relationship between prefixes and suffixes in a single language (Greek), outside Optimality theory, is assumed in Efthymiou, this volume.

Syntactic approaches rely on the so-called Mirror Principle: "Morphological derivations must directly reflect syntactic derivations (and vice versa)" (Baker 1985, p. 375). Manova and Aronoff (2010b) make a clear distinction between syntactic and semantic affix ordering (contra Rice 2000), as syntactic ordering relies on syntactic information such as subject and object, whereas semantic ordering depends on information that involves semantic categories like 'human,' or 'animate.' The most frequently cited example of semantic ordering in the literature is the so-called scopal ordering, as illustrated for the whole language family, the Athapaskan, in Rice (2000). According to the principle of scopal semantic ordering, an affix with a broader scope follows an affix with a narrower scope in the word. This is, however, not always the case and scope can be overridden by other factors (see the examples in Rice 2000 and the discussion on scopal ordering in Letuchiy, this volume). Another popular instance of semantic ordering is the relevance principle postulated in Bybee (1985). According to this principle affixes that are more relevant to (affect stronger) the semantics of the root are closer to the root.

Psycholinguistic ordering reflects the degrees of ease/difficulty of processing of morphological structure; for example, more parsable affixes are outside less parsable ones because this order is easier to process by the human brain (Hay 2003, Plag & Hay 2004, Plag & Baayen 2009).

Cognitive affix ordering (cf. Manova 2011b) relies on information that is cognitive in nature; that is, it is based on the belief that general cognitive principles are also operative in grammar. Cognitive ordering, as the term implies, follows the assumptions of Cognitive grammar (Langacker 1987, Taylor 2002, among many others), where entrenchment of grammatical structure through frequency of use plays a major role; additionally, cognitive ordering can be linked to theories of cognitive (Fillmore 1982) and conceptual semantics (Jackendoff 1990), see Manova, this volume. Thus cognitive ordering can also be seen as related to semantic ordering, but the former differs from the latter as cognitive ordering does not control for scope or relevance but orders affixes in terms of (a limited number of) combinations of semantic concepts.

Templatic ordering means that the affixes in a language can be allotted to slots in a template that does not necessarily have a grammatical motivation. If only two affixes depend on each other the ordering is morphological; recall the definition of morphological ordering given above. In templatic morphology affixation should happen at once, while morphological ordering implies a step-by-step, layered derivation (see the explanation of templatic and layered morphology in Manova and Aronoff 2010b). In this volume, in order not to split the papers into too many sections, templates are under morphological ordering (see Gündoğdu's paper), which is in line with research in affix order that does not differentiate between templatic and morphological ordering (see, e.g., Hyman 2003 and Rice 2011). However, it should also be mentioned that the template Gündoğdu postulates is not a prototypical instance of a template but exhibits some elements of layered morphology, as some of the affix combinations it accounts for can be seen as derived step-by-step.

Thus the majority of the papers in the present volume are distributed into three groups that form the first three parts of the book: Part I: *Syntactic and Semantic Ordering*, Part II: *Phonological and Morphological Ordering*, and Part III: *Psycholinguistic and Cognitive Ordering*. In this way, the volume covers the major approaches to affix order. The last part of the volume, Part IV, is titled *Description and Acquisition of Affix Order*, as the two papers it contains focus on issues related to the description and acquisition of the order of affixes (which does not mean that those papers do not use semantic and morphological information for the ordering of affixes they discuss).

Finally, as Manova and Aronoff (2010b) underline, it is seldom the case that the order of all affixes in a language depends on a single factor, so the assignment of a paper to a particular part of the book does not mean that it relies on a single type or on two types of information with respect to affix ordering but that the major argument is of a particular type.

2. The Papers in This Volume

In addition to this introductory article, the present volume includes ten papers that, as already mentioned, illustrate the major approaches to affix order. I provide short descriptions of the contributions to the volume here.

Three papers are included in Part I of the book that is on syntactic and semantic ordering. In "Recursive Passivization: A Causative Coercion Account," Ekaterina Lyutikova and Sergei Tatevosov discuss double passivization, a phenomenon that is hard to account for in many theories of grammar. The data come from Karachay-Balkar (Altaic, Turkic). In this language the affixes associated with the expression of passive can be repeated. The authors explain this situation with covert causativization; that is, they maintain that in the course of derivation of double passives a new causing subevent and its individual participants are introduced to the event structure, which thus makes possible the derivation of what looks like a passivization of already passive predicates. Causativization that feeds passivization occurs covertly, as a coercion operation.

"Scope Versus Ordering of Operations: Causativization and Ordering of Valency-Changing Operations in Adyghe" by Alexander Letuchiy stays on the topic of causativization. The paper tackles the ordering of valency-changing operations in the verb of the North Caucasian language Adyghe. The author claims that the suffixal part of the Adyghe verb is a case of layered morphology, while the prefixal part appears templatically organized and thus hard to explain with the Mirror principle. As is typical of template morphology, the linear order of prefixes reflects neither their semantic scope nor the ordering of the valency-changing operations they mark, which may, as the author shows, give rise to ambiguity of reading. Nevertheless, there is a tendency for the causative operation to apply after all other valency-changing operations, which is, according to the author, explicable with the placement of the causative marker—a prefix that immediately precedes the verb root.

"Modern Greek Parasynthetic Verbs: A Hierarchical Relationship Between Prefixes and Suffixes?" by Angeliki Efthymiou contributes to the discussion of the role of prefixes and suffixes in affix ordering. The study is based on 308 parasynthetic verbs collected from the *Reverse Dictionary of Modern Greek* (2002). A parasynthetic verb possesses two affixes, a prefix and a suffix, and it is often hard to decide in which order the two affixes attach. The author claims that the linear ordering of prefixes and suffixes reflects the information content of each affix and its contribution to the morphological structure of the parasynthetic verb. Specifically, prefixes in parasynthetic verbs determine the core semantics of the verb. It is shown that affixation is also influenced by register and frequency.

Part II of the volume is on phonological and morphological ordering and consists of the following contributions: Kim's paper "Mobile Affixation Within a Modular Approach to the Morphology–Phonology Interface" discusses a rare case of mobile affixation in Huave, a language isolate spoken in Oaxaca State,

Mexico. In this language affixes end up as either prefixes or suffixes, depending on the phonological properties of the base. The author discusses possible alternative analyses of the data such as: (1) epenthesis avoidance in an optimality theoretic approach that assumes a P >> M architecture of grammar with phonological epenthesis-avoidance constraints outranking morphological constraints that require mobile affixes to be suffixal, (2) feature-floating analysis, and (3) a suppletive allomorphy analysis with subcategorization frames. Comparing these three alternative solutions to the same problem, Kim also tries to answer the question of whether there is phonological affix ordering.

Like one of the alternative solutions in Kim's paper, Eva Zimmermann's "Hierarchy-Governed Affix Order in Eastern Kiranti" provides an Optimality theory analysis. However, Zimmermann claims to offer an analysis that is superior to the subcategorization approach (another alternative discussed in Kim), as Zimmermann's analysis does not rely on affix-specific constraints in terms of subcategorization frames but instead uses morphosyntactic feature classes (i.e., abstract morphemes in the sense of Distributed morphology, Halle and Marantz 1993), which allows for generalizations and typological comparisons across languages. Zimmerman discusses the linear order of the inflectional suffixes in the Kiranti language Athpare and explains it with the help of a language-specific hierarchy of morphosyntactic feature classes, ALIGNMENT constraints, and a markedness constraint that requires an unambiguous marking of the agent argument. The analysis also involves a comparison with other closely related Eastern Kiranti languages. The Eastern Kiranti languages appear peculiar from a typological point of view, since they do not follow the established in Trommer (2003) Person >> Number ordering with respect to person and number agreement markers.

Songül Gündoğdu's paper "Negation in Kurmanji" investigates the negation morphology of Kurmanji Kurdish verbs. The author, a Turkish-Kurmanji bilingual speaker, provides a fairly detailed description of that language's verbal morphology and argues that Kurmanji is a case of templatic affix ordering as the affixes that occupy the same slot do not co-occur, which is also how negation functions in Kurmanji verbs: the attachment of a negation prefix to a verb stem blocks the appearance of certain tense/aspect/mood prefixes. However, the Kurmanji verb morphology also shows elements of layered morphological organization.

Part III of the book, on psycholinguistic and cognitive ordering, includes two papers: In "Suffix Combinations in Italian: Selectional Restrictions and Processing Constraints" Luigi Talamo discusses the combinations of some Italian derivational suffixes. The analysis tests whether the ordering of the suffixes follows psycholinguistic factors—that is, the so-called Complexity-Based Ordering (CBO). CBO assumes that affix ordering depends on processing constraints, and, based on the degrees of parsability of individual affixes, establishes a hierarchy of affixes. With the help of the hierarchy, one can predict possible affix combinations and restrict the impossible ones. The paper also tackles the correlation between suffix productivity and suffix position in the CBO hierarchy. Additionally, the role of selectional restrictions in suffix combinations is discussed.

Stela Manova's paper "Affix Order and the Structure of the Slavic Word" investigates the structural properties of the Slavic word in terms of affix ordering. The data come from Bulgarian (South Slavic), Russian (East Slavic), and Polish (West Slavic), and thus cover all three subgroups of the Slavic family. After a brief discussion of why CBO fails to account for Slavic data, an alternative approach is applied. The latter is domain-specific and cognitively oriented. It is argued that the Slavic word consists of different domains and subdomains defined through the affix-order peculiarities they exhibit. There are prefixational and suffixational domains; the suffixational domain is further divided into derivational and inflectional domains, and within the derivational domain there are evaluative and nonevaluative subdomains. The paper focuses on the order of the suffixes in the nonevaluative derivational domain where the lexical-category specification of a suffix, suffix-particular semantics, and the notion of default are the factors governing suffix combinability. All suffix combinations are either fixed or predictable and entrenchment, in the sense of Cognitive grammar, plays an important role in suffixation. The structure of the Slavic word is also compared with that of the English word.

The Part IV of the book is devoted to issues related to the description and first language acquisition of affix order and comprises two papers: "Suffix Sets in Polish De-nominal Derivatives" by Iwona Burkacka discusses suffix sequences in denominal derivations in contemporary Polish. The analysis relies on the data of the newest resource on Polish word-formation *Słownik gniazd słowotwórczych współczesnego języka ogólnopolskiego*, which is a dictionary of word-formation nests. The author also follows the methodology of this dictionary. The article shows that most suffix sequences in Polish tend to be short (i.e., they usually involve two suffixes) and homogenous (i.e., addition of suffixes is seldom interrupted by other derivation techniques such as conversion).

In "Reduplication, Repetition, Hypercharacterization, and Other Affix-Doubling in Child Language," Dressler, Dziubalska-Kołaczyk, Gagarina, and Kilani-Schoch deal with cases of affix ordering in first language acquisition that entail repetition. According to the authors, a repetition in the ordering of affixes may be total or partial, and may affect both form and meaning or only meaning. Repetitions are classified into subtypes such as traditional onomatopoetic repetitions, innovative onomatopoetic repetitions, and repetitions of specific suffixes, especially those involved in the formation of diminutives and plurals. The article shows the importance of early language acquisition for the better understanding of affix order and recursivity, since early repetitive operations seem to serve as trainings for morphological (de)composition and affix stacking.

3. Discussion and Conclusions

As already mentioned, this volume comprises original recent papers that explore different approaches to affix ordering and provide analyses within various theoretical frameworks. The data come from well-studied and lesser-studied languages alike. Thus the volume appears indicative about the recent developments in the research on affix ordering.

The papers in the present book reveal a number of novel, positive tendencies in the current research on affix ordering:

- *Looking for facts in more than one language*: While in the previous collections (Manova & Aronoff 2010a, Aronoff & Manova 2010, and Manova 2011) almost all papers were on affix ordering in a single language, six out of the ten contributions in this volume compare their findings with affix ordering in other languages (Efthymiou, Kim, Zimmermann, Talamo, Manova, and Dressler et al.).
- *Consideration of complete inflectional paradigms*: The analyses of affix order patterns in lesser-studied languages no longer involve only pieces of data, but the whole inflectional paradigms (Zimmermann and Gündoğdu). Moreover, there is already research on affix order in lesser-studied languages done by native speakers (Gündoğdu).
- *Discussion of virtually the same affix ordering phenomena*:
    - *The behavior of prefixes and suffixes* and where these two types of affixes should be analyzed as staying in some relationship or as two different domains of affixation (Letuchiy, Efthymiou, Kim, and Manova). In this book both perspectives are taken: Letuchiy and Manova argue for two separate domains, because prefixes and suffixes exhibit two different types of ordering, while Kim's and Efthymiou's analyses rely on specific interactions between the prefixes and suffixes in a particular language.
    - *Repetition and permutation of affixes* (Lyutikova & Tatevosov, Zimmermann, Talamo, Manova, Burkacka, and Dressler et al.). These two types of peculiar affix order patterns are not only typical of lesser-studied languages but quite normal in well-studied languages as well; they challenge many theories of grammar but are much more frequent in the languages of the world than it has been assumed so far.
    - *Hierarchical ordering* of affixes, features, or modules, depending on the theory (Lyutikova & Tatevosov, Efthymiou, Kim, Zimmerman, Gündoğdu, Talamo, and Manova).
    - *Length of affix sequences* (Manova and Burkacka). This issue is still underrated in the literature on affix ordering, but it is directly related, for example, to hierarchical ordering. It raises the important question of what should be described in terms of a hierarchy: that is, whether

a theory should put into a hierarchical relationship all elements of a system, be they real or abstract morphemes, if they all never co-occur in a word.

- *Facts from first language acquisition*: Affix ordering phenomena involving repetition of affixes, which are a real challenge to linguistic theories (Lyutikova and Tatevosov, Manova, and Burkacka), are most probably used by children as training strategies in the early stages of the first language acquisition (Dressler et al.).

Although there has been much interest in affix ordering in recent years and every paper on the topic advances our understanding of the mechanisms behind affix ordering, there are still issues that call for explanation. In what follows, I briefly discuss the difficult issues in the current research on affix ordering.

- *Relation between data and theory*: With respect to this issue, papers on affix ordering are of two types: from-theory-to-data (i.e., top-down) and from-data-to-theory (i.e., bottom-up), which does not mean that the second type of papers does not formulate hypotheses. The belonging of a paper to one of the two groups seems to depend mostly on the language analyzed and contributions on lesser-studied languages are, as a rule, top-down. This is perhaps due to the fact that researchers of lesser-studied languages do not have access to resources such as electronic corpora, to which researchers of well-studied languages do. Consequently, research on lesser-studied languages is based on selected examples and concentrates on peculiar phenomena that challenge a particular theory. This situation, however, seems to heavily influence the way affix ordering is analyzed as well as the conclusions drawn. For example, while syntactic categories such as subject and object are quite common in the description of affix order patterns of lesser-studied languages, they are not used for the description of well-studied languages, though such an analysis is possible. Let me exemplify syntactic ordering with data from English:

  (1) *employ-er*
  employ-subject

  (2) *employ-ee*
  employ-object

  Moreover, affixes that attach to verbal stems and mark subjects and objects are classified as verbal in the analyses of lesser-studied languages and as nominal in the analysis of well-studied languages: that is, *-er* in *employer* (1) and *-ee* in *employee* (2) are nominalizing suffixes in English, but they would have been word-class-preserving verbal affixes if English were a lesser-studied language (see the papers in the first part of the volume).[2] In general, research on affix ordering in verbs usually involves argument structure (i.e., syntactic ordering) whereas the derivation of nouns is, as a rule, explained as due to other factors. The overall picture blurs even more if one considers that research on lesser-studied languages accounts for affix ordering in verbs, whereas affix ordering in well-known languages such as Germanic, Romance, and Slavic (see Talamo, Manova, Burkacka, Dressler et al. in this volume) pays more attention to the derivation of nouns, since in Germanic, Romance, and Slavic the majority of suffixes are nominalizing. Additionally, while Manova argues that lexical categories such as nouns, adjectives, and verbs play important roles in affix order in Slavic languages and English, Letuchiy (this volume), based on data from Adyghe (North Caucasian), claims that there is no principle difference between nouns and verbs. Letuchiy writes that he uses the term 'verb' to refer to forms that typically bear predicative markers, such as tense/aspect/mood markers or markers of valency change, but actually in Adyghe both nouns and verbs can take tense/aspect/mood markers. Note that in well-studied languages we also have instances where markers of categories that are prototypically verbal occur on nouns (Manova 2011c: Chapter 1). Nominalizations from secondary imperfective verbs in Slavic illustrate the point as they exhibit an aspectual suffix. However, Slavic deverbal nominalizations are not considered bearing verbal markers but are seen as derived from verbal stems.

The above facts seem to imply that we are used to approach well-studied and lesser-studied languages in two different ways, which favors more establishing of differences than of commonalities.

- *Alternative analyses are still seldom*: Contributions focused on theoretical issues seldom try alternative analyses, and if they do, it seems that the theoretical framework selected could influence the conclusions drawn. See Kim's paper for an explicit discussion of this issue with respect to morphology-phonology interface and the (non)existence of phonological affix ordering.
- *Form–meaning relations*: Following different theoretical frameworks, the papers also assume different relations between form and meaning in affix ordering. Such issues are seldom explicitly discussed in a paper on affix ordering, but according to the form–meaning relation assumed, there are three types of papers. The first type follows the traditional view that an affix, since a morpheme, is a pairing of meaning and form (see Gündoğdu's, Talamo's, and Burkacka's papers). However, in word- and lexeme-based generative morphology (e.g., Realizational morphology, Anderson 1992, Aronoff 1994, Stump 2001; and Construction

[2] Rice (2011) is on word-class-preserving affix ordering in verbs in lesser-studied languages, and can be used as another source of examples.

morphology, Booij 2010) affixes do not have semantics but only phonological form, and receive semantic interpretation in words and constructions of words. Put differently, form and meaning are treated separately and then linked; the best illustration of this approach is the principle of paradigm linkage in Stump (2001). For various reasons, including traditional education, it is hard to imagine that affixes are semantically empty, but an analysis that sees affixes as just markings also appears difficult from a technical point of view, if one does not adopt Stump's formalisms. This could be the explanation for why none of the papers in this volume explicitly relates to Stump (2001), though Gündoğdu's analysis of Kurmanji could be seen as being in this direction, and Zimmermann's analysis of Athpare also has some such elements. The third type of papers follows Distributed morphology, where affix is an abstract unit and refers to a syntactic terminal node and its content, not to the phonological expression of that terminal. Thus in Distributed morphology, affixes with the same meaning, such as the plural *-s* in *books* and *-en* in *oxen*, are the same affix because they both express the same feature [pl] (Embick and Noyer 2007). Because of the late insertion of phonological material (realization), which means separation of meaning and form, Distributed morphology is also considered an instance of Realizational morphology. Zimmermann's paper is an illustration of a Distributed morphology analysis. It should be noted that the majority of the papers in this volume seem to provide evidence in favor of form–meaning correspondence as that in Distributed morphology, but not all of the papers are compatible with an organization of grammar as the one assumed in Distributed morphology (see the discussion in Manova's paper). Additionally, the alternative analyses of the phonology–morphology interface in Kim's paper could also be related to the meaning–form issue: If in the adopted model of grammar vocabulary insertion—that is, assignment of phonological form or exponence—takes place as the last operation, then morphologically driven phonology is the only possibility; if phonology is not the last in a series of operations, phonologically driven morphology is an eligible alternative option.

In conclusion, due to the above listed problems, it is hard to say what exactly the existing research on affix ordering tells us about the general architecture of grammar. It is still not clear how much of what we believe to know from affix ordering studies about the architecture of grammar represents its real architecture, and how much is just the side effects of our theoretical assumptions. Further research is necessary to assess the role of lexical categories such as nouns and verbs as well as the role of entrenchment of grammatical structure in affix ordering. Plag and Baayen (2009) refer to the latter issue as the role of memory. Nevertheless, this volume provides clear evidence against the most recent proposal for affix ordering in English, the CBO (cf. Hay 2003, Hay & Plag 2004, Plag & Baayen 2009, among others). The majority of the affix order patterns discussed in the different papers of the volume involve permutation and repetition of affixes and are thus incompatible with a hierarchy of affixes of the CBO type. (For an illustration of a hierarchy of the CBO type, see Talamo, this volume.)

As the papers in this volume apply different approaches and provide analyses within different theoretical frameworks, including the East European linguistic tradition, virtually every linguist could find a contribution that would appeal to his or her own theoretical persuasion. This book also has things to offer to linguists interested in the particular languages and language families that provide the data for the analyses in the different papers. Last but not least, as the volume illustrates all major approaches to affix order, it is a good starting point for anybody interested in the topic of affix ordering.

References

Anderson, Stephen R. 1992. *A-Morphous Morphology*. Cambridge: Cambridge University Press.

Aronoff, Mark. 1994. *Morphology By Itself*. Cambridge, MA: MIT Press.

Aronoff, Mark, and Manova, Stela, eds., 2010. "Affix Combinations (Part Two)." *Morphology* 20(2). Special issue.

Baker, Mark. 1985. "The Mirror Principle and Morphosyntactic Explanation." *Linguistic Inquiry* 16:373–415.

Booij, Geert. 2010. *Construction Morphology*. New York: Oxford University Press.

Bybee, Joan L. 1985. *Morphology: A Study of the Relation Between Meaning and Form*. Amsterdam: John Benjamins.

Embick, David, and Rolf Noyer. 2007. "Distributed Morphology and the Syntax/Morphology Interface." In *The Oxford Handbook of Linguistic Interfaces*, edited by G. Ramchand and C. Reiss, 289–324. Oxford: Oxford University Press.

Fillmore, Charles. 1982. "Frame Semantics." In *Linguistics in the Morning Calm*, edited by I.-S. Yang, 111–137. Seoul, Hanshin Publishing Co.

Halle, Morris, and Alec Marantz. 1993. "Distributed Morphology and the Pieces of Inflection." In *The View from Building 20*, edited by K. Hale and S. J. Keyser, 111–176. Cambridge, MA: MIT Press.

Hay, Jennifer. 2003. *Causes and Consequences of Word Structure*. London: Routledge.

Hay, Jennifer, and Ingo Plag. 2004. "What Constrains Possible Suffix Combinations? On the Interaction of Grammatical and Processing Restrictions in Derivational Morphology." *Natural Language and Linguistic Theory* 22:565–596.

Hyman, Larry. 2003. "Suffix Ordering in Bantu: A Morphocentric Approach." In *Yearbook of Morphology* 2002, edited by G. E. Booij and Jaap van Marle, 245–281. Dordrecht: Kluwer.

Jackendoff, Ray. 1990. *Semantic Structures*. Cambridge, MA: MIT Press.

Kim, Yuni. 2010. "Phonological and Morphological Conditions on Affix Order in Huave." *Morphology* 20:133–163.

Langacker, Ronald. 1987. *Foundations of Cognitive Grammar, Volume I, Theoretical Prerequisites*. Redwood City, CA: Stanford University Press.

Manova, Stela, and Aronoff, Mark, eds. 2010a. *Affix Combinations. Morphology* 20(1). Special issue.

Manova, Stela, and Aronoff, Mark. 2010b. "Modeling Affix Order." *Morphology* 20:109–131.

Manova, Stela, ed. 2011a. *Affixes and Bases. Word Structure* 4(2). Thematic issue.

Manova, Stela. 2011b. "A Cognitive Approach to SUFF1-SUFF2 Combinations: A Tribute to Carl Friedrich Gauss." *Word Structure* 4:272–300.

Manova, Stela. 2011c. *Understanding Morphological Rules: With Special Emphasis on Conversion and Subtraction in Bulgarian, Russian and Serbo-Croatian*. Dordrecht: Springer.

McCarthy, John, and Alan Prince. 1993. "Generalized Alignment." In *Yearbook of Morphology* 1993, edited by Geert Booij and Jaap van Marle, 79–153. Dordrecht: Kluwer.

Muysken, Pieter. 1986. "Approaches to Affix Order." *Linguistics* 24:629–643.

Paster, Mary. 2009. "Explaining Phonological Conditions on Affixation: Evidence from Suppletive Allomorphy and Affix Ordering." *Word Structure* 2:18–47.

Plag, Ingo, and Harald Baayen. 2009. "Suffix Ordering and Morphological Processing." *Language* 85:109–152.

Rice, Keren. 2000. *Morpheme Order and Semantic Scope*. Cambridge: Cambridge University Press.

Rice, Keren. 2011. "Principles of Affix Ordering: An Overview." *Word Structure* 4:169–200.

Stump, Gregory T. 2001. *Inflectional Morphology*. Cambridge: Cambridge University Press.

Taylor, John R. 2002. *Cognitive Grammar*. New York: Oxford University Press.

Trommer, Jochen. 2003. "The Interaction of Morphology and Syntax in Affix Order." In *Yearbook of Morphology 2002*, 283–324. Dordrecht: Kluwer.

PART I:

Syntactic and Semantic Ordering

2

Recursive Passivization: A Causative Coercion Account

Ekaterina Lyutikova and Sergei Tatevosov

1. Introduction

Various theories of affix ordering, be they articulated within formal (e.g., Baker's 1985 Mirror principle) or functional (e.g., Bybee's 1985 relevance theory) frameworks, aim at reducing the observed ordering patterns to mechanisms underlying syntactic or semantic scope-taking.[1] For instance, the universal tendency for tense grammatical morphemes to occur outside aspectual morphemes can be accounted for on the assumption that it is temporal operators (which locate the reference time with respect to the speech time) that take scope over aspectual operators (which establish temporal relations between the reference time and event time), but not vice versa.

Relative ordering of different pieces of morphology is thus thought of as reflecting hierarchical relations between linguistic categories, apparent deviations from this pattern being subject to various intricate restrictions (see, e.g., den Dikken 2003 on Atapaskan verbal morphology).

A trivial consequence of this view is that the output of an operation merging a piece of morphology must be a suitable input to a subsequent merge of the next piece of morphology. But the matter becomes less trivial when it comes to a recursive application of the same grammatical morpheme. The prediction is that we only find recursive derivations, where the structure created by the first instance of a morpheme can feed an application of a second instance of the same morpheme. Consider a double causative in Japanese:

[1]We are grateful to the audience at the Workshop on Affix Order (University of Vienna, 2011) and the Languages of the Caucasus conference (MPI for Evolutionary Anthropology, Leipzig, 2008), especially to Stela Manova, Wolfgang Dressler, Andrei Malchukov, and Bernard Comrie for their insightful comments and discussion. The study has been supported by Russian Science Foundation (grant #14-18-03270 at MGGU) and Russian Foundation for Basic Research (grant #13–06–00884–a).

(1) Intyoo-sensei-wa Taroo-ni tuuzyoo-yori nagaku byooin-ni
chief physician-TOP Taro-DAT usual-than longer hospital-at
hahaoya-o i-***sase-sase***-ta.
mother-ACC be-**CAUSE-CAUSE**-PAST
'The chief physician caused Taro to cause his mother to stay in the hospital longer than usual.' (Miyagawa 2012)

Abstracting away from a few details about Japanese causativization irrelevant for our illustrative purposes, one can assume with Pylkkänen (2002) that the causative adds a causing subevent to the event structure. If so, its first application to the predicate 'the mother stay in the hospital' gives rise to a more complex configuration 'Taro cause his mother to stay in the hospital.' This configuration is a suitable input to further causativization: merging the second causative morpheme one gets one more causing subevent, and the whole sentence ends up describing a complex event where the chief physician causes Taro to cause his mother to stay in the hospital. Recursive causativization can be constrained by extralinguistic factors (e.g., the increasing morphological complexity of the verb brings about processing problems), but linguistically, there seems to be nothing wrong about causativizing (1) even further and getting a triple causative, quadruple causative, and so on.

Things are different for such operations as passivization. Theories of the passive (Chomsky 1981, Jaeggli 1986, Baker et al. 1989, Collins 2005, Gehrke & Grillo 2009) differ as to what exactly they assume about the syntactic mechanism involved. There is, however, a fundamental agreement that passivization deprives the external argument of the ability to be projected as a sentential subject. The prediction about recursive passivization, then, is clearly negative: a suitable input for the passive is a structure where the external argument is syntactically active, but as soon as the passive applies, we get a configuration with no external argument available for further manipulations. Any application of the passive should block further passivization.

If this reasoning, even stated loosely, is correct, we do not expect to find anything like (2) from Karachay-Balkar (Altaic, Turkic).

(2) kölek zırt-**ıl-ın**-dı.
shirt.NOM tear-**PASS-PASS**-PST.3SG
'A/the shirt was torn.'

What we see in (2) are the two instances of the passive *-l* morpheme occurring in combination with the transitive stem *zırt* 'tear,' illustrated in (3).[2]

---

[2]The passive morpheme in Balkar, as well as in other Turkic languages, allows for two phonological realizations, *-l* and *-n*. Their distribution is phonologically conditioned: *-n* occurs if the final consonant of the stem (no matter whether lexical or derived) is *-l*, *-l* is found elsewhere. (i) illustrates this for a lexical *-l* stem, for a derived *-l* stem, and for two stems with no stem-final *-l*.

(i) a. al al-ın *al-ıl
buy buy-PASS buy-PASS

(3) alim kölek zırt-tı.
A.NOM shirt tear-PST.3SG
'Alim tore a/the shirt.'

Any transitive verb in Balkar is effectively passivized by exactly one application of the *-l* morpheme, as illustrated in (4):

(4) kölek zırt-**ıl**-dı.
shirt.NOM tear-**PASS**-PST.3SG
'A/the shirt was torn.'

Given (4), the very possibility of (2) looks like a mystery: how can the passive apply to (4), which is passive already, to yield (2)?

A closer look at the differences between (2) and (4) reveals more issues. Compare (5) and (6), corresponding to (4) and (2), respectively, where a postpositional phrase (PP) refers to the external argument.

(5) kölek kerim-ni küc-ü ble zırt-**ıl**-dı[3].
shirt.NOM K.-GEN force-3 with tear-**PASS**-PST.3SG
'A/the shirt was torn by Kerim.'

(6) kölek kerim-ni küc-ü ble zırt-**ıl**-**ın**-dı.
shirt.NOM K.-GEN force-3 with tear-**PASS**-**PASS**-PST.3SG
1. 'A/the shirt was torn by Kerim (on someone's order).'
2. 'A/the shirt was torn (by someone) on Kerim's order.'

(5) is unambiguous, and the PP *kerimni kücü ble* 'by (the force of) Kerim' functions in exactly the same way as the *by*-phrase in English, referring to the external argument of the verb 'tear.' (6) is different. Here we have three participants of the event, the theme (shirt), the agent (the one who tears the shirt), and

---

b. zırt-ıl zırt-ıl-ın *zırt-ıl-ıl
tear-PASS tear-PASS-PASS tear-PASS-PASS

c. qıjına *qıjına-n qıjına-l
torture torture-PASS torture-PASS

d. al-ın *al-ın-ın al-ın-ıl
buy-PASS buy-PASS-PASS buy-PASS-PASS

[3]The literal meaning of the Karachay-Balkar counterpart of English *by*-phrases is 'by X's force.' Here the postposition *bla*/*ble* 'by, with' takes the DP 'X's force' as a complement, where X is a Genitive possessor that triggers agreement on the possessee *küc* 'force.'

the causer (the one who makes the agent tear the shirt). The PP *kerimni kücü ble* in (6) can either refer to the agent, (6.1), or to the causer, (6.2).[4]

Examples in (5) and (6) suggest that the second application of the passive does not result in passivization—not surprisingly, given that (4) is passive already. What is surprising is that (6) contains one more argument (even if implicit) as compared to (5), the causer who makes the agent tear a shirt. Introducing arguments is not what we expect from the passive morphology.

Argument-introducing capacity of the passive morpheme does not only manifest itself in passive configurations like (6). The same happens if the passive is combined with intransitives, either unaccusatives or unergatives, as illustrated in (7) and (8).

(7) a. illew (*alim-ni küc-ü ble) sın-dı.
toy.NOM A.-GEN force-3 with break-PST.3SG
'The toy broke (*by Alim).'

b. illew alim-ni küc-ü ble sın-**ıl**-dı.
toy.NOM A.-GEN force-3 with break-PASS-PST.3SG
'The toy was broken by Alim.'

(8) a. alim (*farida-nı küc-ü ble) sekir-di.
A.NOM F.-GEN force-3 with jump-PST.3SG
'Alim jumped (*by Farida).'

b. alim farida-nı küc-ü ble sekir-**il**-di.
A.NOM F.-GEN force-3 with jump-**PASS**-PST.3SG
'Alim jumped (because something was done by Farida).'

The *by*-phrase, ungrammatical in the (a) examples in (7) and (8), is readily available in the (b) examples, where the passive morpheme occurs. In both cases, the *by*-phrase refers to an argument that brings about an event denoted by the nonderived verbal predicate, the breaking event in (7b) and jumping event in (8b).

Note, crucially, that *sın* 'break, intr.' in Karachay-Balkar is a genuine intransitive, and *zırt* 'tear' is a genuine transitive verb. Neither license causative-inchoative alternation, as evidenced by examples in (9) and (10):

(9) *kölek zırt-tı.
shirt.NOM tear-PST.3SG
'The shirt tore.'

---

[4]Most native speakers find configurations with two *by*-phases like (i) degraded.

(i) ??kölek kerim-ni küc-ü ble Alim-ni küc-ü ble zırt-**ıl**-**ın**-dı.
shirt.NOM K.-GEN force-3 with K.-GEN force-3 with tear-**PASS**-**PASS**-PST.3SG
'A/the shirt was torn by Kerim on Alim's order.'

It is not completely clear whether speakers' judgments point toward genuine ungrammaticality or merely represent processing difficulties. We believe that the latter option is more plausible, given that some speakers marginally accept (i) and can assign the intended meaning ('by X on Y's order') to it.

(10) *alim illew-nü sın-dı.
A.NOM toy-ACC break-PST.3SG
'Alim broke the toy.'

As (9) and (10) show, 'tear, tr.' cannot, and 'break, intr.' must project an intransitive clause.

Had (10) been grammatical, *sın* 'break' would have been a counterpart of *break* in English, and examples like (7a–b) would have never been problematic: one would say that while (7a) represents an unaccusative/inchoative/anticausative construal, (7b) is a product of passivization of the transitive in (10). But given that (10) is irreparably ungrammatical, we have every reason to conclude that 'break' in Balkar is intransitive. As a consequence, we see no way of making sense of the idea that (7a) is a product of morphologically unmarked anticausativization, and (7b) results from overt passivization of the transitive 'break,' respectively.[5]

(7) and (8), together with (6), is a challenge for our understanding of the passive. In each case, the passive morphology appears in a configuration which is not a suitable input for passivization, either because it is intransitive in the first place, (7) and (8), or because passivization has already occurred, (6). This does not prevent the *-l* morpheme from applying in such configurations, however. Moreover, what this morpheme does in these examples is right the opposite from what the canonical passive is supposed to do: instead of demoting an argument, it introduces a new one.

Finally, the "intransitive passive" in (7) and (8) can be combined with one more occurrence of the passive morpheme. (11a–b) show this for the unergative 'jump' (for the sake of space we leave out corresponding examples for 'break'):

(11) a. alim (kerim-ni küc-ü ble) sekir-**il**-di.
A.NOM K.-GEN force-3 with jump-**PASS**-PST.3SG
'Alim jumped (because something was done) by Kerim.'

b. alim (kerim-ni küc-ü ble) sekir-**il**-**in**-di.
A.NOM K.-GEN force-3 with jump-**PASS**-**PASS**-PST.3SG
1. 'Alim jumped (because something was done by Kerim on someone's order).'
2. 'Alim jumped (because something was done by someone on Kerim's order).'

---

[5]More generally, verbs like *break* and *open* in English that allow for the causative-inchoative alternation whereby a verb can project both transitive and inchoative/anticausative clauses without visible consequences for the morphology are remarkably absent from Karachay-Balkar and, to the best of our knowledge, from most of other Turkic languages. Some lexical equivalents of English alternating verbs (e.g., *sın* 'break, intr.') are intransitive, their causative counterparts being derived through causativization. Others (e.g., *zırt* 'tear') are transitive, and their inchoative/anticausative/unaccusative counterparts require anticausativization (see, e.g., Göksel 1993, Göksel and Kerslake 2011 on Turkish, and Vinokurova 2005 on Sakha). We are grateful to the anonymous reviewer for encouraging us to discuss this issue.

Native speakers' judgments about (11b) are entirely consistent: (11b) describes a jumping event and two implicit causing events, with one causing participant each. That is, (11b) is ambiguous in a way that (11a) is not. In (11b.1), the *by*-phrase 'by Kerim' can refer to the immediate causer of Alim's jumping. On this reading, there is also an implicit remote causer whose causal contribution is to make Kerim make Alim jump. On the other reading, (11b.2), it is Kerim who is involved as such a remote causer. The immediate causer, the one who makes Alim jump, is not specified.

Here is a summary of the distribution of the passive morpheme in Karachay-Balkar. If the verb stem is transitive, the first occurrence of the passive morpheme creates a regular passive clause (see (4)). The second occurrence introduces the meaning of external causation, as in (2), which will be called *the causal passive* from now on (the term is coined in Lyutikova & Bonch-Osmolovskaya 2006). If, on the other hand, the verb is intransitive, both instances of the passive (cf. (7)–(8) and (11)) do the same thing: they introduce causal semantics.

Therefore, the puzzle is: where does the causal passive come from? How can the passive morpheme introduce a causing subevent and a causing participant of that subevent, even if implicit? Or, as Matthias Gerner (personal communication) puts it, given that morphological passives are understood to decrease valency and morphological causatives to increase valency, how can a marker perform both at the same time?

Two possible ways of solving this puzzle seem to be open at this juncture. First, one can say that there are in fact two distinct operations that happened to be realized by the same phonological exponent. The *-l* marker is ambiguous between 'passive' and 'causal passive' meanings. This view may look appealing given evidence from languages like Korean, where "the causative construction can potentially express both passive and causative senses" (Washio 1993: 61) and causative and passive morphology is literally identical (see also Kim & Pires 2003):

(12) a. totwuk-i swunkyeng-eykey cap-hi-ess-ta.
thief-NOM policeman-BY catch-PASS-PST-PLAIN
'The thief was arrested by a policeman.' (Washio 1993: 48)

b. na-nun sensayngnim-eykey chayk-ul ilk-hi-ess-ta.
I-TOP teacher-BY book-ACC read-VOICE-PST-PLAIN
'I had the teacher read a/the/his book.' (Washio 1993: 60)

Various connections between causative and passive semantics and morphosyntax, both synchronic and diachronic, have also been discussed in Keenan 1985, Santorini and Heycock 1988, Haspelmath 1990, Aoyagi 2007, and McIntyre 2011, among others.

However, for Balkar, assuming that the *-l-* morpheme is ambiguous between the two readings (or that it has some general abstract meaning comprising both,

Washio 1993) more creates problems than solves them. The crucial difference between Korean and Balkar is: the two readings of *-hi-* are available in the same environment, the two 'readings' of *-l-* are complementarily distributed.

In Korean, *-hi-* readily derives both causatives and passives from transitive verbs, as in (12) and (13), and this is exactly what one would expect from a truly ambiguous grammatical morpheme. We find such morphemes across different grammatical domains, an example being Imperfective in Russian, ambiguous between at least progressive and habitual readings (e.g., Grønn 2003, among many others):

(13) Volodja gulja-et v park-e.
V. walk.IPFV-3SG in park-PREP
1. 'Volodja is walking in the park.'
2. 'Volodja walks in the park.'

The *-l-* morpheme in Balkar is different. If it were ambiguous between 'passive' and 'causal passive' readings, we would expect to find at least some transitive verbs that show both. This never happens, however: in the transitive environment, *-l-* must be interpreted as passive, cf. (14), where the intended causative construal is blatantly ungrammatical:

(14) kitap alim-ni küc-ü ble oqu-l-du.
book A.-GEN force-3 with read-PASS-PST.3SG
1. 'The book was read by Alim.'
2. *'The book was read by Alim (on someone's order).'
3. *'The book was read (by someone) on Alim's order.'

In (14), no external causation of the reading event in which Alim is the agent and the book is the theme can be detected, and the readings in (14.2) and (14.3) corresponding to the alleged causal passive construal, are completely unavailable. Specifically, unlike what happens with true 'causal passives' in (6), (7b), (8b), and (11), the *by*-phrase can only refer to the agent of reading, not to a remote causer that initiates the reading event without participating in it.[6]

Given this evidence, we argue that another account for the 'causal passive' is more appealing, whereby the *-l* morpheme receives a uniform analysis. In what follows we will argue that *-l* always has the same meaning, and this is the passive meaning in a strict sense. The apparent differences between 'passive' and 'causal

[6]Another argument against treating the *-l-* morpheme as ambiguous between passive and causal passive readings can be based on diachronic observations from Haspelmath (1990: 46–49), who argues for the existence of the grammaticalization path from the causative to the passive but not the other way around. If this generalization is correct, it must be the 'causal passive' reading of the *-l-* morpheme that gives rise to the passive reading, not the other way around. This is unlikely to be the case, however: in other Turkic languages, *-l-* is invariably a passive marker, while causal passive uses are, to the best of our knowledge, only productive in Balkar, suggesting the "passive → causal passive" development.

passive' emerge as an outcome of the interaction between this single general meaning of *-l* and its lexico–syntactic environment. 'Causal passives' do involve causation, but it is not a semantic contribution of the *-l-* morpheme. Rather, they are a product of the *causative coercion*, a process of covert causativization that feeds the application of the passive morpheme whenever the initial configuration fails to satisfy its semantic requirements.

We argue that the *-l* morpheme always applies to a transitive predicate. What it does is the usual job of the passive: it existentially binds the agent/causer argument and induces promotion of the theme to the subject position. If a verb provides a suitable argument structure, the *-l* morpheme produces a regular passive. That is what we see in examples like (4). But if the verb fails to have a transitive argument structure, which is the case in (2), (6), (7b), (8b), and (11), it undergoes coercion through covert causativization. Causativization introduces a new agent/causer subevent, as usually, creating a transitive structure, and it is this structure that the *-l* morpheme applies to.[7] This is how the causal passive is derived.

The rest of the paper is organized as follows. In section 2, we briefly review coercion phenomena discussed in the literature and make a case for a new type of such an operation, the causative coercion. In sections 3 and 4 we establish an argument supporting our hypothesis that the *causative coercion* underlies what superficially looks like a causal passive. The argument is based on striking parallelism between causal passives and overt causatives. In section 3, we observe that *causal passives of intransitives* are identical event-structurally to corresponding *causatives* derived from the same verbs, abstracting away from the difference in the active/passive voice. This identity is preserved across distinct verb classes. We show that the hypothesis that passives of intransitives involve covert causativization correctly predicts this pattern. In section 4, we extend the argument to *double passives*: their event-structural characteristics are best accounted for if covert causativization happens in between two applications of the passive morpheme. Main results and wider implications of the analysis are summarized in the conclusion.

2. Coercion

Coercion has been a topic of quite a number of studies, including Moens and Steedman 1988, Jackendoff 1997, Michaelis and Lambrecht 1996, de Swart 1998, Zucchi 1998, Michaelis 2004, 2006, Deal 2007, and Koontz-Garboden 2007, among many others. Zucchi (1998) emphasizes that coercion is essentially a conflict resolution:

> Let d be a syntactic constituent of form $[_a$ b c$]$ and R the semantic rule which forms the meanings of the constituents of type a. If combining the meanings

[7] The same idea can be found in the draft version of Malchukov 2011, who hypothesizes that the verb can undergo "a shift to a causative/transitive meaning in the presence of the passive morphology."

of b and c via R leads to trouble (for example, leads to inconsistency or prevents from assigning an interpretation to d), the speaker has the option of reanalyzing the meaning of b or c (coercing the meaning of b or c into a different meaning) in order to apply R to form the meaning of d. (Zucchi 1998)

Coercion occurs if two elements we want to combine do not have matching meanings, and one of them gets reanalyzed. It follows, then, that coercion cannot occur without motivation; it has always to be activated by a certain trigger.

Different types of coercion have been discussed in the literature so far, including type coercion, domain structure coercion, and aspectual coercion.

Examples of type coercion are (15a–b), where proper names are used as property-type DPs, not as referential DPs.

(15) a. This is not *the Paris* I know.
b. There is *a Santa Claus*.

In examples like these, arguments of the type of individuals, *Paris* and *Santa Claus*, are converted into predicates; that is, into the type expected by a function in the denotation of the definite and indefinite articles.

Another example of type coercion are expressions like *a stone lion, a fake gun, a chocolate teapot* (Coulson & Fauconnier 1999, Pustejovsky 1995, Kamp & Partee 1995, Partee 2010). Since the intersection of lions and stone things is empty, the predicate *stone lion* must have an empty extension. As a matter of fact, it does not. A possible reason for this is that the adjective forces the noun to be interpreted as denoting images of things in its original extension. As soon as this happens, the mismatch disappears. Levinson (1983: 110) makes a similar point about examples like *Queen Victoria was made of iron*: "The straightforward interpretation, he indicates, is that since Queen Victoria in fact lacked the definitional properties of iron, she merely had some of the incidental properties like hardness, resilience, non-flexibility."

An ontological type shift is also attested in examples like (16) (Anick & Bergler 1992, Pustejovsky 1991, Godard & Jayez 1993), where DP *the book* based on a predicate of individuals is arguably coerced to a phrase of type of events.

(16) a. to begin *reading the book*
b. to begin *the book*

Analyzing examples like (16b) in terms of individual-event coercion, one can explain how verbs like *begin* are combined with complements of different syntactic types without positing multiple lexical entries for *begin*.

Examples in (17) and (18) represent another type of coercion known as a domain structure coercion.

(17) I had *a tea*.

(18) Give me *some blanket*.

Mass nouns like *tea* can receive "an individuated construal when paired with the indefinite article, and a count noun receives a mass construal when paired with unstressed *some*" (Michaelis 2006), as in (17) and (18). If domains from which count and mass nouns take their denotations are structured in different ways, the latter being nonatomic (Link 1983, and much subsequent literature), in (17) and (18) domains of *tea* and *blanket* must be restructured before combining with determiners.

Intensively studied in the past few years is aspectual coercion, which is attested if an eventuality description does not meet the input requirements of an aspectual operator (e.g., Pustejovsky & Bouillon 1995, de Swart 1998). If this happens, as in examples in (19), where the progressive combines with a state description, the state is coerced into a dynamic eventuality:

(19) a. Susan *is liking* this play a great deal.
b. Peter *is believing* in ghosts these days.
c. Charles *is being silly*.

Aspectual coercion can also be induced by adverbials, as in (20a–b). In (20a), *suddenly* triggers coercion of a state into an inception of state; in (20b), the accomplishment predicate acquires the iterative interpretation after combining with the durative adverbial *for three hours*.

(20) a. Suddenly, I *knew* the answer.
b. She *played* sonata for three hours.

Now that we have seen a wide variety of coercion operations, we are in a position of introducing the main ingredient of the proposal.

We argue that one of the coercion operations available in natural languages is a causative coercion, represented in (17).

(21) $Op_{trans}(V_{intrans}) \rightarrow Op_{trans}(CAUS(V_{intrans}))$
where Op_{trans} is an operator that selects for a transitive verbs, and CAUS is a covert causative operator

According to (21), if a semantic operator that wants a transitive predicate as its input combines with an intransitive predicate, in order to resolve this transitivity mismatch the latter can be coerced by means of covert causativization.

To the best of our knowledge, coercion phenomena involving causativization have not been much discussed in the literature so far. Sag and Pollard (1991: 83) indicate that in languages like English a semantic shift is forced in nonagentive clauses when they occur in an agentive environment; for example in the imperative: *Be optimistic!* means, roughly, 'make yourself optimistic' or 'put effort in maintaining an optimism.' Here the addressee of the imperative is interpreted as exercising control over the state 'be optimistic'—by virtue of causative coercion, Sag and Pollard suggest. Sag and Pollard (1991) and Pollard and Sag (1994) argue that the same type of coercion is observed in control configurations like *John promised Mary to be allowed to attend the reception* (Sag & Pollard 1994: 312) where the infinitival complement is interpreted as 'cause X to be allowed to attend the reception.' In both cases, the coercion seems to be triggered by the requirement that a relevant argument must have a certain properties of the agent for the whole structure to receive a coherent interpretation, but the input (the holder of states like 'optimistic' or the internal argument of *allow)* fails to meet this requirement.

However, what happens in Sag and Pollard's examples is not exactly what (21) is aimed to account for. (21) represents a situation where the causative coercion is induced by the operator what expects a *transitive* input, not just an input where the agent is present. (21) predicts that an intransitive configuration, be it agentive or not, can be causativized if some Op_{trans} wants to merge with it. Below we will develop an argument that this is precisely what happens in Karachay-Balkar when the passive morpheme attaches to an intransitive configuration, as represented in (22).

(22) *-l-*$_{PASS}$ ($V_{intrans}$) → *-l-*$_{PASS}$ (CAUS($V_{intrans}$))

The argument is based on the following reasoning. (22) says that an intransitive predicate undergoes covert causativization. Therefore, it predicts a strict parallelism between causal passives and overt causatives. Whatever properties overt causatives have, their coerced counterparts must have the same, because causativization is part of the derivation of both. Besides, if (22) is essentially a coercion phenomenon, it cannot go without a trigger. We should not find a coerced intransitive predicate in any environment other than in combination with the passive morpheme.

In what follows we will test both of these predictions. First, we will show that causal passives resemble overt causatives in all relevant respects, and this would be a crucial piece of evidence that causal passives involve covert causativization. Specifically, evidence for the analysis in (22) comes from the observations that intransitives combined with the passive pattern with overt causatives with respect to the semantic type of causation, interpretation of agent-oriented, time-span and rate adverbials, and the scope of negation.

Secondly, we will show that coercion can only occur if an intransitive predicate is an argument of the passive operator. This generalization will be crucial for determining the order of operations in the derivation: any two instances of overt passivization are to be mediated by covert causativization, and an output of covert causativization cannot be subject to another covert causativization.

3. Coerced Intransitives

In this section, we present evidence for a strict parallelism between the causative and the causal passive. We will take causatives of different types of intransitives and compare them with corresponding causal passives, and what we will see is their identity in all relevant respects. Then we will show that generalizations drawn from these observations extend naturally to recursive passivization. This will provide us with clues to solving the problem we started with in the introduction: how to treat recursive derivations where one application of an operator destroys the environment for another application.

Table 2.1 summarizes semantic characteristics of overt causatives and their causal passive counterparts derived from unaccusative and unergative verbs.

In a nutshell, what we see is that there is a huge difference between unaccusative-based and unergative-based configurations. However, there is no difference at all between causatives and causal passives derived from the same predicate, be it unergative or unaccusative. Differences between the causative applied to unaccusatives and unergatives are mirrored by the causal passive.

Let us review these data in more detail. The first observation has to do with the semantic type of causation.

(23) alim butaq-nı sın-dır-dı.
Alim.NOM branch-ACC break.INTR-CAUS-PST.3SG
1. 'Alim broke the branch.'
2. *'Alim caused the branch to break.'

The causative of unaccusative in (23) involves direct causation whereby the agent's activity is an immediate cause of the change of state of the theme (e.g., Kratzer 2005). (18) is illicit if the activity and the change of state are mediated by other events. For instance, Alim asking his friend to break a branch, his friend taking a hammer and hitting the branch, and the branch finally breaking would be an impossible scenario for (23).

Compare the causal passive in (24).

(24) butaq (alim-ni küc-ü ble) sın-ıl-dı.
branch.NOM Alim-GEN force-3 with break.INTR-PASS-PST.3SG
1. 'The branch was broken (by Alim).'
2. *'The branch was caused to break by Alim.'

TABLE 2.1
Causatives and Causal Passives of Intransitives

	Unaccusative		Unergative	
	Causative	Causal passive	Causative	Causal passive
Type of causation	Direct	Direct	Indirect	Indirect
Scope of time-span and rate adverbials	Unambiguous	Unambiguous	Three-way Ambiguous	Three-way Ambiguous
Interpretation of agent-oriented adverbials	Unambiguous	Unambiguous	Two-way Ambiguous	Two-way Ambiguous
Scope of negation	Two-way Ambiguous	Two-way Ambiguous	Three-way Ambiguous	Three-way Ambiguous

Native speakers' judgments are very consistent: abstracting away from transitivity, (24) has precisely the same meaning as (23), the one of direct causation with no intermediate events in a causal chain connecting the agent's activity and the change of state.

Unergatives are different. For unergative-based causatives, the indirect causation construal is readily available. In (25), there can be quite a long causal chain connecting the causer's activity and the causee's action. A possible scenario can, for instance, include Alim asking his friend to convince Kerim's father to remind Kerim that he has to run.

(25) alim kerim-ni cap-tır-dı.
Alim.NOM Kerim-ACC run-CAUS-PST.3SG
'Alim caused || allowed || asked || convinced Kerim to run.'

The same is true of the causal passive in (26).

(26) kerim cab-ıl-dı.
Kerim.NOM run-PASS-PST.3SG
'Kerim was caused || allowed || asked || convinced to run.'

Where the difference between unergatives and unaccusatives with respect to causativization comes from is irrelevant for our argument. The direct/indirect distinction has been a constant topic in the studies of causativization phenomena since late 1960s and one of the central issues surrounding the debate on lexical and syntactic causatives (Lakoff 1965, 1971, Fodor 1970, McCawley 1971, 1972, Cruse 1972, Shibatani 1973, Yang 1976, Haiman 1983, only mentioning the early literature on the topic; see a recent discussion in Kratzer 2005, and Miyagawa 2012). In this paper we do not have to add much to this discussion; see Lyutikova and Tatevosov (to appear) for more details.[8] What we are interested in here is a simple generalization that can be drawn from these examples: the causative and the causal passive never show differences in terms of the semantic type of causation. Whenever the indirect interpretation is available for the causative, the causal passive will have it, too.

[8]Lyutikova and Tatevosov (to appear) assume, in line with many proposals found in the literature (e.g., Pylkkänen 2002, Harley 2008, Travis 2010, Tubino Blanco 2011) that the directness of causation corresponds to the syntactic location of the causative morpheme. Within the functional framework, in contrast, the directness is frequently related to the type of grammatical marking associated with the causative. Haiman (1983), for instance, claims that lexical causatives correspond to more direct, and analytic causatives to a less direct causation. For languages like Balkar, however, this view is problematic given that the same piece of morphology—the *-tır-* morpheme—is associated with all types of causatives.

The same pattern—semantic contrast between unaccusatives and unergatives, no contrast between causatives and causal passives—is attested with temporal adverbials. Consider the distribution of time span adverbials as an example:

(27) alim beš minut-xa cojun-nu tol-dur-du.
Alim.NOM five minute-DAT pot-ACC fill.INTR-CAUS-PST.3SG
1. 'Alim filled the pot in five minutes.'
2. *'What Alim did in five minutes was make a/the pot fill.'
3. *'What Alim did was make a/the pot fill in five minutes.'

(28) cojun beš minut-xa tol-un-du.
pot.NOM five minute-DAT fill.INTR-PASS-PST.3SG
1. 'The pot was filled in five minutes.'
2. *'Some event that happened in five minutes made the pot fill.'
3. *'Some event made the pot fill in five minutes.'

Causatives and causal passives again pattern together: in both cases the adverbial indicates that it takes five minutes for the whole filling event to occur, as in (27.1) and (28.1). Interpretations where the adverbial only takes scope over a causing subevent, (27.2) and (28.2), or over a caused subevent, (27.3) and (28.3), are impossible.

Interpretation of time span adverbials combined with causatives and causal passives of unergatives is different. Consider the causative first:

(29) alim beš minut-xa kerim-ni škol-ʁa cap-tır-dı.
Alim.NOM five minute-DAT Kerim-ACC school-DAT run-CAUS-PST.3SG
1. 'In five minutes, Alim made Kerim run to the school.'
2. 'What Alim did in five minutes was make Kerim run to the school.'
3. 'What Alim did was make Kerim run to the school in five minutes.'

Here the adverbial can take scope over the causing subevent, over the caused subevent or over both. On the first reading, both subevents occur within a five-minute interval, (29.1). On the second reading, the adverbial identifies duration of the causing subevent without indicating how long the running subevent lasts, (29.2). The other way around, in (29.3), it takes five minutes for the causee to reach the school, but the duration of the causing subevent is left unspecified.

The empirical reality of the readings in (29.2) and (29.3) is easy to see by specifying an appropriate context. For (29.2), this would be (30), where the duration of the caused subevent can be asserted to be different from five minutes without yielding a contradiction:

(30) alim beš minut-xa kerim-ni škol-ʁa cap-tır-dı.
Alim.NOM five minute-DAT Kerim-ACC school-DAT run-CAUS-PST.3SG
'What Alim did in five minutes was make Kerim run to the school.'
andan sora kerim on minut-xa škol-ʁa cap-tı.
this-ABL after K. ten minute-DAT school-DAT run-PST.3SG
'Then Kerim ran to the school in ten minutes.'

Similarly, a coherent piece of discourse obtains in (31), corresponding to (29.3):

(31) alim beš minut-xa kerim-ni škol-ʁa cap-tır-dı.
Alim.NOM five minute-DAT Kerim-ACC school-DAT run-CAUS-PST.3SG
'What Alim did was make Kerim run to the school in five minutes.'
on minut alim-ge bol-du.
ten minute A.-DAT be-PST.3SG
'Ten minutes were enough for Alim (to do so).'

Had the scope of the adverbial in (31) necessarily included the causing subevent, the second clause would have leaded to a contradiction. The discourse in (31) is fully consistent, however, suggesting that 'in five minutes' modifies the running, but not the causing subevent.

What is crucial is that the range of interpretations of the causative and the causal passive is again identical. The causative in (29) is three-way ambiguous, and so is the causal passive in (32), where (32.1) corresponds to (29.1), (32.2) to (29.2), and (32.3) to (29.3):

(32) kerim beš minut-xa škol-ʁa cab-ıl-dı.
Kerim.NOM five minute-DAT school-DAT run-PASS-PST.3SG
1. 'In five minutes, Kerim was made to run to the school.'
2. 'Some event that happened in five minutes made Kerim run to the school.'
3. 'Some event made Kerim run to the school in five minutes.'

The next piece of evidence supporting the proposal comes from the distribution of purpose adjuncts. Purpose adjuncts, unless used metaphorically, are strictly agent-oriented, since only agents can act volitionally and exercise control over their actions. Causatives of unaccusatives only have one agent, hence purpose adjuncts do not yield ambiguity. The same holds for causal passives and their implicit agents, as illustrated in (33a–b).

(33) a. fatima lejla-ʁa acuuʁa illew-nü sın-dır-dı.
Fatima.NOM Leila-DAT to.spite toy-ACC break.intr-CAUS-PST.3SG
'Fatima broke the toy to spite Leila.'

b. illew lejla-ʁa acuuʁa sın-ıl-dı.
toy.NOM Leila-DAT to.spite break.intr-PASS-PST.3SG
'The toy was broken to spite Leila.'

Most unergatives are agentive, so causativization creates a structure that can contain two agents—the agent of a causing subevent and the agent of a caused subevent. As a result, purpose adjuncts can be associated with either of these two agents. This is what (34a) illustrates for causatives and (34b) for causal passives. Again, no difference can be detected as to the range of interpretations available for (34a) and (34b).

(34) a. fatima lejla-ʁa acuuʁa alim-ni züz-dür-dü.
Fatima.NOM Leila-DAT to.spite Alim-ACC swim-CAUS-PST.3SG
1. 'Fatima let Alim swim to spite Leila. (It was Alim who wanted to spite Leila.)'
2. 'In order to spite Leila, Fatima made Alim swim. (It was Fatima who wanted to spite Leila.)'

b. alim lejla-ʁa acuuʁa fatima-nı küc-ü
Alim.NOM Leila-DAT to.spite fatima-GEN force-3
ble züz-ül-dü.
with swim-PASS-PST.3SG
1. 'Alim was allowed by Fatima to swim to spite Leila. (It was Alim who wanted to spite Leila.)'
2. 'Alim was allowed to swim by Fatima, who wanted to spite Leila.'

Finally, consider the scope of negation. Causativized unaccusatives license two interpretations:

(35) alim cojun-nu tol-dur-ma-dı.
Alim.NOM pot-ACC fill.INTR-CAUS-NEG-PST.3SG
'Alim did not fill the pot.'
1. No filling activity has occurred.
2. Alim started filling the pot but has not finished.

Being derived accomplishments, causatives like *fill* provide the negation with two scope possibilities (Dowty 1979, and much further work). In (35.1), the negation takes wide scope that includes both activity and change of state. In (35.2), the change of state only falls under the scope of negation. Again, the same range of interpretations is available to the causal passive in (36).

(36) cojun tol-un-ma-dı.
pot.NOM fill.INTR-PASS-NEG-PST.3SG
'The pot was not filled.'
1. No filling activity has occurred.
2. Someone started filling the pot but has not finished.

Causatives of unergatives are ambiguous under negation, too, but in a considerably different way. The first interpretation of (37) corresponds to (35.1): neither causing nor caused subevents occur. The second interpretation involves the causer's activity that fails to bring about the caused subevent. It can be thought of as being scopally equivalent to (35.2). However, causatives of unergatives allow for one more option not available for causatives of unaccusatives. On the third reading, (37) means that the agent was caused to not run:

(37) alim kerim-ni škol-ʁa cap-tır-ma-dı.
Alim.NOM Kerim-ACC school-DAT run-CAUS-NEG-PST.3SG
1. 'Alim did not make Kerim run to the school.' (Neither causing nor running events occur.)
2. 'Alim did not make Kerim run to the school.' (He attempted but has not succeeded.)
3. 'Alim made Kerim not run to the school.'

The precise analysis of the negation in Karachay-Balkar is not our immediate focus. Rather, we use it as a means for discerning the difference between unaccusatives and unergatives and making sure that causatives and causal passives are alike. We see that the causal passive behaves in exactly the same way as the corresponding causative: (38), like (37), licenses the 'cause to not run' reading in addition to the other two.

(38) kerim škol-ʁa cab-ıl-ma-dı.
Kerim.NOM school-DAT run-PASS-NEG-PST.3SG
1. 'Kerim was not made to run to the school.' (Neither causing nor running events occur.)
2. 'Kerim was not made to run to the school.' (He attempted, but has not succeeded.)
3. 'Someone made Kerim not run to the school.'

Again, the data from the scope of negation lend support to the view that causal passives and causatives involve the same derivation.

What the evidence from causativization of unaccusatives and unergatives teaches us is that semantic characteristics of causatives and causal passives are strictly parallel. The hypothesis that the derivation of causal passives involves causativization, even if covert, accounts for this parallelism in a principled way.

With this in mind, we have everything we need to approach recursive passivization phenomena in the next section.

4. Recursive Passivization

4.1. DOUBLE PASSIVES OF TRANSITIVES

In the previous section, we have seen that combining intransitives with the *-l* morpheme involves covert causativization. Now we are going to show that the same happens if the *-l* morpheme combines with a predicate that is passive already. The relevant example is repeated as (39):

(39) kölek kerim-ni küc-ü ble zırt-**ıl-ın**-dı.
shirt.NOM K.-GEN force-3 with tear-**PASS-PASS**-PST.3SG
1. 'A/the shirt was torn by Kerim (on someone's order).'
2. 'A/the shirt was torn (by someone) on Kerim's order.'

One complication that emerges at this point is: we cannot use the same strategy as before, namely, compare the double passive like *zırtılın* in (39) with a corresponding overt causative of the form **zırtıldır* and show that they are identical. In Balkar, as in many other Turkic languages, the causative *-tır* can only be merged inside the passive morpheme, not outside, which possibly reflects the hierarchy of functional projections these morphemes are inserted into (Enç 2003). Since causatives like **zırtıldır* do not exist, we propose another, slightly less direct way of identifying effects of covert causativization.

We show that double passives like (39) are to single passives like (4) as causatives of transitives are to ordinary transitives, as represented in Scheme 1:

Scheme 1. Transitives, causatives, passives, and double passives

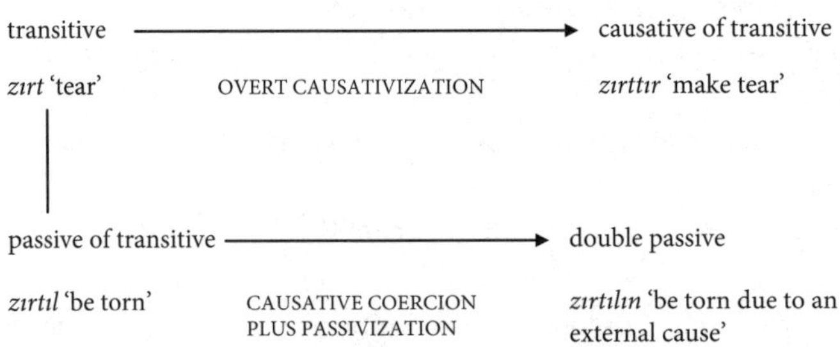

We will look at the difference between transitive verbs like *zırt* and their causatives like *zırttır* and make sure that the event structure of the latter contains an additional causing subevent, as expected. The presence of this subevent can

be detected due to additional scope possibilities it creates. Relevant scope taking operators will be the same as before: various types of adverbials and negation. Then we will see that the difference between the passive like *zırtıl* 'be torn' and the double passive *zırtılın* 'be torn due to an external cause' is exactly the same, and this will be our crucial argument for the claim that the derivation of the latter involves causativization through causative coercion.

Let us first look at adverbial modification data. In terms of the scope of adverbials, causatives of transitives pattern together with causatives of unergatives discussed above. Unlike their nonderived counterparts, causatives allow for multiple scope ambiguities, whereby adverbials modify causing and caused subevents independently from each other. Examples in (40b) and (41b) illustrate this for time-span and agent-oriented adverbials, respectively. Nonderived verbs in (40a) and (41a) are unambiguous.

(40) a. farida beš minut-xa xalı zırt-tı.
Farida.NOM five minute-DAT thread tear-PST.3SG
'Farida tore a thread in five minutes.'

b. alim beš minut-xa farida-ʁa xalı zırt-tır-dı.
Alim.NOM five minute-DAT Farida-DAT thread tear-CAUS-PST.3SG
1. 'Alim made Farida to tear a thread in five minutes.'
2. 'What Alim did in five minutes was make Farida tear a thread.'
3. 'What Alim did was make Farida tear a thread in five minutes.'

(41) a. farida lejla-ʁa acuuʁa xalı zırt-tı.
Farida.NOM Leila-DAT to.spite thread tear-PST.3SG
'Farida tore a thread in order to spite Leila.'

b. alim lejla-ʁa acuuʁa farida-ʁa xalı zırt-tır-dı.
Alim.NOM Leila-DAT to.spite Farida-DAT thread tear-CAUS-PST.3SG
1. 'Alim let Farida tear a thread to spite Leila (it was Farida who wanted to spite Leila).'
2. 'In order to spite Leila, Alim made Farida tear a thread (it was Alim who wanted to spite Leila).'

When it comes to the negation, one can observe that nonderived transitives are two-way ambiguous (cf. (35) above) whereas the range of interpretations of causatives is much wider.

(42) a. farida xalı zırt-ma-dı.
Farida.NOM thread tear-NEG-PST.3SG
'Farida did not tear a thread.'
1. Farida did not do anything.
2. Farida tried to tear a thread, but did not succeed.

b. alim farida-ʁa xalı zırt-tır-ma-dı.
Alim.NOM Farida-DAT thread tear-CAUS-NEG-PST.3SG
'Alim did not make Farida tear a thread.'
1. Neither Alim nor Farida did anything.
2. Alim failed to make Farida tear a thread.
3. Farida tried to tear a thread on Alim's order but did not succeed.
4. Alim made Farida not tear a thread.

On the widest scope reading, (42b.1), the sentence conveys that no subevent of the whole complex event has occurred. In (42b.2), the causing subevent escapes from the scope of negation, and Alim performs some causing activity that, however, does not bring about tearing of the thread. In (42b.3), both causing subevent and tearing activity occur, and what is negated is a change of state of the theme. One more reading is (42b.4), where the causal relation is established between Alim's activity and Fatima's not tearing the thread.

Observations from (40)–(42) are summarized in Table 2.2.

The literature can provide us with various ideas about where the range of interpretations we observe in (40)–(42) comes from. For instance, the scope of adverbials can be accounted for if one assumes an event-based semantic analysis of the causative in Pylkkänen's (2002) style coupled with the view that syntactically, the causative morpheme is an exponent of the *v* head (Folli & Harley 2007, Harley 2008, Travis 2010, Miyagawa 2012). If causatives of transitives involve a configuration with two *v*Ps (Lidz 2004, Harley 2008, Ramchand 2008), if semantic contribution of *v* is a causing subevent, and if time-span and rate adverbials are *v*P adjuncts, the pattern in (40) is easily explained:

(43) a. [… [$_{vP}$ Alim [*v* -tır] [$_{vP}$ in five minutes [$_{vP}$ farida tear a thread]]]]
b. [… [$_{vP}$ in five minutes [$_{vP}$ Alim [*v* -tır] [$_{vP}$ farida tear a thread]]]]

When the adverbial adjoins to the lower *v*P, (43a), the narrow scope reading in (40.3) obtains. If the adverbial merges at the higher *v*P level, (43b), two options are available. First, it can only modify a causing subevent introduced by *v* within the same *v*P, in which case the reading in (40.2) is derived. If, secondly, it scopes over the whole complex *v*P, including the lower one, one gets the reading in (40.1). See Lyutikova and Tatevosov (to appear) for a more detailed discussion.

TABLE 2.2
Transitives Vis-à-Vis Causatives of Transitives

	Transitive *zırt* 'tear'	**Causative of transitive** *zırt-tır* 'tear-CAUS'
Time-span and rate adverbials	Unambiguous	Three-way ambiguous
Agent-oriented adverbials	Unambiguous	Two-way ambiguous
Negation	Two-way ambiguous	Four-way ambiguous

TABLE 2.3

Passives Vis-à-Vis Double Passives

	Passive of transitive *zırt-ıl* 'tear-PASS'	**Double passive of transitive** *zırt-ıl-ın* 'tear-PASS-PASS'
Scope of time-span and rate adverbials	Unambiguous	Three-way ambiguous
Agent-oriented adverbials	Unambiguous	Two-way ambiguous
Scope of negation	Two-way ambiguous	Four-way ambiguous

However, in the case at hand we do not need to know why the scope of negation and scope of adverbials pattern the way they do in (40)–(42). We simply take them as a signature of causativization, and use them as a diagnostic when we need to make sure that a certain configuration has undergone the causative derivation. For instance, if we see that a structure based on a transitive verb stem is four-way ambiguous in the same way as the causative in (42b), we have every reason to suggest that its derivation involves causativization even if no causative morphology shows up on the surface. If this reasoning is correct, and if the recursive passivization in (39) does indeed involve causative coercion, the prediction is straightforward: double passives must differ from plain passives in exactly the same way as causatives in (40b)–(42b) differ from their nonderived counterparts in (40a)–(42a). The prediction is borne out, as the data in Table 2.3 show.

Table 2.3 summarizes differences between simple passives and double passives of transitives: these differences turn out to be identical to those between transitives and their causatives in Table 2.2. Like causatives, double passives offer more scope possibilities than simple passives. Let us see a few illustrations.

(44) kölek eki minut-xa kerim-ni küc-ü ble zırt-ıl-dı.
shirt.NOM two minute-DAT K.-GEN force-3 with tear-PASS-PST.3SG
'The shirt was torn by Kerim in two minutes.'

(44) illustrates a simple passive, where the time span adverbial is unambiguous: the tearing event occurs in two minutes, in exactly the same way as with its active counterpart in (40a). This is what happens to passives in languages like English, too. Now consider the double passive in (45):

(45) kölek eki minut-xa kerim-ni küc-ü ble zırt-ıl-ın-dı.
shirt.NOM two minute-DAT K.-GEN force-3 with tear-PASS-PASS-PST.3SG
1. 'The shirt was torn by Kerim on someone's order in two minutes.'
2. 'The shirt was torn by Kerim because someone convinced him in two minutes to do so.'
3. 'The shirt was torn in two minutes by Kerim, because someone convinced him to do so.'

In contrast with (44), (45) licenses three readings. On one reading, (45.1), the adverbial takes scope over both causing and tearing subevents. On the second reading, (45.2), the causing subevent only falls under the scope of the adverbial, and the sentence means that it took two minutes for someone to make Kerim tear his shirt. The temporal extent of the tearing event is not specified on this reading. Finally, the third reading, (45.3), is where the adverbial scopes over the caused event, and it is the tearing event that occupies two minutes, the duration of the causing event being left indeterminate. Therefore, time-span adverbials provide us with evidence that the event structure of the double passive is more complex than that of a regular passive. Moreover, the difference between (44) and (45) is identical to that between (40a) and (40b). If the derivation of the double passive involves causativization, this is what we expect.

Consider agent-oriented adverbials:

(46) kölek lejla-ʁa acuuʁa kerim-ni küc-ü ble zɪrt-**ɪl-ɪn**-dɪ.
shirt.NOM Leila-DAT to.spite K.-GEN force-3 with tear-**PASS-PASS**-PST.3SG

1. 'A/the shirt was torn by Kerim (on the order or by permission of some person X) to spite Leila.'
   a. It was Kerim who wanted to spite Leila.
   b. It was X who wanted to spite Leila.
2. 'A/the shirt was torn (by some person X) on Kerim's order (or by Kerim's permission) in order to spite Leila.'
   a. It was X who wanted to spite Leila.
   b. It was Kerim who wanted to spite Leila.

As we have seen in (6), the *by*-phrase 'by (the force of) Kerim' can refer to either the agent of a tearing event or to the external causer, the one who causes the agent to tear a shirt. No matter which option is realized, the adverbial can be anteceded by either of these two arguments, which creates four possible combinations, represented in Table 2.4:

For the native speakers, readings in (46.1a) and (46.2b), where the antecedent of the adverbial is identical to the referent of the *by*-phrase, are the first choice. However, the other two readings are by no means excluded: an appropriate scenario provided, these readings are not difficult for the speakers to process. What this means is: like the causative in (41b), the double passive in (46) always creates

TABLE 2.4
Possible Readings of (46)

	Antecedent of the adverbial is	
By-phrase refers to	Agent	Causer
Agent	(46.1a)	(46.1b)
Causer	(46.2a)	(46.2b)

two possible antecedents for a purpose adjunct, one of which is the agent of the tearing event, the other one being a participant who makes or lets him perform this event. Again, if the causative coercion occurs in between two applications of the passive, this result falls out for free.

Under negation, the double passive shows the same four interpretations as the corresponding causative in (42b):

(47) kölek kerim-ni küc-ü bile zırt-**ıl-ın**-dı.
shirt.NOM K.-GEN force-3 with tear-**PASS-PASS**-PST.3SG
'The shirt was not torn by Kerim on somebody's order.'
1. Neither the causer nor Kerim did anything.
2. The causer failed to make Kerim tear a thread.
3. Kerim tried to tear a thread on the causer's order but failed to do so.
4. The causer made Kerim not tear a thread.

As in (42b.1), in (47.1) the negation has the widest possible scope including the causing subevent and the whole tearing event. In (47.2), the occurrence of the tearing event is only negated. In (47.3), the scope of negation is narrower: it only includes the change of state of the shirt, cf. (42b.3). Finally, (47.4) corresponds to (42b.4), where the causer's action prevents the agent from tearing a shirt.

4.2. DOUBLE PASSIVE OF INSTRANSITIVES

If the above line of reasoning is correct, and the second instance of the passive morpheme induces causative coercion when applies to the passive of a transitive verb, our predictions about the double passive of intransitives is straightforward. The relevant example is repeated in (48):

(48) alim (kerim-ni küc-ü ble) sekir-**il-in**-di.
A.NOM K.-GEN force-3 with jump-**PASS-PASS**-PST.3SG
1. 'Alim jumped (because something was done by Kerim on someone's order).'
2. 'Alim jumped (because something was done by someone on Kerim's order).'

(48) shows the type of ambiguity characteristic of a configuration that involves three subevents, Alim's jumping, X's making Alim jump, and Y's making X make Alim jump. Depending on whether Kerim, the causer introduced by the *by*-phrase, is identified with X or with Y, two readings of (48) obtain.

As we know from section 3, the passive morpheme, when combined with intransitives, induces causative coercion, (22), repeated as (49):

(49) $\textit{-l-}_{PASS}\,(\mathrm{V}_{intrans}) \rightarrow \textit{-l-}_{PASS}\,(\mathrm{CAUS}(\mathrm{V}_{intrans}))$

In (49), covert causativization creates a transitive structure $[\mathrm{CAUS}(\mathrm{V}_{intrans})]_{trans}$ required by the passive. Merging the passive with $[\mathrm{CAUS}(\mathrm{V}_{intrans})]_{trans}$ produces a configuration which is intransitive again: [-*l*-PASS ([CAUS($V_{intrans}$)]$_{trans}$)] $_{intrans}$. What happens when a second instance of the passive appears in the derivation, as in (48), is now clear: like the first instance, it induces causative coercion of the derived intransitive:

(50) -*l*-PASS (-*l*-PASS (CAUS ($V_{intrans}$))) → -*l*-PASS (CAUS (-*l*-PASS (CAUS($V_{intrans}$))))

The same logic as above tells us that since the double passive of an intransitive involves two instances of covert causativization, (48) should pattern event-structurally with the double causative of the same verb. The double causative counterpart of (48) is illustrated in (51):

(51) marat kerim-ge alim-ni sekir-**t**-**tir**-di.
M.NOM K.-DAT A-ACC jump-**CAUS**-**CAUS**-PST.3SG
'Marat ordered Kerim to make Alim jump.'

The expected parallelism of (48) and (51) should manifest itself in the whole range of phenomena sensitive to the event structure of a complex predicate (scope of adverbials, scope of negation, etc.), and this is indeed the case. For the sake of space, we limit ourselves to showing this for the rate adverbial 'quickly' in (52); other diagnostics work in a similar way.

(52) marat kerim-ge alim-ni terk sekir-**t**-**tir**-di.
M.NOM K.-DAT A-ACC quickly jump-**CAUS**-**CAUS**-PST.3SG
1. 'Marat ordered Kerim to make Alim jump quickly.'
2. 'Marat ordered Kerim to quickly make Alim jump.'
3. 'Marat quickly ordered Kerim to make Alim jump.'

(53) alim kerim-ni küc-ü ble terk sekir-**il**-**in**-di.
A.NOM K.-GEN force-3 with quickly jump-**PASS**-**PASS**-PST.3SG
1. 'Alim quickly jumped because something was done by Kerim (on someone's order).'
2. 'Alim jumped because something was quickly done by Kerim (on someone's order).'
3. ? 'Alim jumped because something was done by Kerim (on someone's quick order).'
4. 'Alim quickly jumped (because something was done by someone) on Kerim's order.'
5. ?'Alim jumped (because something was quickly done by someone) on Kerim's order.'
6. 'Alim jumped (because something was done by someone) on Kerim's quick order.'

(52) illustrates three readings of the rate adverbial *terk* 'quickly' where it modifies three subevents of the whole complex event. In (52.1), it is Alim's jumping that happens quickly; on this reading, Kerim making him jump and Marat giving an order are underspecified as to how quickly they occur. On the second reading, (52.2), the sentence indicates that it did not take much time for Kerim to make Alim jump, and the adverbial takes scope of the causing subevent in which Kerim is a causer. What happens quickly in (53.3) is Marat's delivering an order: here the scope of the adverbial includes the most external causing subevent.

(53) is strictly parallel, as expected. Since (11b), on which (52) is based, is ambiguous as to whether the *by*-phrase refers to an immediate or remote causer of Alim's jumping, we get six possible readings, not three. But crucially, the scopal possibilities available for 'quickly' are exactly the same as in (52). 'Quickly' takes scope either over the jumping subevent, (53.1) and (53.4), the subevent of making Alim jump, (53.2) and (53.5), or the subevent of giving an order, (53.3) and (53.6).

For some speakers it may be somewhat difficult to get readings where the adverbial modifies a subevent not associated with an overtly realized participant. These are (53.3) and (53.5): in (53.3), the *by*-phrase refers to an immediate causer, but the adverbial modifies the remote causing subevent, and in (53.5) things are the other way around. However, taken together, examples in (53) suggest clearly that all the three subevents are 'visible' for the adverbial and can be modified independently from each other. This is exactly what happens with the double causative in (52), too, hence (52) and (53) provide us with significant evidence that event-structurally, double causatives and double passives of intransitives are identical. Given our coercion analysis, this is again what we predict: as (50) indicates, double passives pattern with double causatives because *they are* double causatives, even if both instances of causativization occur covertly.

The coercion analysis makes one more (less immediately obvious) prediction about the structure of the double passive. It requires the order of covert causativization and overt passivization operations be exactly as in (50), and not, say, as in (54):

(54) *-l-*$_{PASS}$ (*-l-*$_{PASS}$ ($V_{intrans}$))) → *-l-*$_{PASS}$ (*-l-*$_{PASS}$ (CAUS (CAUS($V_{intrans}$))))

In (54), both instances of causative coercion precede the application of both instances of the passive morpheme. However, there is a reason to believe that (54) cannot be a correct analysis of the double passive of intransitive, since (54) is clearly incompatible with the coercion view of the phenomenon in question.

What is problematic about (54) is the second application of CAUS. At the first stage of derivation, where the first instance of the passive morpheme merges, the need of the passive to meet a transitive verb is satisfied by exactly one instance of CAUS. There is no trigger for the second CAUS at this stage, which means, by definition of coercion, that it cannot occur. The right environment for the second CAUS is only created after the second application of the passive, which means that there is no way for CAUS to be on top of another CAUS. This necessitates an analysis along the lines of (50), where each application of the passive is mediated by exactly one instance of causative coercion. Assuming something like (54) would inevitably lead us to abandoning the coercion analysis altogether. We doubt, however, that this is what one may want: this analysis, as we have argued above, gives us a number of empirical advantages, predicts correctly a wide range of data, and requires the minimal amount of stipulations.

Now we are in the position of summarizing our observations and generalizations. In all relevant respects, as evidence from the scope of adverbials and negation suggests, the double passive of a transitive verb differs from the single passive in the same way as the causative differs from a nonderived transitive in the active voice. If the derivation of the double passive involves covert causativization, this identity falls out naturally. The double passive possesses an additional causing subevent visible for adverbials and an implicit participant of this subevent that can be identified by the *by*-phrase. On the proposed analysis, this follows from the fact that the simple passive is essentially intransitive; hence the second occurrence of the passive morpheme induces causative coercion. As we know from section 3, exactly the same mechanism is at work if intransitives get passivized.[9]

The double passive of intransitives involves two instances of cover causativization (one per each occurrence of the passive morpheme), and this type of structure is correctly predicted to pattern together with double causatives.

Now that we have established an argument that recursive passivization involves causative coercion, we can complete the analysis by developing semantic derivations for double passives and other configurations we have discussed so far. The purpose of the next section is to argue that the causative coercion triggered by the passive is essentially a way of repairing the logical type mismatch between the passive morpheme and its intransitive complement.

5. Derivations

In this final section, we will present a model-theoretic fragment that makes the analysis presented above fully explicit. The two essential components of the

---

[9]The anonymous reviewer asks if the passive of ditransitives exhibit any peculiarities relevant to the present discussion (in view of the behavior of ditransitives in English, where two options are available: *Peter was given the book, The book was given to Peter)*. Karachay-Balkar is a language that lacks the dative alternation altogether, and it is always the theme, not the recipient that appears in the subject position in the passive:

(i) kitap kerim-ge ber-il-di
book K.-DAT give-PASS-PST.3SG
'The book was given to Kerim.'

(ii) *kerim kitap ber-il-di.
K. book give-PASS-PST.3SG
'Kerim was given the book.'

Further passivilzation of (i) derives a clause identical to the double passive of transitives, already discussed, in all relevant respects.

analysis, semantics for the coercion operator and for the passive, are represented in (40) and (41):[10]

(55) Causative coercion:
$||\text{CAUS}|| = \lambda P_{<s,t>}.\lambda x.\lambda e.\exists e'[P(e') \wedge agent(x)(e) \wedge cause(e')(e)]$

(56) Passive:
$||\ \textit{-l-}\ || = \lambda R_{<e,<s,t>>}.\lambda e.\exists x[R(x)(e)]$

Both operators do exactly what the causative and passive operators are supposed to do. The causative coercion operator takes a predicate of events and returns a two-place relation between individuals and events. It existentially binds an event from the original extension of a predicate and introduces a causing event as well as a participant of that event.[11] The passive morpheme takes a relation between individuals and events and returns an event predicate, existentially binding the individual variable. Crucially, as we will see in a moment, the outcome of causative coercion is always a suitable input for passivization.

The simplest case is the passive of transitives repeated in (57) that does not require any coercion: the verb phrase is of logical type <e, <s,t>>, a relation between individuals and events (or, equivalently, a function from individuals to functions from events to truth-values), which serves as an argument of the passive morpheme:

(57) kölek zırt-ıl-dı.
shirt.NOM tear-PASS-PST.3SG
'The shirt was torn.'

(58)
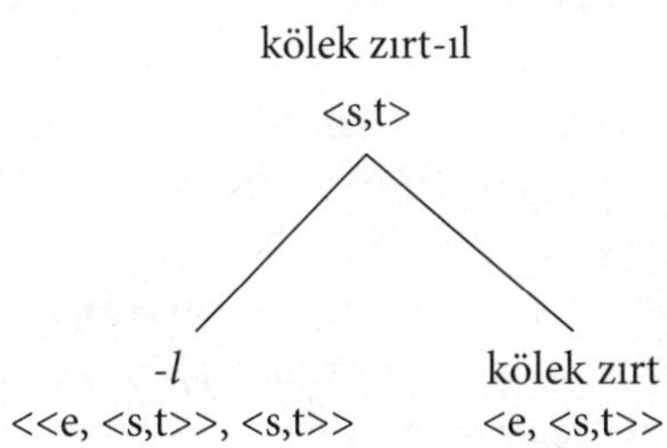

---

[10]In what follows, we assume an extensional λ-calculus with basic types e (individuals), s (events) and t (truth values). Complex types are recursively defined in the standard manner. The only rule of semantic composition used below is Functional Application. We focus on the derivation of complex event predicates where all individual argument positions are saturated; for simplicity DPs are taken to denote individual constants. Irrelevant details about the unaccusative/unergative distinction are ignored. For the sake of space, we leave integration of adverbials, *by*-phrases and negation into the picture for another occasion. We believe that this simplification does not affect the argument we are establishing throughout the paper.

[11]We assume with Pylkkänen 2002, Ramchand 2008, and many others that the causee is not an argument of the causative. See Alsina 1992 for the alternative view.

The relevant part of the derivation of (57) is shown in (59).

(59) a. **Transitive verb:**
$|| \text{zırt} || = \lambda y.\lambda x.\lambda e.\exists e'[\text{tear}(y)(e') \wedge \text{agent}(x)(e) \wedge \text{cause}(e')(e)]$

b. **Saturation:**
$|| \text{kölek zırt} || = \lambda y.\lambda x.\lambda e.\exists e'[\text{tear}(y)(e') \wedge \text{agent}(x)(e) \wedge \text{cause}(e')(e)](\text{shirt})$
$= \lambda x.\lambda e.\exists e'[\text{tear}(\text{shirt})(e') \wedge \text{agent}(x)(e) \wedge \text{cause}(e')(e)]$

c. **Passive:**
$|| \textit{-l-} || = \lambda R.\lambda e.\exists x[R(x)(e)]$

d. **Passivization:**
$|| \textit{-l-} \text{ [kölek zırt]} || = \lambda R.\lambda e.\exists x[R(x)(e)](\lambda x.\lambda e \exists e'[\text{tear}(\text{shirt})(e') \wedge \text{agent}(x)(e) \wedge \text{cause}(e')(e)]) = \lambda e.\exists x \exists e'[\text{tear}(\text{shirt})(e') \wedge \text{agent}(x)(e) \wedge \text{cause}(e')(e)]$

The denotation of the transitive verb in (59a) is a three-place relation between two individuals and events. After saturation of the internal argument position in (59b), the passive, repeated in (59c), applies. The outcome is an event predicate in (59d) that denotes tearing events in which the shirt is torn. These events are brought about by some individual with the agent properties.

Let us consider a more complex case, the passive of intransitives discussed in section 3; an example of this type of the passive derived from an unaccusative verb *sın* 'break' is repeated in (60) (the account for unergatives would be similar, possibly with minor technical adjustments).

(60) illew (alim-ni küc-ü ble) sın-ıl-dı.
toy.NOM Alim-GEN force-3 with break-PASS-PST.3SG
'The toy was broken (by Alim.)'

Unlike in (57), in (60) the causative coercion is necessary to obtain an interpretable configuration. Indeed, the passive morpheme in (56) requires a relation between individuals and events to apply to. But the intransitive predicate fails to provide the passive with a suitable input: as soon as its single argument position is saturated we get an event predicate, and a type mismatch results, as shown in (61a). This mismatch is repaired if the event predicate is coerced by means of covert causativization into a relation between individuals and events, as required. This is shown in (61b).

(61) a. b. illew sın-ıl

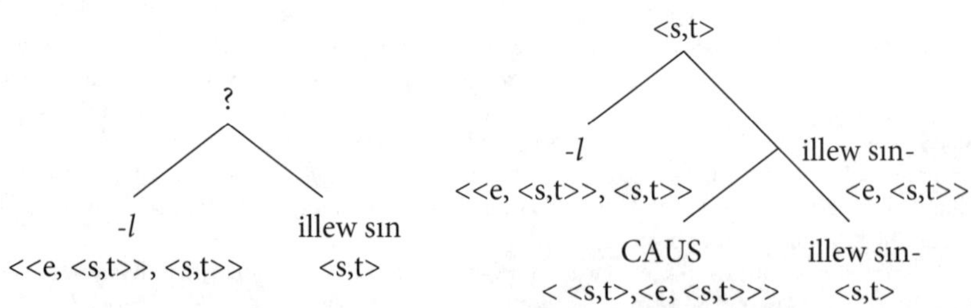

(62) presents the derivation of (60).

(62) a. **Unaccusative verb:**
|| sın || = $\lambda x.\lambda e.break(x)(e)$

b. **Saturation:**
|| illew sın || = $\lambda e.break(toy)(e)$

c. **Coercion operator:**
|| CAUS|| = $\lambda P.\lambda x.\lambda e.\exists e'[P(e') \wedge agent(x)(e) \wedge cause(e')(e)]$

d. **Coercion:**
|| CAUS [illew sın] || = $\lambda x.\lambda e.\exists e'[break(toy)(e') \wedge agent(x)(e) \wedge cause(e')(e)]$

e. **Passive:**
|| *-l-* || = $\lambda R.\lambda e.\exists x[R(x)(e)]$

f. **Passivization:**
||*-l-* [CAUS [illew sın]] || = $\lambda e.\exists x\exists e'[break(toy)(e') \wedge agent(x)(e) \wedge cause(e')(e)]$

The most significant step of the derivation is (62d): application of the coercion operator to the event predicate in (62b) creates a relation between causing events and agent individuals in (62d). This relation is a legitimate input for the passive operator in (62e). The output of passivization is shown in (62f): it is a predicate that denotes events that bring about a change of state of the toy. The individual variable ranging over agent individuals is existentially bound.

Now the final component of the proposal comes: semantic derivation of recursive passivization. We are going to see that a well-formed semantic configuration can only be derived through the causative coercion. Essentially, we will show that this type of passivization can be reduced to the derivation in (62).

Recall that the double passives of transitives involve an extra causing event and its agentive participant, as in (63).

(63) kölek (kerim-ni küc-ü ble) zırt-ıl-ın-dı.
shirt.NOM K.-GEN force-3 with tear-PASS-PASS-PST.3SG
'The shirt was torn by Kerim (on someone's order).'

The informal idea we have discussed in section 1 is that the first instance of the passive morpheme yields an intransitive configuration, which must be coerced to be combined with the second one. Specifically, after the first application of the passive morpheme we get a predicate of events, similar to the nonderived one in (62b), and the causative coercion is called for the same reason as in (62d), that is, to resolve a type mismatch:

(64) a. b.

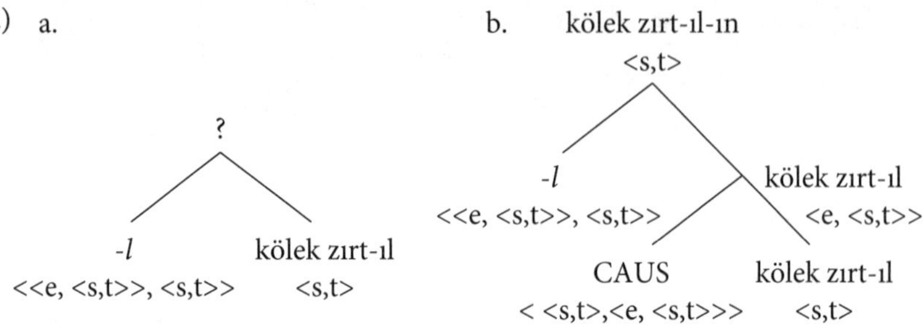

In (64a), the result of combining the transitive verb 'tear' with the passive is of type <s,t>. This has been shown in (59d) above. As a consequence, this predicate cannot be merged with the second instance of the passive morpheme (of type <<e, <s,t>>, <s,t>>) due to the familiar type mismatch. To resolve the mismatch, CAUS applies in (64b) in precisely the same way as in case of lexical intransitives in (60)–(62). After the causative coercion, the second instance of the passive morpheme can be merged, yielding an interpretable configuration.

The derivation of the double passive is shown in (65).

(65) a. **Passive (=59d):**
$||\text{-}l\text{- [kölek zırt]}|| = \lambda e.\exists y \exists e'[\text{tear(shirt)}(e') \wedge \text{agent}(y)(e) \wedge \text{cause}(e')(e)]$

b. **Coercion operator:**
$||\text{CAUS}|| = \lambda P.\lambda x.\lambda e.\exists e'[P(e') \wedge \text{agent}(x)(e) \wedge \text{cause}(e')(e)]$

c. **Coercion:**
$||\text{CAUS (kölek zırt-ıl)}|| = \lambda P.\lambda x.\lambda e.\exists e'[P(e') \wedge \text{agent}(x)(e) \wedge \text{cause}(e')(e)](\lambda e.\exists y \exists e'[\text{tear(shirt)}(e') \wedge \text{agent}(y)(e) \wedge \text{cause}(e')(e)]) = \lambda x.\lambda e.\exists y \exists e' \exists e''\ [\text{tear(shirt)}(e'') \wedge \text{agent}(y)(e') \wedge \text{cause}(e'')(e') \wedge \text{agent}(x)(e) \wedge \text{cause}(e')(e)]$

d. **Passive:**
$||\ \text{-}l\text{-}\ || = \lambda R.\lambda.e \exists x[R(x)(e)]$

e. **Passivization:**
$||\text{-}l\text{- (CAUS (kölek zırt-ıl))}|| = \lambda e.\exists x \exists y \exists e' \exists e''\ [\text{tear(shirt)}(e'') \wedge \text{agent}(y)(e') \wedge \text{cause}(e'')(e') \wedge \text{agent}(x)(e) \wedge \text{cause}(e')(e)]$

The product of causative coercion is a relation between causing events and individuals in (65c). This relation combines with the passive morpheme creating an event predicate in (65e). This predicate involves a complex event structure consisting of three subevents, with one participant each. One of the subevents is a change of state where the shirt gets torn. The second subevent is a causing subevent, with a corresponding individual variable existentially bound. The third subevent is the one that causes the second subevent; its individual participant receives an existential interpretation, too. It is not difficult to check that (65e) is exactly what the sentence in (48) means.

The derivation of double passive of intransitives in (6) is straightforward. Up to the stage where the first occurrence of the passive morpheme is merged, the derivation proceeds exactly as shown in (62). Two extra steps are the second application of the causative coercion operator followed by the second occurrence of the passive. This results in an event predicate in (67).

(66) alim (kerim-ni küc-ü ble) sekir-**il**-**in**-di.
A.NOM K.-GEN force-3 with jump-**PASS**-**PASS**-PST.3SG
'Alim jumped (because something was done by Kerim on someone's order).'

(67) $\|$-*l*- [CAUS [-*l*- [CAUS [alim sekir]]] $\| = \lambda e.\exists x \exists y \exists e' \exists e''$[jump(alim)(e'') $\wedge$ agent(x)(e') $\wedge$ cause(e'')(e') $\wedge$ agent(y)(e) $\wedge$ cause(e')(e)]

This completes the analysis of recursive passivization. In the concluding section, we will summarize main results of this study and its implications for a theory accounting for multiple occurrences of the same piece of derivational morphology.

6. Conclusion

We started this paper by asking the following question: what constrains recursive application of the same morphosyntactic operation, which results in multiple occurrences of the same piece of morphology? If one assumes a theory where the main tool of semantic composition is functional application (Heim & Kratzer 1998, von Fintel & Heim 2010), and if grammatical morphemes are functions that take their complements as arguments, an obvious expectation is that we only find true recursion if an operation *O* is essentially a modification—that is, a function of logical type $<\sigma, \sigma>$, where σ is a logical type. Whenever the type of the input is not identical to the type of the output, something else should happen for the operation to apply more than once. Specifically, if *O* is of type $<<\sigma,\tau>, \tau>$, the configuration of the form (68) is predicted to be uninterpretable, for the obvious reason (see (69)): whenever β is semantically well-formed, it would be of type τ, not of type $<\sigma,\tau>$ required by *O*. α, then, would fail to have a denotation.

(68) $[_{\alpha}$ M(O) $[_{\beta}$ M(O) $[_{\gamma} \ldots]\,]\,]$, where M(O) is a morphological exponent of O

(69)

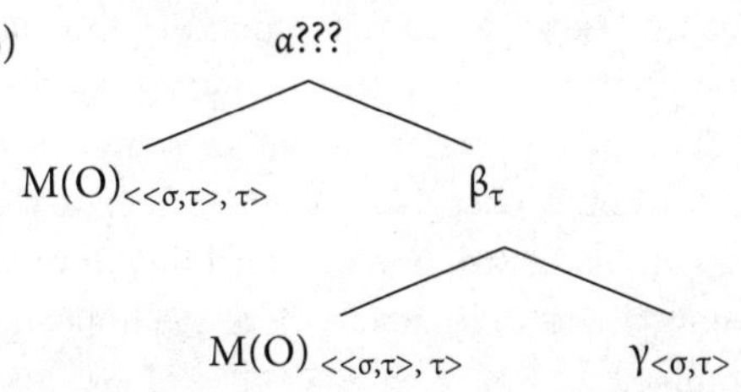

It is not difficult to see what we need to repair (69): we have to turn β into an expression of type $<\sigma,\tau>$. One way of doing so is to mediate two applications of *O* by a function of type $<\tau, <\sigma,\tau>>$:

(70) $[_{\alpha}\ M(O)\ [_{\beta'}\ f_{<\tau,<\sigma,\tau>>}\ [_{\beta}\ M(O)\ [_{\gamma} \ldots]\ ]$

We can think of different logical types in place of σ and τ, and of different means of morphosyntactic realization of the *f* function. In this paper we have examined one specific case showing that the scenario outlined in (68)–(70) is empirically well-motivated.

In this case, *O* is a passivization operation of type $<<e,<s,t>>,<s,t>>$ (so that *e* corresponds to σ in (68)-(70), and $<s,t>$ corresponds to τ). The *f* function is thus of type $<<s,t>,<e,<s,t>>>$, the one the causative morpheme has in event-based systems like Pylkkänen 2002.

We have seen that there are good reasons to believe that causativization is indeed called for to mediate between two applications of the passive. Evidence from the scope of adverbials and negation suggests clearly that in the course of derivation of double passives a new causing subevent and its individual participant are introduced to the event structure. The distribution of the passive combined with intransitives points toward the same conclusion: since intransitives fail to provide a suitable input to passivization in the first place, they need to be causativized to be able to merge with the passive morpheme. Causativization, which maps event predicates of type $<s,t>$ to relations between events and individuals of type $<e,<s,t>>$, is invoked whenever the type mismatch occurs between the verbal predicate and the passive morpheme. It enables the derivation of what superficially looks like a passivization of intransitive predicates and predicates that are passive already.

The only peculiarity of Karachay-Balkar is therefore as follows: causativization that feeds passivization can occur covertly, as a coercion operation. This type of causativization has indeed all the hallmarks of coercion; crucially, it only occurs whenever the verbal predicate fails to meet the type requirement of the passive morpheme. It is not available otherwise: we do not find, say, lexically intransitive verbs that acquire new arguments through the free application of covert causativization.

Our narrative about causative coercion comes to the end. We believe we managed to make a case for causative coercion that enables the passive morpheme to apply more than once in languages like Karachay-Balkar. Much more work still has to be done, though. One obvious question we did not address in this paper is what makes languages like Karachay-Balkar different from languages like English where the causative coercion cannot repair ungrammaticality of sentences like *John was was killeded*. For a complete and comprehensive answer we need to know much more about parameters of cross-linguistic variation in availability of various coercion operators, as well as about constraints on that variation. We hope, however, that this task will be soon accomplished.

References

Alsina, Alex. 1992. "On the Argument Structure of Causatives." *Linguistic Inquiry* 23: 517–555.

Anick, Peter, and Sabine Bergler. 1992. "Lexical Structures for Linguistic Inference." In *Lexical Semantics and Knowledge Representation*, edited by James Pustejovsky and Sabine Bergler, 121–135. Berlin: Springer Verlag.

Aoyagi, Hiroshi. 2007. "Toward a Unified Account of Morphological Causatives and Passives in Korean." In *Locality and Minimalism: Proceedings of the 9th Seoul International Conference on Generative Grammar*, edited by Doo-Won Lee, 9–27. Hankuk Publishing.

Baker, Mark. 1985. "The Mirror Principle and Morphosyntactic Explanation." *Linguistic Inquiry* 16: 373–415.

Baker, Mark, Kyle Johnson, and Ian Roberts. 1989. "Passive Argument Raised." *Linguistic Inquiry* 20: 219–251.

Bybee, Joan L. 1985. *Morphology: A Study of the Relation Between Meaning and Form*. Amsterdam: John Benjamins.

Chomsky, Noam. 1981. *Lectures on Government and Binding*. Berlin: Mouton de Gruyter.

Collins, Chris. 2005. "A Smuggling Approach to the Passive in English." *Linguistic Inquiry* 36: 289–297.

Coulson, Seana, and Gilles Fauconnier. 1999. "Fake Guns and Stone Lions: Conceptual Blending and Privative Adjectives." In *Cognition and Function in Language*, edited by Barbara Fox, Dan Jurafsky, and Laura Michaelis, 143–158. Stanford: *Center for the Study of Language and Information*.

Cruse, D. Aalan. 1972. "A Note on English Causatives." *Linguistic Inquiry* 3: 520–528.

De Swart, Henriette. 1998. "Aspect Shift and Coercion." *Natural Language and Linguistic Theory* 16: 347–385.

Deal, Amy R. 2007. "The Asymmetry of Argument Structure: A View from Coercion." Unpublished Ms., University of Massachusetts, Amherst. Accessed in May 2011, Portable Document Format.

den Dikken, Marcel. 2003. "Lexical Integrity, Checking, and the Mirror: A Checking Approach to Syntactic Word Formation." *Journal of Comparative Germanic Linguistics* 6: 169–225.

Dowty, David R. 1979. *Word Meaning and Montague Grammar*. Dordrecht: Reidel.

Enç, Murvet. 2003. "Functional Categories in Turkish." Paper presented at the First Workshop on Altaic Formal Linguistics, May 2003, MIT.

Fodor, Jerry A. 1970. "Three Reasons for Not Deriving 'Kill' from 'Cause to Die.'" *Linguistic Inquiry* 1: 429–438.

Folli, Rafaella, and Heidi Harley. 2007. "Causation, Obligation and Argument Structure: On the Nature of Little *v*." *Linguistic Inquiry* 38: 197–238.

Gehrke, Berit, and Nino Grillo. 2009. "How to Become Passive." In *Explorations of Phase Theory: Features, Arguments, and Interpretation at the Interfaces*, edited by Kleanthes Grohmann, 231–268. Berlin: Mouton de Gruyter.

Göksel, Aslı. 1993. "*Levels of Representation and Argument Structure in Turkish*." PhD diss., University of London.

Göksel, Aslı, and Celia Kerslake. 2011. *Turkish, An Essential Grammar*. London: Routledge.

Grønn, Atle. 2003. "*The Semantics and Pragmatics of the Russian Factual Imperfective*." PhD diss., University of Oslo.

Haiman, John. 1983. "Iconic and Economic Motivation." *Language* 59: 781–819.
Harley, Heidi. 2008. "On the Causative Construction." In *Oxford Handbook of Japanese Linguistics*, edited by Shigeru Miyagawa and Mamoru Saito, 20–53. Oxford: Oxford University Press.
Haspelmath, Martin. 1990. "The Grammaticization of Passive Morphology." *Studies in Language* 14: 25–71.
Heim, Irene, and Angelika Kratzer. 1998. *Semantics in Generative Grammar*. Oxford: Blackwell.
Jackendoff, Ray. 1997. *The Architecture of the Language Faculty*. Cambridge, MA: MIT Press.
Jaeggli, Osvaldo A. 1986. "Passive." *Linguistic Inquiry* 17: 587–622.
Godard, Danièle, and Jacques Jayez. 1993. "Towards a Proper Treatment of Coercion Phenomena." In *Proceedings of the Sixth European ACL Conference*, 168–177. Somerset: Association for Computational Linguistics.
Kamp, Hans, and Barbara Partee. 1995. "Prototype Theory and Compositionality." *Cognition* 57: 129–191.
Keenan, Edward. 1985. "Passive in the World's Languages." In *Language Typology and Syntactic Description* Vol. 1, edited by Timothy Shopen, 243–281. Cambridge: Cambridge University Press.
Kim, Hee-Soo, and Acrision Pires. 2003. "Ambiguity in the Korean Morphological Causative/Passive." In *Japanese/ Korean Linguistics 12*, edited by William McClure, 255–266. Stanford: CSLI Publications.
Koontz-Garboden, Andrew. 2007. "*Aspectual Coercion and the Typology of Change of State Predicates*." PhD diss., Stanford University.
Kratzer, Angelika. 2005. "Building Resultatives." In *Event Arguments: Foundations and Applications*, edited by Claudia Maienborn, and Angelika Wöllstein, 177–212. Tbingen: Niemeyer.
Lakoff, George. 1965. "*On the Nature of Syntactic Irregularity*." PhD diss., Indiana University.
Lakoff, George. 1971. "On Generative Semantics." In *Semantics. An Interdisciplinary Reader in Philosophy, Linguistics, and Psychology*, edited by Danny Steinberg and Leon A. Jakobovits, 232–252, edited by Danny Steinberg and Leon A. Jakobovits. Cambridge: Cambridge University Press.
Larson, Richard. 1998. "On the Double Object Construction." *Linguistic Inquiry* 19: 335–391.
Levinson, Stephen. 1983. *Pragmatics*. Cambridge: Cambridge University Press.
Lidz, Jeffrey. 2004. "Causation and Reflexivity in Kannada." In *Clause Structure in South Asian Languages*, edited by Veneeta Dayal and Anoop Mahajan, 93–130. Dordrecht: Kluwer.
Link, Godehard. 1983. "The Logical Analysis of Plurals and Mass Terms." In *Meaning, Use, and Interpretation of Language*, edited by Rainer Bauerle, Christoph Schwarze, and Arnim von Stechow, 302–323. Berlin: De Gruyter.
Lyutikova, Ekaterina, and Anastasia Bonch-Osmolovskaya.2006. "A very active passive". In *Case, Valency and Transitivity*, edited by Leonid Kulikov, Andrej Malchukov, and Peter de Swart, 393–416. Amsterdam: John Benjamins.
Lyutikova, Ekaterina, and Sergei Tatevosov (to appear). "Causativization and Event Structure." In *Causation in Grammatical Structures*, edited by Bridget Copley and Fabienne Martin. Oxford: Oxford University Press.
Malchukov, Andrey. 2011. "Interaction of Verbal Categories: Resolution of Infelicitous Grammeme Combinations." *Linguistics* 49: 229–282.

McCawley, James D. 1971. "Pre-lexical Syntax." In *Report Of The Twenty-Second Annual Round Table Meeting on Linguistics and Language Studies. Linguistics: Developments of the Sixties—Viewpoints for the Seventies*, edited by Richard J. O'Brien, 19–34. Washington, DC: Georgetown University Press.

McCawley, James D. 1972. *Syntactic and Logical Arguments for Semantic Structures*. Bloomington: Indiana University Linguistic Society Club.

McIntyre, Andrew. 2011. "English Get-Passives, Middle Voice and Causative-Passive Ambiguities." Paper presented at the Workshop on Non-Canonical Passives, Annual Meeting of Deutsche Gesellschaft für Sprachwissenschaft, University of Göttingen.

Michaelis, Laura. 2004. "Type Shifting in Construction Grammar: An Integrated Approach to Aspectual Coercion." *Cognitive Linguistics* 15: 1–67.

Michaelis, Laura. 2006. "Construction Grammar." In *The Encyclopedia of Language and Linguistics*, Second edition, volume 3, edited by Keith Brown, 73–84. Oxford: Elsevier.

Michaelis, Laura, and Knud Lambrecht. 1996. "Toward a Construction-Based Model of Language Function: The Case of Nominal Extraposition." *Language* 72: 215–247.

Miyagawa, Shigeru. 2012. *Case, Argument Structure, and Word Order*. New York: Routledge.

Moens, Marc, and Mark Steedman. 1988. "Temporal Ontology and Temporal Reference." *Computational Linguistics* 14: 15–28.

Partee, Barbara H. 2010. "Privative Adjectives: Subsective Plus Coercion." In *Presuppositions and Discourse: Essays offered to Hans Kamp*, edited by Rainer Bauerle, Uwe Reyle, and Thomas E. Zimmermann, 273–285. Bingley: Emerald Group Publishing.

Pollard, Carl, and Ivan Sag. 1994. *Head-Driven Phrase Structure Grammar*. Chicago: University of Chicago Press.

Pustejovsky, James. 1991. "The Generative Lexicon." *Computational Linguistics* 17: 409–441.

Pustejovsky, James. 1995. *The Generative Lexicon: A Theory of Computational Lexical Semantics*. Cambridge, MA: MIT Press.

Pustejovsky, James, and Pierrette Bouillon. 1995. "Aspectual Coercion and Logical Polysemy." *Journal of Semantics* 12: 133–162.

Pylkkänen, Liina. 2002. "*Introducing Arguments*." PhD diss., MIT.

Ramchand, Gillian. 2008. *Verb Meaning and the Lexicon: A First Phase Syntax*. Cambridge: Cambridge University Press.

Ritter, Elizabeth, and Sara Thomas Rosen. 1993. "Deriving Causation." *Natural Language and Linguistic Theory* 11: 519–555.

Sag, Ivan, and Carl Pollard. 1991. "An Integrated Theory of Complement Control." *Language* 67: 63–113.

Santorini, Beatrice, and Caroline Heycock. 1988. "Remarks On Causatives And Passive." Technical report. Department of Computer and Information Science, School of Engineering and Applied Science, University of Pennsylvania.

Shibatani, Masayoshi, ed. 1976. *The Grammar of Causative Constructions*. New York: Academic Press.

Shibatani, Masayoshi. 1973. "Semantics of Japanese Causatives." *Foundations of Language* 9: 327–373.

Travis, Lisa. 2010. *Inner Aspect: The Articulation of VP*. Dordrecht: Springer.

Tubino Blanco, Mercedes. 2011. *Causatives in Minimalism*. Amsterdam: John Benjamins.

Vinokurova, Nadja. 2005. "*Lexical Categories and Argument Structure: A Study with Reference to Sakha*." PhD diss., University of Utrecht.

Von Fintel, Kai, and Irene Heim. 2010. "*Intensional Semantics*." MIT lecture notes.

Washio, Ryuichi. 1993. "When Causatives Mean Passive: A Cross-Linguistic Perspective." *Journal of East Asian Linguistics* 2: 45–90.

Yang, In-Seok. 1976. "Semantics of Korean Causation." *Foundations of Language* 14: 55–87.

Zucchi, Alessandro. 1998. "Aspect Shift." In *Events and Grammar*, edited by Susan Rothstein, 349–370. Dordrecht: Kluwer.

3

Scope Versus Ordering of Operations: Causativization and Ordering of Valency-Changing Operations in Adyghe

Alexander Letuchiy

1. Introduction

This paper tackles the interaction of valency-changing operations in Adyghe, a polysynthetic language of the Caucasus region.[1] In what follows, I will consider the problem of the order of morphosyntactic operations in Adyghe—namely, the peculiar properties of the causative morpheme. I will show that even in cases where different orderings of the operations would yield the same semantics of the derived verb, morphological tests show that causativization usually occurs after any other valency-changing operation.

Another remarkable feature of Adyghe examined throughout the article is the mismatch between the ordering of morphosyntactic operations and the ordering of the affixes themselves in the verb form. The causative prefix has a fixed position after all other prefixes; in other words, it immediately precedes the verb root. The Mirror Principle formulated by Baker (1985) and discussed in 4.1 below says that roughly speaking, the closer the marker is to the stem, the earlier the derivation occurs with respect to other operations. Given this position, it is unexpected that causativization takes place after, and not before, all other operations.

There are two important theoretical questions that arise as soon as valency change in Adyghe is considered. First, can the relative ordering of operations be tested on any reliable ground? I will show that this is possible for some combinations of the causativization and other operations, but impossible for other ones.

[1]Adyghe belongs to the North-West Caucasian language family. Native speakers of Adyghe live in Russian Federation, as well as in Turkey and some other countries. The language can be considered to be well-described, including grammars and other types of descriptions, such as a grammar by Rogava and Kerasheva (1966), a grammatical sketch of the Abzakh dialect by Paris (1989), descriptions of Adyghe morphology by Kumakhov (1971), and of Adyghe morphophonology by Smeets (1996), the recent collection of papers on different aspects of grammar in Testelets et al. (eds.) (2009), and so on.

Second, it is worth testing whether the scope of the valency-changing operators (e.g., the causative operator, modal operators) and the relative ordering of operations are the same. The answer should be negative for Adyghe, though mismatches between the scope and the ordering are hard to find. Note that in Hyman 2003, for instance, the relative scope and the relative ordering are taken to be the same.

The article is structured as follows. Section 2 describes some relevant typological features of Adyghe. In section 3, the makeup of Adyghe verbal form is considered in some detail. Section 4 represents the crux of the analysis and considers the characteristic features of combinations containing the causative prefix and other valency-changing markers. Finally, in the last section, some explanations and conclusions are proposed.

The data used in this paper come from fieldwork materials on the Temirgoy dialect of Adyghe collected in the village Heqᵂərənehabl (Хакуринохабль), Republic Adygeya, in 2003–2006. These fieldtrips were organized by the Russian State University for the Humanities. Most of the speakers we worked with are Adyghe-Russian bilinguals.

2. Typological Characterization of Adyghe

Let me list some typologically unusual features of Adyghe. First of all, in Adyghe we find (almost) no distinction between nouns and verbs (Lander & Testelets 2006). In fact, dictionaries (e.g., Shaov 1975, Tharkakho 1991, Paris & Batouka 2005) translate Adyghe words with Russian or French verbs and others with nouns; in reality, both verbs and nouns can appear in predicative as well argument syntactic positions without any dedicated marking. The morphosyntactic criteria that distinguish the two major classes are subtle.

Note that throughout the paper, the term *verb* is nevertheless used to mark the form bearing typically predicative markers, such as TAM markers or markers of valency change. This traditional term seems to be more convenient than the term *predicate*, though in reality, both nouns and verbs can take TAM markers. Some (not all) markers of valency change (i.e., the causative, malefactive, and some other prefixes) are also applicable to (base) nouns and adjectives: for instance, the noun *ṣ̂ʷənč̣'* 'darkness' can bear the malefactive prefix *ṣ̂ʷe-*, introducing in this case the subject of opinion: the combination *ṣ̂ʷe-ṣ̂ʷənč̣'* means 'it is dark for someone.'

Second, Adyghe is a highly polysynthetic language, which means that the verb form contains pronominal affixes expressing all syntactic arguments of the predicate. In what follows, I will call *arguments* only A of transitive verbs, S of intransitive verbs, DO and IO. These arguments are always cross-referenced in the verb form. The language also has a rich system of valency-increasing operations (Smeets 1992, Letuchiy 2009a, 2009b), including causative, which is the main topic of the present paper.

The Adyghe verb form is organized in slots. For the reader's convenience, I provide here a complete list of slots showing in general words in which grammatical meanings are encoded in each prefixal or suffixal slot.

Adyghe is also a pro-drop language. In other words, all pronominal argument NPs can be omitted. In this case the pronominal arguments are only expressed by cross-reference markers in the verb form.

The following example from the article by Lander and Letuchiy (2010) illustrates the richness of Adyghe verb forms:

(1) *Sə-qə-t-de-p-fə-Ø-r-a-ʁa-ǯe-š̕tə-ʁ.*
1SG.ABS-DIR-1PL.IO-COM-2SG.IO-BEN-3SG.IO-DAT-3PL.A-CAUS-read-AUX-PST
'They were making me read it to you together with us.'

The verb form in (1) contains an absolutive 1SG cross-reference marker *sə-* (slot -9) of Figure 3.1; a directional prefix *qə-* (slot -9); an oblique 1PL IO cross-reference marker *t-* introduced by the comitative applicative marker *de-* (-6); an oblique 2SG IO cross-reference marker *p-* introduced by the benefactive

Prefixes							
Argument structure zone						Pre-stem elements[i]	
Absolutive	Directional prefix	Prefix referring to the temporal argument	Applicatives	Indirect object	Agent	'Dynamic' prefix	
						Optative	Negation
−9	−8	−7	−6	−5	−4	−3	−2

Prefixes	Root	Suffixes				
Stem				'Endings'[ii]		
Causative	Root	Directional, inceptive, antipassive	Propositional operators	Plural	'Dynamic' suffix	Negation, illocutionary force, epistemic modality
−1	0	+1	+2	+3	+4	+5

FIGURE 3.1 The structure of the Adyghe verbal complex (based on Lander & Letuchiy 2010: 264–266).

applicative marker *fə-* (-6); a zero 3SG IO cross-reference marker (-5), introduced by the dative (default applicative) marker *r*-; an oblique agent 3PL cross-reference marker *a-* (-4); the causative marker *ʁa-* (-1); the root *ǯe-* 'read'; and a composed imperfective suffix *ŝ'tə-ʁ* (+2).

Third, Adyghe has an ergative alignment that manifests itself both in case-marking and cross-reference system (see Smeets 1992 and Kumakhov & Vamling 2006 for details). However, Adyghe is not a canonical ergative language in any sense: the set of functions of the ergative case are wider than the functions of the absolutive and ergative in many other ergative languages. While the suffix *-r* (absolutive), typically for the absolutive case, marks intransitive subjects (3) and direct objects (2), the suffix *-m* (ergative) is used not only for transitive subjects (agents), as in (2), but also for indirect objects, certain adverbials, and adnominal possessors. In example (4), it marks an indirect object of the intransitive verb *bewən* 'kiss,'[2] and in (5), it marks a possessor. This is why, throughout the article, we term what has been traditionally called 'ergative' 'oblique.'

(2) *Ç'ale-m pŝaŝe-r Ø-ə-λeʁ^{W}ə-ʁ.*
boy-OBL girl-ABS 3SG.ABS-3SG.A-see-PST
'The boy saw the girl.' (transitive verb)

(3) *Pŝaŝe-r Ø-ma-ķ^{W}e.*
girl-ABS 3SG.ABS-DYN-go
'The girl goes.' (intransitive verb)

(4) *Ç'ale-r pŝaŝe-m Ø-Ø-je-bewə-ʁ.*
boy-ABS girl-OBL 3SG.ABS-3SG.IO-DAT-kiss-PST
'The boy (intransitive subject, absolutive) kissed the girl (IO, oblique).'

(5) *çəfə-m jə-č'etəw*
man-OBL 3SG.PR+POSS-cat
'The man's (possessor, oblique) cat'

Fourth, in terms of the characterization presented by Nichols et al. (2004), Adyghe is a "transitivizing" language; that is, it has a lot of morphosyntactic devices that allow to add an argument to the valency structure of the verb, including causative and a variety of applicatives. However, it does not have as many

[2]Note that the notions of transitive and intransitive verb classes are interpreted in a nonstandard way in Adyghe. See section 4.2, where I show that the two classes can mainly be distinguished on the morphological (cross-reference) basis. In (4), it is crucial that the patient (the 'indirect object') of 'kiss' is not marked in the same way as the patient of verbs with high semantic transitivity such as 'kill'. In contrast, the 'direct object' of 'see' in (2) is marked in the same way (by the absolutive case) well as the patient of verbs with high semantic transitivity such as 'kill' (see section 4.2 for details).

grammaticalized means of valency decrease. For instance, anticausative is compatible with only a handful of verbs, such as *feṣ̂ən* 'close,' and is marked with the reflexive marker *ze-*.[3] Antipassive is more productive but most frequently co-occurs with the unmarked omission of the object; thus, its compatibility is not so wide as in many other languages. In addition, Adyghe possesses a rich system of locational preverbs (Paris 1989). All of them have a double function: semantically, they denote the type of localization of the situation with respect to some spatial reference point, whereas syntactically they add an indirect object to the valency structure of the verb.

3. Structure of Verbal Form in Adyghe

As said above, the verbal form in Adyghe consists of both prefixes and suffixes; Figure 3.1 above represents the overall structure in terms of "positions" or "slots" (cf. Smeets 1996: ch. 2; Paris 1989: 196–198; Testelets et al. (eds.) 2009: 42). The "slots" in Figure 3.1 are given for convenience, and do not always imply that any position can be filled by only one morpheme at a time (see Lander & Letuchiy 2010). Thus the morphological pattern employed in the prefixal part of the Adyghe verb form seems to be templatic rather than layered (see Hyman 2003, for the notion of templates and templatic morphology). This means that each morpheme has its fixed slot in all the verb forms it is appropriate for—Hyman, dealing with Bantu, states that "there is no evidence that Bantu suffix ordering is driven by semantic compositionality or by a general Mirror Principle." The order of prefixes does not quite depend on the scope of each morpheme or the operation that is included by this morpheme. For instance, in (6), the verb only has an absolutive argument introduced by a 1SG prefix *se-* and an oblique IO argument cross-referenced by the 2SG marker *w-*, but it has no oblique A argument—thus the -4 slot reserved for agent cross-referencing prefixes remains empty (as does the -6 slot used for applicative markers):

(6) *Se* *we* *se-w-e-žə.*
boy-ABS girl-OBL 1SG.ABS-2SG.IO-DYN-wait
'I (intransitive subject, absolutive) am waiting for you (IO, oblique).'

---

[3]Moreover, it is doubtful that *ze-* marks valency decrease in examples where it is semantically interpreted as an anticausative marker. In spite of the verb form being interpreted as an anticausative ('close (intransitive) by itself'), it contains two valency slots and can be read literally as 'to close itself,' which is more similar to reflexive than to anticausative: cf. *z-jə-wəfa-ʁ* [REFL.ABS-3SG.A-bend-PST] 'it bent (lit. 'it bent itself').' In this example, both the 3SG agent prefix *ə-* and the reflexive marker *z(ə)-* are present. Thus, the verb 'bend (by itself)' is a transitive verb with two agreement slots, only that one of them (the absolutive slot) is occupied by the reflexive marker, and the verb form is homonymous to 'it/(s)he bent itself.'

Usually each prefixal slot is reserved for marking one grammatical category or group of meanings: for instance, the slot -9 is filled by cross-referencing absolutive arguments. Note that the suffixal morphology is not as strict as the prefixal one: for instance, one slot can be filled by several suffixes; for some pairs of suffixes, both theoretically possible orders exist (see Korotkova & Lander 2010, Arkadiev & Letuchiy 2008 and section 4 below for details).

The arguments of the verb are cross-referenced in the prefixal part of the verb form. The elements cross-referenced include all absolutive arguments (if we accept that the 3rd-person absolutive markers are zero, which is usually done in the existing grammatical descriptions) and most oblique participants:

(7)(=1) *Sə-qə-t-de-p-fə-Ø-r-a-ʁa-ǯe-š̓tə-ʁ.*
1SG.ABS-DIR-1PL.IO-COM-2SG.IO-BEN-3SG.IO-DAT-3PL.A-CAUS-read:
AP-AUX-PST
'They were making me read it to you together with us.'

In this example, the 1SG absolutive DO, the 1PL oblique IO, the 2SG oblique IO, the 3SG oblique IO, and the 3PL agent are cross-referenced in the verb form. Note that almost any combination of persons is grammatically correct in Adyghe, even if it is semantically strange, as in (7).

The most intriguing situation is observed for temporal arguments. The slot -7 can only be filled by the prefix *zə-*, referring to the temporal argument in struc tures like 'when he came' (lit. 'in the time when he came'), analogous to temporal embedded clauses in SAE languages:

(8) *Qə-<u>zə</u>-ḳ^{W}a-ʁe-m,* *s-λeʁ^{w}ə-ʁe-p.*
DIR-<u>REL.TEMP</u>-go-PST-ERG 1SG.A-see-PST-NEG
'When he came, I didn't see him.'

In this example, the temporal argument is cross-referenced by *zə-* because the first verb form in (8) is embedded: *zə-* expresses the meaning like 'the moment when (he came).'

In all other cases, the -7 slot reserved for relativization of temporal argument remains empty. For instance, in examples like 'He (has) lived in Majkop for three weeks,' the temporal argument 'three weeks' is not cross-referenced in the verb form by a 3rd-person plural marker.

A few words should be said on the morphological glossing of verbal arguments. Throughout the article, labels 'OBL' (oblique) and 'ABS' (absolutive) will be used for nominal case endings. For cross-reference prefixes, 'ABS' will be used for all absolutive prefixes (whether they mark S of intransitive verbs or DO of transitive verbs), while among the oblique prefixes, 'A' is used for prefixes cross-referencing the actor of transitive verbs and 'IO' for prefixes cross-referencing the IO of intransitive or transitive verbs. Zero absolutive prefix of the 3rd person is not written down and not glossed.

4. Causativization

The causative meaning is coded by the prefix *ʁe-*. This marker occupies the position immediately before the verb root.

Note first that the position of the causative affix is different from that of applicative affixes that add an indirect object. All the applicative affixes are situated in the -6 position together with the cross-reference marker of the indirect object introduced by the operation.

The historical sources of this difference remain unknown. However, it is obviously related to the fact that the causative operation is in a sense (more precisely, in two senses) distinct from all applicative derivations.

The first difference lies in the semantic domain. As Ramchand (2008) and others point out, the causative operation creates a new subevent (i.e., 'causation'), while various types of applicative do not. At least this is what is often supposed in studies on the typology of causatives (such as Comrie 1976, Shibatani 1976, Shibatani & Pardeshi 2002), though some authors (including Pylkkänen 2002) consider causativization to be rather an operation of adding an agentive participant, and not a subevent. They only introduce a new participant (or a participant that used to be irrelevant for the sense of the base verb) into the situation.

The second difference is a purely syntactic/morphosyntactic one. While the causative not only adds a participant to the subject position, but also demotes the status of the former subject, the applicative valency increase in Adyghe does not change the syntactic status of the arguments of the base verb (though it can be the case in other languages; see Peterson 2007 for Kinyarwanda).

The syntactic mechanism of causativization follows a typologically widespread pattern, described by Comrie (1976) in his influential work: the new argument occupies the subject position. The former subject occupies the highest of the vacant syntactic positions: namely, it becomes an absolutive direct object if the base verb is intransitive, and it becomes an oblique indirect object if the base verb is transitive, and the direct object position is occupied.[4]

---

[4]Note that for Adyghe, Comrie's account is not unproblematic. According to Lander (2009) and Letuchiy (2012), almost no tests for subjecthood and syntactic priority are applicable to Adyghe. One can, however, see that the case marking in the causative construction does not directly contradict to Comrie's rule. For instance, in (9), the causative construction is built on the base of an intransitive construction, and the causee bears the marking characteristic of what is similar to standard average European languages direct objects (absolutive patientive argument). In (10), the causative construction derived from a transitive construction, the causee bears the same marking as indirect objects (e.g., recipients) do, though it is also possible to say that in (10b) the causee (= the initial subject, e.g., 'boy') just retains its oblique marking (cf. (10a)).

Intransitive verb:

(9) a. *Pŝaŝe-r Ø-ma-ḳ^{W}e.*
girl-ABS 3SG.ABS-DYN-go
'The girl goes.'

b. *Č̦'ale-m pŝaŝe-r Ø-j-e-ʁa-ḳ^{W}e.*
boy-OBL girl-ABS 3SG.ABS-3SG.A-DYN-CAUS-go
'The boy makes the girl go.'

Transitive verb:

(10) a. *Č̦'ale-m g^{W}əč̦ə-r Ø-j-e-wəfe.*
boy-OBL iron-ABS 3SG.ABS-3SG.A-DYN-bend
'The boy bends the iron.'

b. *Pŝaŝe-m č̦'ale-m g^{W}əč̦ə-r Ø-Ø-r-j-e-ʁe-wəfe.*
girl-OBL boy-OBL iron-ABS 3SG.ABS-3SG.IO-DAT-3SG.A-DYN-CAUS-bend
'The girl makes the boy bend the iron.'

In (9a), the base verb is monovalent intransitive ('go'). The sole argument (S in Dixon's terms) is marked by the absolutive case and cross-referenced with a zero absolutive marker. The causative in (9b) is transitive, and an oblique causer ('the boy') is introduced and cross-referenced with an oblique agent prefix *j*-, while the initial S becomes a DO and retains its absolutive marking and zero cross-reference, though it is used in another function. In (10a), the base verb is transitive ('bend something'): the absolutive argument is cross-referenced with a zero prefix, and the agent is marked with *j*-. Its causative counterpart in (10b) is ditransitive, the initial A ('the boy') becoming an IO (causee) and retaining its oblique marking, but now the oblique case and the zero cross-reference marker marks an IO. A new argument (A, causer)—namely, 'the girl'—is introduced, which is cross-referenced by means of the agent marker *j*-.

4.1 ORDERING AND SCOPE OF DERIVATIONS

In what follows, I will try to find out what the semantic scope of the causative marker with respect to other valency-changing mechanisms is.

I use the following notions throughout the article: 'scope of operator' and 'order of operations.' The choice between them is not always trivial, as I will show.

The scope is interpreted semantically here. Saying that the scope of the marker A includes the marker B denotes that the meaning of A applies to B, among other components of the expression. For instance, consider the following example: there are two operators: a causative *make*-like verb and an adverbial meaning 'on purpose.'

The two operators can be applied to a verb meaning 'fall.' The meaning of the resulting construction can be different. If the resulting construction means 'Peter made Jane fall *on purpose*,' where 'on purpose' refers to what Peter did, this means that the causative operator is in the scope of the purpose one, but not vice versa. The causative operator refers to the situation 'fall,' but not to 'fall on purpose.'

Let us now imagine that the construction means 'Peter made Jane [fall on purpose]' where 'on purpose' refers to Jane's falling. Here the causative operator includes the purpose one in its scope. The causation refers not to falling but to falling on purpose, because the purposive component is caused by Peter, as well as the act of falling itself. In contrast, the scope of the purpose operator does not include the causative operator.

The notion of 'ordering of derivations' is used in the following way. According to Baker's (1985) Mirror Principle, the morphological order reflects the order of syntactic operations. Operation A precedes operation B if some criteria allow us to say that B modified the verb or the expression that was modified already by A. For instance, consider a situation when A is a detransitivizer and B is a habitual marker that only can modify intransitive verbs. If we see a base transitive verb 'kill,' modified by both A and B, we can easily conclude that the operation marked by A took place before the operation marked by B, but not the other way around. Operation B could not precede A, because B can only modify intransitive verbs, and before operation A is applied, the verb 'kill' is transitive.

Let me remind that the order of derivations and the order of affixes do not correspond to each other in Adyghe in the same way as they do in many European languages. It is well-known that in many languages the linear order of affixes corresponds to the order of derivations. This is what the Mirror Principle formulated by Baker (1985) and Grimshaw (1986) and formalized by, for instance, Kobele (2002) tells us. In other words, affixes that are situated closer to the root often mark derivations that occur earlier than others.

For instance, the Russian adjective *učitel'skij* 'teacher's, designed for teachers and so on,' obeys this rule. The verbal stem is *uči-* 'teach.' Then, the noun *učitel'* 'teacher, person who teaches' is formed by means of the suffix *-tel.'* Finally, the adjectivizer *-sk* forms the adjective *učitel'skij* 'teacher's' ('which belongs to or is related to the teacher'). Here, the linear order is obviously iconic. The agentive suffix *-tel'* is situated closer to the root than the adjectivizer *-sk*, and, correspondingly, the agentive noun is formed first, and then the adjective is built on its base.

In Adyghe, the suffixal part of the word obeys roughly the same rules (see Korotkova and Lander 2010): the suffix that is situated closer to the verbal root marks the derivation that took place earlier—in general, the Mirror Principle is observed here. For instance, in some pairs of affixes, the order can be dependent on which affix has the widest scope.

(11) a. *Waŝʷe-m ẑʷaʁʷe qə-tje-s-xə-**ŝʷə-ŝʷe**.*
sky-OBL star DIR-LOC-1SG.A-take-HBL-SML
'I pretend as if I could take a star from the sky.' (simulative > habilitive)

b. *Waŝʷe-m ẑʷaʁʷe qə-tje-s-xə-**ŝʷe-ŝʷə**.*
sky-OBL star DIR-LOC-1SG.A-take-SML-HBL
'I can pretend as if I were taking a star from the sky.' (habilitive > simulative) (Korotkova & Lander 2010)

In (11a), the scope of the simulative operator ('to pretend to') includes the habilitive one ('to be able'): what the subject is pretending is **to be able** to do something. Correspondingly, the suffix with the widest scope occupies the rightmost position in the verb form. In (11), in contrast, the scope of the habilitive operator includes the simulative one: what the subject is able to do is **to pretend** to do something. Not surprisingly, the respective order of the morphemes changes, the habilitive being in the rightmost position. Thus, in terms of Simpson and Withgott (1986), Mithun (2000), Paster (2005), and so on, the suffixal part is organized in the spirit of layered morphology. The order of morphemes depends on their relative scope and the order of the applying grammatical operations.

However, the situation changes as soon as the prefixal part of the verb is considered.

In the prefixal part of the verb, most morphemes have their own rigid morphological positions (see Figure 3.1), which they occupy irrespective of the order of derivations and scope of affixes. In other words, the linear position of the affix does not reflect its semantic scope with respect to other affixes. The prefixal part organization is based on the templatic morphology principles (see above). Of course, this yields many cases of scope ambiguities, where the same linear order of prefixes can have more than one reading, as in (12), borrowed from Arkadiev and Letuchiy (2008):

(12) *Jate ç̣'ale-m meqʷə s-fə-r-jə-ʁe-wəpç̣'a-ʁ.*
father boy-OBL hay 1SG.IO-BEN-3SG.IO-3SG.A-CAUS-mow-PST
i. 'The father made the boy [mow the grass for me].' (causative > benefactive)
ii. '[The father made the boy mow the grass] for me.' (benefactive > causative)

Under the first interpretation, the scope of the causative marker is wider than that of the benefactive marker. In other words, the benefactive operation took place earlier, yielding the meaning 'He mows the grass for me.' Then causativization took place. Under the second interpretation, causativization took place earlier (the meaning was something like 'The father made the boy mow the grass'), and then the benefactive meaning was added. However, the linear order of affixes—the benefactive marker *fe-* and the causative prefix *ʁe-*—is identical in both cases. The scope of the benefactive operator under the second interpretation is wider than the scope of the causative one. Under the first interpretation, causative scopes over benefactive (what is caused is an action carried out for me, i.e., the benefactive component is included into the scope of the causative operator). Thus, the difference in the semantic scope of derivations does not presuppose any difference in the linear order of prefixes.[5] The Mirror Principle does not seem to be at work in the prefixal part of examples like (12), otherwise the two readings would yield two different linear orders of the causative versus benefactive morpheme.

Therefore, I argue that the organization of the prefixal part of Adyghe verb (at least the position of the causative marker with respect to other ones) is based on the principles of templatic morphology. Each morpheme (except applicative markers inside position -6) has its fixed position, irrespective of the scope of operation it marks.

In what follows, I will focus on the combinations of the causative morpheme with other derivational morphemes. The issue of the semantic scope of valency-changing operations will be central to the present study. The problem of the linear order of morphemes will only be considered in the concluding section. In general, the causative morpheme always occupies the position immediately before the verbal root (the -1 slot in terms of Smeets 1996), and the only challenge is to try to explain how this fixed position is related to the function of the causative morpheme.

Since the position of the causative prefix is fixed and does not depend on the semantic scope of derivations, generally the main problem is to find some reliable criteria defining the semantic scope of derivations. In what follows, I propose these criteria.

4.2 CAUSATIVIZATION AND REFLEXIVIZATION

In Adyghe, the causative prefix *ʁe-* can be combined with the reflexive prefix *ze-/zə-* within one verb form. In other words, the same verb can be causativized and reflexivized, and the main issue is the scope of the derivational markers and the order of derivations).

First, a few words about the reflexive marker should be said. This marker (as well as the reciprocal markers discussed in 4.3) always occupies one of the cross-reference slots reserved for one of arguments it binds (including all examples discussed below). This allows us to conclude that the very reflexive and reciprocal

---

[5]Here I do not distinguish the semantic ordering from the scope of operators. Later on, I will show that the scope and the semantic ordering do not always match (see the comment after example (21)).

markers are types of cross-reference markers that only differ from usual cross-reference markers in that the entities they refer to are coreferential with another argument and are never expressed with a separate NP.

Let us consider the case in which the base verb is bivalent. In this case, the base verb can theoretically be reflexivized, yielding a structure like 'Peter wounded himself,' and then causativized, resulting in 'I let/made Peter wound himself.' In this case the semantic scope of the causative operator would be wider than that of the reflexive marker—in other words, the reflexivization would occur before the causativization, and the scope of the causative marker would include the reflexive marker.

At the same time, the reverse relative scope of causative and reflexive is equally possible. Theoretically, the reflexive morpheme can scope above the causative one—that is, the verb that has undergone causativization can further be reflexivized. If the base verb is bivalent transitive (and the derived causative is thus trivalent), there are three theoretically possible variants of reflexivization. However, the first of them is used very rarely by native speakers:

(i) The causer is coreferential with the causee. The base verb being 'X wounded Y,' the causative is 'Z made X wound Y,' and the reflexive derivative will mean 'Z made him/herself wound Y.' The direct object of the base verb does not take part in the reflexive relation. This variant occurs only rarely with causatives of transitive verbs. Though the variant like 'I made myself read this book' is semantically perfect, it is rarely expressed with a combination of causative and reflexive.
(ii) The causer is coreferential with the direct object of the base verb. In this case, the reflexive means something like 'Z made X see him (Z).' The causee (X) does not participate in the reflexive relation.

(13) *Zə-s-jə-ʁe-λeʁ^{W}ə-ʁ.*
REFL.ABS-1SG.IO-3SG.A-CAUS-see-PST
'He showed himself to me.' (lit. 'He made me see himself').

In (13), the reflexive marker applied to the trivalent transitive verb 'show' occupies the absolutive position, while in the oblique IO and the oblique agent position, the 'usual' cross-reference markers *s-* (1SG) and *jə-* (3SG) are retained.

(iii) Finally, the causee can be coreferential with the direct object of the base verb, the reflexive having the meaning 'Z made X wound himself (X).' In this case, the causer does not take part in the reflexive relation.

In sum, variants (i) and (ii) presuppose unequivocally that the reflexive operation follows the causative one. It is only after causativization that the causer is introduced into the semantics of the situation. Thus, reflexivization follows causativization. The scope of the reflexivization includes causativization: the causative operator modifies the initial situation like 'I saw you,' yielding something like 'He made me see you.' Then, the reflexive operator applies to the causative situation 'He made me saw you,' yielding 'He made himself see you' in variant (i) or 'He made me see himself' in variant (ii).

However, the situation gets more complicated in variant (iii). It is obvious that the meaning of the resulting verb in variant (iii) ('I made Peter wound himself'), where the causative verb is reflexivized, is the same as in the reverse case in which causativization follows reflexivization. This means that if we see in a language L the verb form containing a causative and a reflexive marker with the meaning of 'I made Peter wound himself,' it is not always the case that we can decide, based on the purely morphological criteria, which of the two markers has a wider scope.

However, I will show that in cases like these in Adyghe, causativization follows reflexivization, and not vice versa. The evidence comes from the distinction between the classes of bivalent transitive and bivalent intransitive verbs.

In Adyghe there are two classes of bivalent verbs (i.e., verbs that cross-reference two arguments in the verb form). From the semantic point of view, these verb classes seem to be analogous to Dixon's (1980) transitive and extended intransitive classes. For instance, one of the classes that we term 'transitive' throughout the paper contains verbs like *wəč'ən* 'kill,' which are transitive in the majority of languages, the other class include *dəʔepəʔen* 'help,' which tends to be intransitive. What also makes the second class similar to the 'usual' intransitive verb class in European languages is that their (indirect) object is marked with the same (oblique) case, as the recipient of trivalent verbs like *jetən* 'give.'

If a bivalent verb is 'transitive' it has an oblique subject (A), which in our case simply means 'the argument situated in the -4 slot' and an absolutive DO in the -9 slot. If a verb is 'intransitive' it has an absolutive subject S), also in the -9 slot, and an oblique IO in the -6 or -5 slot. The former class includes verbs like *wəč'ən* 'kill' or *wəʔen* 'wound,' but also the verb 'see,' which is intransitive in many other Caucasian languages (in other words, the Adyghe 'transitive' and 'intransitive' classes do not fully correspond to their counterparts in Nakh-Daghestanian languages, where perception verbs tend to be intransitive). The latter class contains, among others, lexemes like *jebewən* 'kiss,' *jewen* 'hit,' or *jepλen* 'look at,' which do not presuppose that the patient is strongly affected by the actions of the agent. All lexemes of the latter class include a special prefix *je-*, which marks that the verb has an indirect object.

Both verb classes can be causativized and reflexivized, and for both classes the reflexive and causative markers are compatible within the same verb form. Under variant (iii) above, for the intransitive class, their combination can yield a meaning like 'I made Peter look at himself (e.g., in the mirror),' while for the transitive class the meaning is similar to 'I made Peter see himself (also in the mirror)' or 'I made Peter kill himself/commit a suicide.'

Transitive verb:

(14) *S-ja.te* *ʁʷənӡ̌e-m-ç̣̌ʼe*
1SG.PR-POSS.father mirror-OBL-INS
zə-qə-s-jə-ʁe-λeʁʷə-žʼə-ʁ.
REFL.ABS-DIR-1SG.IO-3SG.A-CAUS-see-RE-PST
The father$_i$ showed me$_j$ myself$_j$ in the mirror. (lit. 'The father$_i$ made me$_j$ see myself$_j$ in the mirror') (Causative > Reflexive)[6]

Intransitive bivalent verb:

(15) *ʁʷənӡ̌e-m-ç̣̌ʼe* *s.-a.ne*
mirror-OBL-INS 1SG.PR-POSS.mother
sə-ze-r-jə-ʁe-pλə-žʼə-ʁ
1SG.ABS-REFL.IO-DAT-3SG.A-CAUS-look-RE-PST
The mother$_i$ made me$_j$ look at myself$_j$ in the mirror. (causative > reflexive)

In (14), a causative of bivalent transitive verb is shown. The resulting verb is trivalent transitive. The reflexive marker occupies the absolutive slot, and the IO and the agent slot are reserved for 'usual' cross-reference markers. In (15), the reflexive marker occupies the IO slot, and the absolutive and the agent slots are occupied by standard cross-reference markers.

I will show now that the morphological structure of the verb form in examples (14) and (15) allow us to suppose that the causative operation occurs after the reflexive one.

What would we expect if reflexivization takes place after causativization? For transitive verbs like 'kill,' the causee, following Comrie's (1976) rule formulated in the beginning of section 3, must occupy the indirect object position, while the direct object of the base verb remains in the direct object position.

For intransitive verbs like 'hit,' the causee must occupy the direct object position, because this position is free, and the patient/theme must remain in its inherent indirect object position.

Afterward, when reflexivization would take place, the reflexive operator would, in both cases, bind the direct and the indirect object. In the transitive verb, the IO position would be occupied by the causee, and the DO position by the base direct object. In the intransitive verb, conversely, the DO position would be occupied by the causee and the IO position by the base indirect object.

When coreference between DO and IO occurs in Adyghe, the reflexive marker should always be situated (except some minor verb classes analyzed separately by Letuchiy 2009b) in the indirect object slot and controlled from the direct object slot.

This means that if reflexivization scoped over causativization, the reflexive marker would have to be situated in the indirect object slot in both examples like

[6]The refactive morpheme *-žʼə* does not bear any meaning of repeatedness here: it just accompanies the reflexive prefix as it often does (though in all examples of this sort it can be dropped).

(14) and (15) above, which would yield the following—in reality, ungrammatical—form for (14):

(14') **S-ja.te* *ʁʷənǯe-m-čʼe*
1SG.PR-POSS.father mirror-OBL-INS
sə-qə-z-jə-ʁe-λeʁʷə-žʼə-ʁ.
1SG.ABS-DIR-REFL.IO-3SG.A-CAUS-see-RE-PST
The father$_i$ showed me$_j$ myself$_j$ in the mirror. (causative > reflexive)

For reflexivization, DO behaves as a more privileged argument than IO. The reflexive marker occupies the IO slot and is controlled from the DO slot.

However, this is not the case. In (15), the reflexive marker is indeed situated in the indirect object slot, while in (14), with a transitive verb, it occupies the direct object (absolutive) slot and is controlled from the IO slot.

What does this mean? Presumably, reflexivization precedes causativization. When reflexivization occurs in (14) and (15), it affects two noncausative verbs, one having an A and DO and the other one an S and IO. In the former case, the reflexive marker is situated in the DO slot, whereas in the latter one it occupies the IO slot. This situation is not surprising: in Adyghe, as in most languages, the reflexive marker tends to be controlled from the subject slot, but not from the direct or indirect object slot. To make it more precise, in both cases the reflexive marker is in both cases controlled from the subject slot, though later, when the verb is causativized, the subject argument ceases to be one.

Importantly, the notion of scope is very opaque and not always useful in the case of the coreference between the causee and the base object. In the case under analysis (the combination of reflexive and causative) it really is the case that the ordering and the scope of operations match each other. Causative is applied after reflexive, and the scope of reflexive in the given type of coreference does not include the causative situation. However, the situation could well be more complicated: even if reflexive was applied after causative, the reflexive operator could well bind the DO to the causee (i.e., the reflexive would not have the widest semantic scope). I will demonstrate a case like this when considering the reciprocal + causative combination.

Note that the combination of the causative and reflexive operators demonstrates one of the possible orders of morphosyntactic operations. This situation, however, does not allow us to choose between the two variants of description of the morphological markers themselves. The markers can either attach one after another, or simultaneously (here I do not say anything on the precise mechanism of attaching the morphemes to the verb form). However, semantically and morphosyntactically the operations have a fixed order—namely, causativization after reflexivization.

4.3 CAUSATIVIZATION AND RECIPROCALIZATION

Roughly the same situation takes place with the combination causative + reciprocal. As said above, the reciprocal markers behave just like the reflexive one in that they always occupy one of the cross-reference slots.

The linear order of causative versus reciprocal, as well as that of causative versus reflexive, is strictly fixed: the causative prefix always follows the reciprocal marker (i.e., the causative marker is closer to the root). However, it is possible to prove that causativization usually takes place after reciprocalization and that, nevertheless, there are some exceptions where causativization precedes reciprocalization.

When the combination of causative and reciprocal is considered, it is also theoretically possible that the initial bivalent verb (for instance, 'wound') is reciprocalized with the meaning 'They wounded each other'; it is then causativized, the resulting meaning being 'I made them wound each other.'

On the other hand, the combination of causative and reciprocal has the same three possible interpretations, as with the combination of causative and reflexive: the reciprocal relation can bind (i) the causer and the causee, (ii) the causer and the initial direct object, or (iii) the causee and the initial direct object. The first two variants are only possible if reciprocalization follows causativization, and the scope of the reciprocal operator includes the causative one.

Variant i:

(16) *Te lə O-te-zere-ʁe-šxə-ʁ.*
we meat 3SG.ABS-1PL.IO-REC.A-CAUS-eat-PST
'We fed each other with meat.' (lit. 'We made each other eat meat').

In (16), the causer ('one who feeds/makes eat') and the causee ('one who eats') are bound by the reciprocal relation. The marker *zere-* occupies the agent slot, and the IO and the absolutive slot are reserved for a usual 1PL cross-reference marker (the absolutive marker is zero in this case).

Variant (ii) is in principle possible, but mostly represented by unnatural sentences like 'We showed each other to the teacher' (lit. 'We$_i$ made the teacher$_j$ see each other$_i$').

In contrast, the third type of coreference (causee = initial object) can be a result of two different scenarios:

(a) reciprocalization precedes causativization:
'X wounded Y' → 'X and Y wounded each other' (reciprocalization) → 'Z made X and Y wound each other' (causativization).
This variant presupposes that two arguments of the initial verb (e.g., the transitive verb 'wound') are first reciprocalized, leading to the meaning 'X and Y wounded each other.' The resulting reciprocal form is then causativized, and a new argument Z is introduced ('Z made X and Y wound each other').

(b) causativization precedes reciprocalization:
'X wounded Y' → 'Z made X wound Y' (causativization) → 'Z made X and Y wound each other' (reciprocalization).

In this variant, the initial verb 'wound' is first causativized, yielding 'Z made X wound Y.' Then the reciprocal operation is applied to the causee and the initial object (direct for the verb 'wound') to make them arguments of the reciprocal relation ('Z made X and Y wound each other').

In the case of reciprocal and causative, as well as in the case of reflexive and causative, there is a test showing that the causativization usually takes place after the reciprocalization. Adyghe has two main reciprocal markers: *zere-* and *ze-/zə-*. The first of them is used if the reciprocal relation binds the subject and the direct object of a transitive verb, as in (17).[7]

In almost all other cases, Adyghe makes use of the second marker; for instance, in (19) it denotes coreference of the A and IO of a trivalent transitive verb. Sometimes, *ze-* is used instead of *zere-*, even in the cases of A = DO correspondence (see (18)), but not vice versa.

(17) *te te-zere-wəʔe-žʼə-ʁ.*
we 1PL.ABS-REC.A-wound-RE-PST
'We wounded each other.' (A and DO of a transitive verb, *zere-* in the Agent slot)

(18) *te ze-te-wəʔe-žʼə-ʁ.*
we REC.IO-1PL.A-wound-RE-PST
'We wounded each other.' (A and DO of a transitive verb, *ze-* in the absolutive)—rarely.

(19) *te ŝwəhaftən-xe-r O-ze-te-tə-žʼə-ʁ.*
we gift-PL-ABS DAT-REC.IO-1PL.A-give-RE-PST
'We gave gifts to each other.' (A and IO of a trivalent verb, *zere-* in the IO slot)

An important feature of the affix *zere-* is that it occupies the -4 position of the agent, not that of the absolutive argument.

Let us consider two variants, (a) and (b), explained above, which yield the same structure and semantics as a result. Which reciprocal marker must be used under scenario (a) and scenario (b)?

It is obvious that if scenario (a) is true—that is, the causative operator has scope above the reciprocal one—the reciprocal marker *zere-* must be used by the general rule. When the reciprocal operator applies, the arguments bound by the

[7] Notably, the prefix *zere-* occupies the agent slot, the absolutive one being reserved for a usual cross-reference marker. However, historically, *re-* in *zere-* is an instrumental applicative marker, and *ze-* occupies the IO slot introduced by the applicative.

reciprocal relation are the subject (A) and the direct object of the transitive verb like 'wound.' Afterward, the causative operator is used.

If Adyghe makes use of the scenario (b), *ze-/zə-* must be used. Under this scenario, when the two arguments are reciprocalized, causativization has already taken place. As I have shown, causativization introduces a new argument, the base subject (A) becoming an indirect object. Thus under scenario (b) the reciprocal operator binds a direct and an indirect object, and there is no reason to expect *zere-* in this case.

In fact, Adyghe usually uses scenario (a):

(20) *čʼetəwə-m-re ha-m-re zere-z-ʁe-λeʁʷə-ʁe-x.*
cat-OBL-COORD dog-OBL-COORD REC.IO-1SG.A-CAUS-see-PERF-PL(ABS)
I_i showed the cat and the dog_j to each $other_j$. (causative after reciprocal)

In (20), the causee and the base direct object of the verb *ʁeλeʁən* 'show' are reciprocalized. Reciprocalization takes place first; it binds the A and DO of the transitive verb (the reciprocal prefix occupies the Agent position, the absolutive one being occupied by the zero absolutive 3rd-person prefix). Afterward, the causative operator adds the causer in the subject position, making the base subject an indirect object.

Note, however, that for reciprocal marking, the second alternative is also employed by some native speakers of Adyghe. The reciprocalization can occur after causativization:

(21) *čʼetəwə-m-re ha-m-re ze-z-ʁe-λeʁʷə-ʁe-x.*
cat-OBL-COORD dog-OBL-COORD REC.IO-1SG.A-CAUS-see-PST-PL(ABS)
I_i showed the cat and the dog_j to each $other_j$. (reciprocal after causative)

In (21), *ze-* is used instead of *zere-*, making it evident that reciprocalization occurs after causativization. First, the causative verb *ʁe-λeʁʷəEn* 'show' is formed. This verb is trivalent: the causer (A) is 'I,' the causee (indirect object) is someone who the patient is shown to; the patient itself occupies the direct object position. Afterward, the reciprocal operator binds the causee and the base direct object, leaving the causer intact.

Note that the existence of variant (21) proves that the ordering of derivations and the scope of markers can sometimes conflict with each other. In (21), the reciprocal operator is used after the causative one. However, the scope of the reciprocal marker is not wider than one of the causative one even in (21). The reciprocal marker binds a direct and indirect object, and does not include the causative component. In other words, if an operation (reciprocalization) FOLLOWS another one (causativization), this does not mean that reciprocalization SCOPES over causativization. However, the variant in (21), where reciprocalization precedes causativization, is preferred by almost all native speakers.

By the way, distinguishing the two notions: semantic scope and ordering of operations could be useful not only for dealing with Adyghe. Recall that for Bantu, according to Hyman (2003), templatic morphology principles should be postulated to account for some unclear suffix orderings. Hyman (2003: 5) points to the act that the applicative suffix *-il* always follows the causative suffix *-its*. His example (3) from Chichewa shows that this suffix ordering represents both scope combinations: (i) those where the applicative operator scopes over the causative one ('The hunters are [making the child cry] with sticks') and (ii) those where the causative operator scopes over the applicative one ('The hunters are making the woman [stir with a spoon]').

In this interpretation, the Chichewa data really seem to show that the combination *-its-il* is templatic; that is, that it does not reflect the scope of the operators. However, it may be the case, though not necessarily, that the causative-applicative ordering in Chichewa reflects the ordering of operations and not the scope. It could be the case that applicative is in both cases applied after causative, but in the case (i) it is applied to the whole causative event, while in (ii) it is applied only to the non-causative subevent 'to stir.' At least it is what takes place in (21) in Adyghe: the choice of the reciprocal marker shows unequivocally that the reciprocal operator is applied after the causative one, but does not takes the whole causative situation into its scope.

4.4 CAUSATIVE AND ANTIPASSIVE/A-LABILE VERBS

Let me consider now the combination of the causative and the antipassive operator. Antipassive (see Cooreman 1994 for details) is a type of valency change that either eliminates the direct object of the transitive verb, making it intransitive, or just demotes the direct object to the status of indirect/oblique object (in this case too, the verb becomes intransitive).

In Adyghe, the antipassive operation is marked by the suffix *-e*, which can attach to certain transitive verb stems ending on *-ə*, making the verb intransitive. For instance, the verb *txən* means 'write something,' and it is transitive. The antipassive counterpart *txen* means 'write (something irrelevant, general, etc.)'; it is intransitive, and no direct object can be added.

(22) a. *ç̣'ale-m pis'me-r j-e-txə.*
boy-OBL letter-ABS 3SG.A-DYN-write
'The boy is writing a letter.'

b. *ç̣'ale-r ma-tx-e.*
boy-ABS DYN-write-AP
'The boy is writing.'

In some other cases, the direct object can be eliminated and the verb can be used intransitively without any morphological marking—this is the case of so-called A-labile verbs.

(23) a. *Xate-r Ø-ə-pč̣'a-ʁ.*
garden-ABS 3SG.ABS-3SG.A-weed -PST
'[(S)he] weeded the garden.'

b. *Ç̣əfə-r mefe.rjenə-m Ø-pč̣'a-ʁe.*
man-ABS day.whole-OBL 3SG.ABS-weed-PST
'The man weeded the whole day.' (Arkadiev & Letuchiy 2008).

For most verbs, no oblique NP can be added to the antipassive construction—as is the case in (23b) and (22b). In other words, the antipassive construction is strictly monovalent. There are, however, several lexemes that form a bivalent intransitive antipassive construction; for instance, *ǯən* 'read smth' (with an antipassive variant *ježen* 'read smth') and *ŝ^{W}ən* 'drink smth' (with an antipassive variant *jeŝ^{W}en* 'drink smth'). For these verbs, bivalent antipassive variants are more frequent and less semantically marked than bivalent transitive ones. The transitive variants are used in special cases when the action is finished—in other words, the transitive versus antipassive opposition conveys here a sort of aspectual semantic correlation.

The causativization can take place after the antipassive operation. In other words, verbs that are made intransitive by means of antipassivization can be further causativized and become transitive again:

(24) a. *Hač̣'e-xe-r ma-šx**e**-x.*
guest-PL-ABS DYN-eat.AP-PL(ABS)
'The guests are eating.'

b. *A-š' hač̣'e-xe-r j-e-ʁ**a**-šx**e**-x.*
DEM-OBL guest-PL-ABS 3SG.A-DYN-CAUS-eat.AP-PL(ABS)
'(S)he is feeding the guests.' (causative > antipassive)

The causativization in Adyghe is not subject to any restrictions related to the transitivity of the base verb. It is not surprising that for many verbs—for instance, *šxən/šxen* 'eat'—both the base (transitive) and the antipassive (intransitive) variants of the verb can be causativized. Causativization of the antipassive variant is illustrated in (24b), while (25b) shows causativization of the base transitive verb:

(25) a. *Hač̣'e-xe-m lə O-a-šxə.*
guest-PL-OBL meat 3SG.ABS 3PL.A-eat
'The guests are eating meat.'

b. *A-š' hač̣'e-xe-m lə O-a-r-j-e-ʁ**a**-šxə.*
DEM-OBL guest-PL-OBL meat 3SG.ABS 3PL.IO-DAT-3SG.A-DYN-CAUS-eat
'(S)he is making the guests eat meat.'

When the base verb is transitive, the derived causative verb, as I have shown before, is trivalent. This is what is shown in (25b). In the derived causative verb, the causer (A) is cross-referenced by means of the affix *j-*, the affix *a-* pointing to the causee occupying the IO position, and the zero prefix that corresponds to the DO *lə* 'meat.'

Thus the causative operation can occur after the antipassive one. Crucially, the opposite case is impossible. The antipassive operation cannot modify verbs causativized by means of the prefix ʁe-.

Consider, for instance, the verb *ʁeẑen* 'bake, fry' (the initial verb *ẑen* means 'be baked, fried'). This verb is a morphological causative from *ẑen* 'bake (be baked), fry (be fried)'. Though it belongs to the semantic group of verbs that are mostly compatible with the antipassive operation (verbs of work and professional activities), this particular lexeme does not take the suffix *-e*:

(26) a. *haləʁ^{w}əʁaẑe-m haləʁ^{w}ə-r j-e-ʁa-ẑe.*
baker-OBL bread-ABS 3SG.A-DYN-CAUS-be.baked
'The baker is baking bread.' (Arkadiev & Letuchiy 2008: 92)

b. **haləʁ^{w}əʁaẑe-r njepe rjen-ew me-ʁa-ẑe.*
baker-ABS today whole-ADV DYN-CAUS-be.baked.AP
intended meaning: 'The baker bakes the whole day.' (Arkadiev & Letuchiy 2008: 92) (intended: antipassive > causative)

The same is true, for instance, for *ʁeẑ^{w}en* 'cook, boil.' For the meaning 'he boils (different things) the whole day,' no monovalent verb can be employed.

It seems that the only reason for the ungrammaticality of (26b) is that the verb is a morphological causative and cannot be antipassivized by general rule. Thus, the scope of the antipassive operation cannot include the causative marker.

4.5 CAUSATIVE AND APPLICATIVES

In the sections 4.2 through 4.4, I showed that in cases where both orderings of causative and other operations are theoretically possible, Adyghe tends to choose one ordering: namely, the variant where causativization follows reflexivization, the antipassive operation, and, in most cases, reciprocalization. In this section and the following one I will consider less obvious cases where no reliable criteria can be used to check the relative scope and the ordering of morphosyntactic operations.

In Adyghe, there is a number of applicatives—derivations that introduce a new object to the valency structure of the verb. All applicative markers occupy the -6 slot of the verb form. Applicatives always introduce an oblique-marked indirect object, which is cross-referenced in the verb form. However, the semantic role of the new argument can be different: for instance, it can be beneficiary (*fe-*), maleficiary (*ṣ̂^{w}e-*), locative participant (*šʼe-* and other markers), and so on.

All applicative operations can be combined with causativization in one verb form. In this case, the restrictions on the scope are not valid. The causative operator can have the scope either above or under applicatives.

The following example provides an illustration. In (12), repeated here as (27), the combination of causative and benefactive can have two meanings:

(27) *Jate* *č̓ale-m* *meqʷə* ***s-fə***-*r-jə*-**ʁe**-*wəpč̓a-ʁ*.
POSS:father boy-OBL hay 1SG.IO-BEN-DAT-3SG.A-CAUS-mow-PST
i. '[The father made the boy mow the grass] for me.' (benefactive > causative)
ii. 'The father made the boy [mow the grass for me].' (causative > benefactive or benefactive > causative)

Again, just as in the case of reflexive and reciprocal, there is a reading in which the relative scoping of benefactive and causative and the ordering of morphosyntactic operations is obvious. Under the first translation (i), the benefactive operator apparently has scope over the whole verb form, including the causative operator. In this meaning, what took place for me (i.e., to my benefit), is everything including the causative component, thus the benefactive operator modifies the causative one. The benefactive operation follows causativization.

In contrast, under the second translation (ii), either of the operations could have occurred after the other one. It may be the case that first the benefactive operation takes place ('mow for someone'), and afterward the causativization occurs. But it may equally be the case that causative modifies the verb 'mow,' yielding the meaning 'make mow,' and afterward the benefactive component is added. Though it is true that the scope of benefactive marker does not include the whole causative situation, this does not help us to choose any of the two orderings (see the comments on examples (20) and (21) above).

In other words, translation (ii) corresponds to two orders of operations.

The order of the benefactive and causative operations cannot be defined under the reading (ii), since the benefactive derivations operates the meaning component ('mow the grass') that is present both before and after causativization. In contrast, under the meaning (i), the benefactive component interacts with a component introduced by causativization ('**make** mow the grass'). The choice between the two variants under (ii) could in principle be done in favor of the variant Causative > Benefactive, if only we were sure that, after the causativization takes place, the caused subevent ('mow grass') becomes inaccessible for other operators once the causing subevent ('make') is added above it. In this case the benefactive operator applied after the causative one could only modify the causing situation or the event as a whole. However, we have no theoretical prerequisites that allow us to make such a conclusion.

4.6. CAUSATIVE AND HABILITIVE

In Adyghe there are two morphological means of marking possibility and potentiality: the prefix *fe-* and the suffix *-ŝʷə*, glossed as HAB (habilitive). While the latter does not change the valency structure of the verb and is not subject to any syntactic restrictions, the former can only attach to transitive verbs and makes them 'inverse' in terms of Smeets (1992, 1996): the most agentive participant is an indirect object, the patient is a subject. With intransitive verbs, it is ungrammatical (see (29)).

(28) *Se pχe-r Ø-s-fe-qʷəte-r-ep.*
I wood-ABS 3SG.ABS-1SG.IO-POT-break-DYN-NEG
'I cannot break the piece of wood.' (transitive verb)

(29) **Se s-fa-k̦ʷe-r-ep.*
I 1SG.IO-POT-go-DYN-NEG
Intended: 'I cannot go.' (intransitive verb)

The prefix *fe-* usually expresses internal possibility when the participant is not able to carry out an action because of his/her own properties or state. The suffix *-ŝʷə* has no restrictions of this sort. It can denote any type of possibility or probability. The precise nature of possibility expressed by these markers is not yet clear and lays outside the scope of the present paper.

The verb marked by the potential *fe-* is syntactically and morphologically intransitive, the absolute S is inanimate, while the oblique IO is animate, so the verb literally means something like 'this is not visible for me':

(30) a. *č̦'ale-xe-m bukve-xe-r O-Ø-a-λeʁʷə-xe-r-ep.*
boy-PL-OBL letter-PL-ABS 3SG.ABS-3PL.A-see-PL(ABS)-DYN-NEG
'The boys do not see the letters.'

b. *č̦'ale-xe-m bukve-Ø-xe-r*
boy-PL-OBL letter-PL-ABS
Ø-a-fe-λeʁʷə-xe-r-ep.
3SG.ABS-3PL.IO-POT-see-PL(ABS)-DYN-NEG
'The boys cannot see the letters.'

In (30), the transitive verb *λeʁwən* 'see' is given. The oblique A is cross-referenced by means of the prefix *a-*, while the absolutive DO is cross-referenced with a zero prefix. The situation in (30b) differs from that in (30a) only by the presence of the potential prefix and the fact that the former A is now cross-referenced with the prefix *a-* and is an argument of the potential prefix (i.e., an IO). The assumption that potential derivatives are intransitive is confirmed by the fact that they no longer take the reciprocal prefix *zere-* marking reciprocity between A and DO. To convey a meaning like 'the boys cannot see each other,' *ze-*, as in (31a), and not *zere-* (31b), is used, proving that the verb is intransitive and its arguments are not related to each other as A and DO.

(31) a. *č̣'ale-xe-m ze-fe-λeʁ^Wə-xe-r-ep.*
boy-PL-OBL REC.IO-POT-see-PL(ABS)-DYN-NEG
'The boys cannot see each other.'

b. **č̣'ale-xe-m zere-fe-λeʁ^Wə-xe-r-ep*
boy-PL-OBL REC.IO-POT-see-PL(ABS)-DYN-NEG
'The boys cannot see each other.'

The potential operation, which bears a modal meaning, acts at the same time as a valency-rearranging operation in terms of Dixon (1980) and Plungian (2000). The operation does not change the number of verbal arguments (i.e., agreement slots in the verb form), but it does change the status of A, making it an indirect object.

The potential operator can scope over the causative one. In other words, a morphological causative can further become an initial point for the potential derivation. For instance, the causative verb *ʁe-stən* 'burn' can be 'potentialized' to yield the potential derivative *fe-ʁestən* 'can burn':

(32) *Čəgə-r pŝaŝe-m O-Ø-fe-ʁe-stə-št-ep.*
tree-ABS girl-OBL 3SG.ABS-3SG.IO-BEN-CAUS-burn-FUT-NEG
'The girl will not be able to burn this tree.'

In (32), it is obvious that the scope of the potential prefix includes the causative one. Without the causative morpheme, the verb *stən* means 'burn'—it can be transitive or intransitive, but the subject cannot be animate. In the transitive use, the subject of *stən* can only be 'fire,' 'flame,' or similar objects. Thus the potential derivation with the meaning as in (32), where the subject is animate, is only possible if the verb is causativized before the potential derivation.

I am not aware of any example of the converse situation where the causative scopes over the potential operator. However, this may be a result of insufficiency of data.

5. Conclusions and Explanations

In this paper, the scope peculiarities of the causative marker in Adyghe have been considered. Throughout the paper I have shown that in some cases, the order of morphosyntactic operations can be checked using syntactic and morphological criteria. In cases such as (14), (15), and (20), two theoretically possible orderings of reflexive versus causative and reciprocal versus causative lead to the same meaning. In both cases, causativization tends to follow, and not precede, the other morphosyntactic operation. Note that the reverse order (causative before reflexive/reciprocal) is realized under other readings of the same combinations of markers, which cannot exist if causativization follows reflexivization/reciprocalization: namely, in those instances where the causer, introduced by the causative operation, is then included into a coreference relation with another argument.

There are stricter cases, like the combination of causative and antipassive where only one ordering is possible, and no exceptions are found. In this case the causative operation also follows the other one, thus confirming the general tendency illustrated by reflexives and reciprocals.

At the same time, some other operations, such as different types of applicatives and potential valency-rearranging operations, do not show any specific preference with respect to the causative marker: under some readings, such as reading (ii) in (27), corresponding operations can act after or before the causative operator to yield the same meaning. Unfortunately, no morphosyntactic tests allow us to find out which operation took place before the other.

How can these facts be accounted for? The data observed show that the combinations of causative and reciprocal or causative and reflexive are regulated in strict ways. Importantly, we are not aware of any restrictions of the same sort that affect combinations of other valency-changing devices—for example, benefactive and reflexive. However, it is more accurate to say that at present we have two possible solutions to choose. Either the causative operation is in a sense unique from any other operation, or there is a system of restrictions that affect each pair of valency-changing operations. Now we rather prefer to choose the first variant.

It seems that the causative derivation is in some respects different from any other valency-changing operation in Adyghe. First of all, as noted above, only causativization changes the semantics of the situation in a crucial way: namely, it adds a new subevent (causation) to the event structure. All other operations, such as applicatives, simply add a new participant to the argument frame of the predicate, but they do not create a new subevent.

Second, the causative morpheme occupies a special place in the verb form. Applicative morphemes are situated in the -6 slot and are immediately preceded by cross-reference prefixes marking the object introduced by the applicative operation. In contrast, the causative prefix is situated immediately before the verbal root, and the cross-reference marker of the causer can be separated from the causative prefix.

There are three combinations where the causative operator tends to occur (though see example (21) for an exception). These are causative + reflexive, causative + reciprocal, and causative + antipassive. They all have one common feature: if the causativization occurred before another valency-changing operation, this other operation (reflexivization, reciprocalization, or antipassive operation) would change the properties of the objects (the causee and the base direct object) of the causative construction: this operation would either make the causee coreferential with the base DO or would eliminate the causee. It seems that operations of this type are avoided in Adyghe. If it is possible to change the argument structure and the properties of some arguments before the verb is causativized, this should be done.

In fact, this makes us suppose that applicative operations (the addition of an indirect object) also occur before causativization under readings like (ii) in (27). However, in these cases we cannot be sure. As said above, no morphological or syntactic tests allow us to know what the relative ordering of operations is. Note also that applicativization, whether it takes place before or after causativization, does not modify any semantic or referential properties of the causer and the causee.

Reflexivization and reciprocalization of the type Causer = IO or Causer = DO can also occur after causativization, simply because there is no causer before causativization. In other words, the language has in these cases no way to order causativization after the other operation.

The Adyghe data are theoretically and typologically relevant. The facts considered throughout the article show that even if a language has a rigid order of markers in a part of the verb form, it is not possible to fully predict the set of possible orders of operations or combinations of semantic scopes. In cases when the derivation affects the objects of the causative construction, this derivation tends to occur before causativization. Morphosyntactic tests show that causativization tends to follow all other valency-changing operations. In this sense, causativization has a special role in the Adyghe system of valency-changing operations.

Another respect in which the Adyghe data can be of theoretic importance is the following: it turns out that sometimes, even in the domain of valency change, we cannot see directly the order in which the valency-changing devices are applied; both or all theoretically possible orders lead to the same meaning of the resulting form. The only possible way for these cases is to use morphosyntactic criteria—for instance, related to transitivity.

Finally, the data under analysis show that the ordering of morphosyntactic operations and the semantic scope of operators cannot be reduced to each other. Though mismatches are hard to find, and examples like (21) are rarely used by native speakers, the very existence of examples like this is important. Though semantically the scope of reflexive does not include the causative component, the morphosyntactic criteria show that reflexivization occurs after causativization.

Abbreviations

1, 2, 3—1st, 2nd, 3rd person; A—agent cross-reference prefixes; ABS—absolutive case, absolutive argument cross-reference prefixes; AUG—augment, prefix inserted for morphophonological reasons; BEN—benefactive; CAUS—causative; COORD—coordination marker; DAT – non-specialized applicative prefix introducing an indirect object; DIR—directional locative marker; DYN—dynamic verb marker; OBL—oblique case; IO—indirect object cross-reference prefixes; POT—potential; PL—plural; POSS—possessive marker; PST—past tense; REC—reciprocal; REFL—reflexive; RE—refactive.

References

Arkadiev, Peter, and Alexander Letuchiy. 2008. "Derivacii antipassivnoj zony v adygejskom jazyke" [Derivations of the antipassive zone in Adyghe]. In *Issledovanija po otglagol'noj derivacii*, edited by Vladimir Plungian and Sergej Tatevosov, 77–102. Moscow: Jazyki slavjanskix kul'tur.

Baker, Mark. 1985. "The mirror principle and morphosyntactic explanation." *Linguistic Inquiry* 16: 373–415.

Comrie, Bernard. 1976. "The syntax of causative constructions: Cross-language similarities and divergences." In *The grammar of causative constructions* [*Syntax and Semantics*, volume 6]., edited by Masayoshi Shibatani, 261–312. New York: Academic Press.

Cooreman, Ann. 1994. "A functional typology of antipassives." In *Voice: Form and Function* [*Typologicl Studies in Language* 27], edited by Barbara A. Fox and Paul J. Hopper, 49–88. Amsterdam: John Benjamins.

Dixon, Robert Malcolm Word. 1980. *Ergativity*. Cambridge: Cambridge University Press.

Grimshaw, Jane. 1986. "A morphosyntactic explanation for the mirror principle." *Linguistic Inquiry* 17: 745–749.

Hyman, Larry. 2003. "Suffix ordering in Bantu. A morphocentric approach." *Yearbook of Morphology 2002*, 245–281.

Kobele, Gregory M. 2002. "Formalizing mirror theory." *Grammars* 5 (3).

Korotkova, Natalia A., and Yury A. Lander. 2010. "Deriving affix ordering in polysynthesis: evidence from Adyghe." *Morphology* 20, 299–319.

Kumakhov, Mukhadin. 1971. *Slovoizmenenie adygskix jazykov* [Inflection of Adyghe languages]. Moscow: Nauka.

Kumakhov, Mukhadin, and Karina Vamling. 2006. Ergativnost' v cherkesskix jazykax. Malmö: School of International Migration and Ethnic Relations.

Lander, Yury A. 2009. "Množestvennaja reljativizacija: podlinnaja i mnimaja" [Multiple relativization: fake and real]. In *Aspekty polisintetizma: Ocherki po grammatike adygejskogo jazyka* [Aspects of polysynthesis: elements of grammar of Adyghe], edited by Yakov G. Testelets, Peter M. Arkadiev, Alexander B. Letuchiy, and Nina R. Sumbatova, 612–653. Moscow: RGGU.

Lander, Yury A., and Alexander B. Letuchiy. 2010. "Kinds of recursion in Adyghe morphology." In *Recursion and Human Language* [*Studies in Generative Grammar* 104], edited by Harry van der Hulst, 263–284. Berlin: Mouton de Gruyter.

Lander, Yury A., and Yakov G. Testelets. 2006. "Nouniness and specificity: Circassian and Wakashan." Paper presented at the conference Universals and Particulars in Parts-of-Speech Systems, Amsterdam. 2006, July 10.

Letuchiy, Alexander B. 2009a. "Kauzativ, dekauzativ i labil'nost" [Causative, anticausative and lability.]. In *Aspekty polisintetizma: Ocherki po grammatike adygejskogo jazyka* [Aspects of polysynthesis: elements of grammar of Adyghe], edited by Yakov G. Testelets et al. Moscow: Russian State University for Humanities. 372–428.

Letuchiy, Alexander B. 2009b. "Affiksy benefaktiva i malefaktiva: sintaksicheskie osobennosti i krug upotreblenij." [Benefactive and malefactive markers: syntactic properties and range of uses.] In *Aspekty polisintetizma: Ocherki po grammatike adygejskogo jazyka* [Aspects of polysynthesis: elements of grammar of Adyghe], edited by Yakov G. Testelets et al. Moscow: Russian State University for Humanities. 329–371.

Letuchiy, Alexander B. 2012. "Ergativity in the Adyghe system of valency-changing operations." In *Ergativity and voice*, edited by Gilles Authier and Katharina Haude, 323–353. Berlin: Mouton de Gruyter.

Mithun, Marianna. 2000. "The reordering of morphemes." In *Reconstructing grammar: comparative linguistics and grammaticalization*, edited by Spike Gildea, 231–258. Amsterdam: John Benjamins.

Nichols, Johanna, David A Peterson, and Jonathan Barnes. 2004. Transitivizing and detransitivizing languages. Linguistic typology 8.2. 149–211.

Paris, Catherine. 1989. "Esquisse grammaticale du dialecte abzakh (tcherkesse occidental)." In *The Indigenous Languages of the Caucasus, Vol. 2. The North West Caucasian Languages*, edited by George B. Hewitt, 155–260. Delmar, NY: Caravan.

Paris, Catherine, and Nijaz Batouka. 2005. *Dictionnaire abzakh (tcherkesse occidental)* [The dictionary of Abdzakh (West Circassian)]. Tomes I–II. Leuven: Peeters.

Paster, Mary. 2005. "Pulaar verbal extensions and phonologically driven affix ordering." *Yearbook of Morphology 2005*, 155–199.

Peterson, David A. 2007. Applicative constructions. Cambridge: CUP.

Plungian, Vladimir P. 2000. *Obščaja morfologija* [General morphology]. Moscow: URSS. There is no full spelling for this publisher.

Pylkkänen, Liina. 2002. "Introducing Arguments." PhD diss., Massachusetts Institute of Technology.

Ramchand, Gillian. 2008. *Verb Meaning and the Lexicon: A First Phase Syntax*. Cambridge: Cambridge University Press.

Rogava, Georgi, and Zejnab Kerasheva. 1966. *Grammatika adygejskogo jazyka* [A grammar of Adyghe]. Krasnodar-Majkop: Krasnodarskoe knizhnoe izdatel'stvo.

Shaov, Zh. A. 1975. *Adygejsko-russkij slovar'* [Adyghe–Russian dictionary]. Majkop: Adygejskij nauchno-issledovatel'skij institut.

Shibatani, Masayoshi. 2002. "Introduction: Some basic issues on the grammar of causation." In *The Grammar of Causation and Interpersonal Manipulation*, edited by Masayoshi Shibatani. Amsterdam: John Benjamins. 1–22.

Shibatani, Masayoshi, and Prashant Pardeshi. "The causative continuum." In *The Grammar of Causation and Interpersonal Manipulation*, edited by Masayoshi Shibatani. Amsterdam: John Benjamins. 85–125.

Simpson, Jane, and Michael Withgott. 1986. "Pronominal clitic clusters and templates." In *The Syntex of Pronominal Cliticcs* [*Syntax and Semantics* 19], edited by Hagit Borer, 149–174. New York: Academic Press.

Smeets, Rieks. 1992. "On valencies, actants and actant coding in Circassian." In *Caucasian Perspectives*, edited by George B. Hewitt, 98–144. München: Lincom Europa.

Smeets, Rieks. 1996. *Studies in West Caucasian Phonology and Morphology*. Leiden: Hakuchi Press.

Testelets, Yakov G. et al. (eds.). 2009. *Aspekty polisintetizma: Ocherki po grammatike adygejskogo jazyka* [Aspects of polysynthesis: elements of grammar of Adyghe]. Moscow: Russian State University for Humanities.

Tharkakho, Junus. 1991. *Adygejsko-russkij slovar'* [Adyghe-Russian dictionary]. Majkop: Adygejskoe knižnoe izdatel'stvo.

4

Modern Greek Parasynthetic Verbs: A Hierarchical Relationship Between Prefixes and Suffixes?

Angeliki Efthymiou

1. Introduction

This paper deals with parasynthetic verbs in Modern Greek, whose structure involves the simultaneous presence of prefixes and suffixes: for example, *apo-cefal-iz-o* 'to decapitate' (DPREF-BASE-DSUFF-ISUFF). After a brief overview of the relevant literature, I will argue in favor of a hierarchical relation among the prefixes and suffixes that participate in these structures. I will also suggest that the ordering of the affixes that participate in Modern Greek parasynthetic verb formation can be explained with reference to their information content; that is, their contribution in the morphological structure of the derived word: the first affix (i.e., the prefix) affects the meaning, the argument structure, and the register of the derivative more strongly than the second affix (i.e., the suffix).[1] More specifically, it will be shown that in the case of Modern Greek parasynthetic verbs (henceforth MGPVs) prefixes behave like internal prefixes (in the sense of Di Sciullo 1997; see also Ralli 2004) and determine the core semantics of the derived verb. I will also demonstrate that all affix combinations are not possible and that the preference for certain suffixes is mainly determined by frequency factors. Finally, it will be argued that the occurrence of multiple derivational affixes in the structure of the MGPVs can be explained in terms of optimal shape, constructional iconicity, transparency, and biuniqueness. My claims will be exemplified by reference to 308 parasynthetic verbs that have been collected from the *Reverse Dictionary of Modern Greek* (2002). The paper is organized as follows. Section 2 presents the structure of parasynthetic verbs and provides a brief overview of the relevant literature. Section 3 explains how the data of this study were selected and offers a brief presentation of the affixes that participate in MGPVs. Section 4 discusses

[1] It should be noted that in Bell and Plag's (2012) study of English compounds the term *informativeness* (or *information content*) is used in a different sense, i.e., to refer to the morphological family sizes of the constituents and the constituents' degree of semantic specificity.

diachronic and synchronic features of Greek prefixes. In sections 5 and 6 evidence is provided that the linear ordering of affixes is not motivated by scopal hierarchy (in the sense of Rice 2000), but reflects the contribution of each affix in the morphological structure of the derived word. In section 7 it is shown that affix combinations are determined by various factors, such as register and frequency. It is also suggested that the preference for certain suffixes is mainly determined by frequency factors. In section 8 it is argued that the occurrence of multiple derivational affixes in the structure of the MGPVs can be explained in terms of optimal shape, constructional iconicity, transparency, and biuniqueness. Section 9 offers a brief summary of the main arguments.

2. The Structure of Parasynthetic Verbs

Parasynthesis is generally described as a process of simultaneous multiple affixation (cf. Lieber 2010: 202, Plag 2003: 40). As the Modern Greek and English examples in (1) illustrate, parasynthetic verbs are formed by an adjectival or nominal base and two derivational affixes, one of which is a prefix and the other is a suffix, which are simultaneously added to the right and the left of the base:

(1) a. *kse-δond-jáz-o* 'to take one's teeth out' (Modern Greek)
DPREF-tooth$_N$-DSUFF-ISUFF
**δondjázo*

b. *kse-fluδ-íz-o* 'to peel'
DPREF-skin$_N$-DSUFF-ISUFF
**fluδ-íz-o*

c. *kse-port-íz-o* 'to go away, to slip out'
DPREF-door$_N$-DSUFF-ISUFF
**port-íz-o*

d. *eks-anθrop-íz-o* 'to humanize'
DPREF-human$_N$-DSUFF-ISUFF
**anθrop-íz-o*

e. *ipo-γram-íz-o* 'to underline'
DPREF-line$_N$-DSUFF-ISUFF
**γram-íz-o*

f. *apo-liθ-ón-o* 'to petrify'
DPREF-stone$_N$-DSUFF-ISUFF
**liθ-ón-o*

g. *apo-vlak-ón-o* 'to make someone stupid, to stupefy'
DPREF-stupid$_A$-DSUFF-ISUFF
**vlak-ón-o*

h. *de-caffein-ate* (English)
DPREF-caffeine$_N$-DSUFF
**caffeinate*

In the literature parasynthetic constructions are usually described as bracketing paradoxes; that is, cases in which the semantic interpretation or the phonological organization of a word seems to conflict with its internal structure (cf. Lieber 2010: 190–193). Although a number of suggestions have been put forward so far in order to account for this paradox, parasynthetic constructions still remain a debatable topic. Most linguists reject the ternary prefix-base-suffix solution (i.e., the simultaneous addition of affixes) and prefer either the addition of an affix (suffix for Scalise 1984, prefix for Corbin 1987) followed by the addition of a second affix, or propose a circumfixal hypothesis (cf. Crocco Galéas & Iacobini 1993). Furthermore, most linguists suggest that parasynthetic verbs have corresponding nonparasynthetic structures that are perfectly legitimate as binary structures and that the unavailability of possible bases should be explained by semantic or pragmatic reasons (cf. Efthymiou 2001, Anastassiadis-Symeonidis & Masoura 2009, Iacobini 2010, Lieber 2010). On the other hand, the circumfixal hypothesis has been subject to severe criticism, because it leads to a certain amount of redundancy (cf. Ralli 2004: 5–6).

With regard to Greek parasynthetic verbs, Thomadaki (1996) and Ralli (2004), following Scalise's analysis of similar cases in Italian (cf. Scalise 1984: 149–150), adopt a binary structure: first, the noun/adjectival base is combined with a derivational suffix in order to produce a possible but nonexisting bound verbal stem. The suffix is responsible for the verbal category of the derivative. Second, the prefix is added to the possible stem. Third, the inflectional suffix *-o* transforms the stem into an actual word.[2] On the other hand, following Corbin's (1987) model, Efthymiou (2001) and Anastassiadis-Symeonidis and Masoura (2009, 2012) suggest that the so-called derivational suffixes of the Modern Greek parasynthetic verbs are class-markers (i.e., verbalizers) and do not carry the semantic content of a suffix, but signal the integration of the derivative into either a grammatical or a semantic category.[3] Interestingly, both solutions imply an intermediate step and accept that the suffix (or the suffix-like ending) is responsible for the grammatical category of the parasynthetic verb.

It should be mentioned here that the adoption of a binary structure, despite its advantages, has been subject to severe criticism (see, among others, Rainer 1989, Crocco Galéas & Iacobini 1993). According to Rainer (1989: 38–39), the principle of binarity is not to be explained necessarily by adopting a solution that implies two steps. He rather suggests that the binarity is a consequence of the hierarchical (i.e., subordinative) relation among the elements that form the word. In addition,

[2] In this paper I will only deal with the ordering of derivational affixes. I will leave the investigation of inflectional suffixes for future research.

[3] According to Corbin's model, class marking is a case of pseudosuffixing. Specifically, a class marker is an item with iconic role and signifier of similar form to the suffix's signifier but without the semantic instruction of the suffix. The ending that serves as the class marker is not selected arbitrarily, but copies the form and the intrinsic properties of the suffix of the corresponding nonprefixed word (see also Anastasiadis-Symeonidis & Masoura 2009, 2012 for discussion).

one could argue that the adoption of an intermediate step is confronted with the problem of determining the meaning of possible bases: are **top-íz-o*, **γram-íz-o*, and **vaθm-íz-o* possible bases for *meta-top-íz-o* 'to displace,' *ipo-γram-íz-o* 'to underline,' and *ipo-vaθm-íz-o* 'to downgrade'? What is their possible meaning? Are these forms avoided because of the well-known phenomenon of blocking (cf. Aronoff 1976)? For example, is **top-íz-o* blocked by the attested compound verb *topo-θet-ó* 'to place'? Can we accept that **port-íz-o* 'to enter (through the door)' is blocked by *béno* 'to enter'? In this case, why isn't the parasynthetic verb *kse-port-íz-o* 'to go away, to slip out' also blocked by *vjéno* 'to go out'? What kind of pragmatic reason could explain the unavailability of **vlak-ón-o* and **atomik-év-o* as bases for the formation of the verbs *apovlakóno* 'to make someone stupid' and *eks-atomik-év-o* 'to individualize'? More recently, the idea of a necessary intermediate step has been challenged by Booij (2010: 44), who proposes a unified schema for the English parasynthetic verbs. He particularly suggests that parasynthetic verbs are derived by means of a generalized schema that is the unification of more basic schemas of word formation as shown in (2):

(2) *de-nuclear-ize*
$[de\ [x_V]]_V + [[x]_{N/A}\ \text{ize}]_V = [de\ [[x]_{N/A}\ ize]_V]_V$

It should be noted that according to Booij (2010), word-formation patterns can be seen as abstract schemas that generalize over sets of existing complex words with a systematic correlation between form and meaning. In this paper I adopt Booij's constructional approach and propose a similar generalized schema for the Modern Greek parasynthetic verbs as follows:

(3) $[pref\ [x_V]]_V + [[x]_{N/A}\ suff]_V = [pref\ [[x]_{N/A}\ suff]_V]_V$

Given that this analysis implies the co-occurrence of two derivational processes (cf. Booij 2010: 43), the nested structure indicates the hierarchical relation between prefixation and suffixation. Interestingly enough, the appearance of parasynthetic verbs in Ancient Greek seems to provide additional evidence for this generalized schema solution:

(4) *apo-cefal-íz(o)* 'to decapitate' (Ancient and Modern Greek)
$[apo\ [[x]_N\ ízo]_V]_V$

(5) *apo-δ ermat-íz(o)* 'to flay' (Ancient Greek)
$[apo\ [[x]_N\ ízo]_V]_V$

As Papanastasiou (2011) suggests, the appearance of Ancient Greek parasynthetic verbs can be explained in terms of reanalysis. According to him, some prefixed verbs with the structure [pref [Noun + suff] $_V]_V$ denoting 'cutting, breaking, splitting' and so on (e.g., *opl-ízo* 'to arm' > *af-opl-ízo* 'to disarm') may have been reanalyzed as pref+Noun+suff, and this gave birth to new formations with the structure: $[pref\ [[x]]_N\ suff]_V]_V$.

In the rest of the paper I will provide evidence in favor of a hierarchical relation among prefixes and suffixes. I will also suggest that the above-mentioned generalized schema has the following specific constructional properties:[4] (a) the first affix (i.e., the prefix, which has an internal character) affects the meaning, the argument structure, and the register of the derivative more strongly than the second affix (i.e., the suffix); and (b) the suffix is responsible for the grammatical category of the parasynthetic verb. Interestingly enough, this suggestion is in line with Efthymiou (2001) and Anastassiadis-Symeonidis and Masoura (2009, 2012), who claim that the suffixes do not determine the semantics of the MGPVs.

3. Methodology, Data, and Preliminary Remarks

Given that electronic corpora of Modern Greek are still relatively small and are not annotated for morphological purposes, I take as a starting point the analysis of 300 parasynthetic verbs in *-ízo*, *-jázo*, *-iázo*, and *-óno* in Tsakou (2010) and a list of eight parasynthetic verbs in *-évo* extracted from the *Reverse Dictionary of Modern Greek* (2002, henceforth *RDMG*). Tsakou's data were also extracted from the *RDMG*. The number of parasynthetic verb types in *-ízo*, *-jázo*, *-iázo*, *-óno*, and *-évo* as attested in the *RDMG* is summarized in Tables 4.1 and 4.2.[5] In Table 4.1 the verbs are ordered according to their suffix 'frequency,' whereas in Table 4.2 the verbs are ranked according to their prefix 'frequency':[6]

TABLE 4.1

Parasynthetic Verbs Extracted from the *RDMG* (2002) Ranked by Suffix 'Frequency'

Verbal suffix	Number of verb types
-ízo	144
-óno	116
-jázo	38
-évo	10
-(i)ázo	2
Total	308

[4] As stated by Booij (2010: 45), unified schemas may possess specific constructional properties of their own.

[5] Note that the resulting list of raw data was cleared, and exclude (a) all forms that did not feature the particular suffixes, and (b) all prefixed forms with an attested base.

[6] In this paragraph the term frequency corresponds to the number of occurrences of each affix in the *RDMG*, i.e., their type frequency. 'Type frequency' is concerned with the number of different words that occur with the morphological category under consideration (cf. Baayen 2008; Plag 2006).

TABLE 4.2

Parasynthetic Verbs Extracted from the *RDMG* (2002) Ranked by Prefix 'Frequency'

Prefix	Number of verb types
kse- 'out of, negative, privative, reversative, excessive'	111
ek- 'out of, privative, intensive, completive'	62
apo- 'away from, privative, reversative, intensive, completive'	40
epi- 'on, over'	20
en- 'in'	15
δia- 'between, across'	14
kata- 'down(wards), towards, excessive'	9
pros- 'towards'	9
ana- 'upwards, again'	8
sin- 'together with, jointly'	4
meta- 'after'	5
ipo- 'under, not sufficiently'	3
para- 'close to, excessive'	2
peri- 'around'	2
pro- 'before, in front of'	2
iper- 'over, excessive'	2
Total	308

The figures in Table 4.1 and Table 4.2 seem to suggest a number of points.[7] First, it appears that *-ízo* and *-óno* are much more frequent than *-jázo*, *-iázo*, and *-évo* in parasynthetic verb formation. Second, Modern Greek prefixes seem to differ considerably in their type frequency (named also 'extent of use' or 'realized productivity'; cf. Baayen 2008, Plag 2006): *kse-* appears to be the most frequent prefix attested in the list, followed by *ek-* and *apo-*. Third, it appears that the 'extent of use' of some prefixes (e.g., *pro-*, *iper-*) is very limited in Modern Greek parasynthetic verb formation.

Another interesting remark about MGPVs is that they prefer attaching to nominal bases, as can be seen in Table 4.3:

TABLE 4.3

Preference for Nominal Bases

Grammatical category of base	Number of verb types
Nominal bases	258
Adjectival bases	50

[7] Note that the attachment of Modern Greek prefixes to a base is very often accompanied by phonological changes. Thus, Modern Greek prefixes have various allomorphic forms, such as *an-* (allomorph of *ana-*), *kat-/ kaθ-* (allomorph of *kata-*), *δi-* (allomorph of *δia-*), *em-* (allomorph of *en-*), *ef-/ep-* (allomorphs of *epi-*), *ap-/af-* (allomorphs of *apo-*), *eks-* (allomorph of *ek-*), *ks-* (allomorph of *kse-*), *sim-/siγ-/sil-/sis-/sir-* (allomorphs of *sin-*), *met-/meθ-* (allomorphs of *meta-*), or *ip-/if-* (allomorphs of *ipo-*).

TABLE 4.4
Occurrence in Nonparasyntetic Verb Formations

(Non-parasynthetic) derivative	Base
aristotel-íz-o 'to imitate Aristotle'	*Aristotélis* 'Aristotle'
ritiδ-jáz-o 'to have wrinkles, to become wizened'	*ritíδa* 'wrinkle'
vutir-ón-o 'to butter'	*vútiro* 'butter'
kse-δén-o 'to untie'	*δéno* 'to tie'
apo-sinδé-o 'to disconnect'	*sinδéo* 'to connect'
ek-pné-o 'to expire'	*pnéo* 'to blow'

In addition, all derivational prefixes and suffixes that participate in the MGPVs can also occur in nonparasynthetic verb formations, as can be seen from the examples in Table 4.4.

The next section presents several diachronic and synchronic features of prefixes that participate in MGPVs and are relevant for the following discussion.

4. Diachronic and Synchronic Features of Greek Prefixes in Parasynthetic Formations

According to Ralli (2005: 42), Modern Greek has twenty well-known prefixes. These are listed, along with examples, in (6)–(25):

(6) *a-*: *á-oplos*$_{\mathrm{A}}$ 'unarmed,' *á-plitos*$_{\mathrm{A}}$ 'unwashed'

(7) *ana-*: *ana-vaptízo*$_{\mathrm{V}}$ 'to rebaptize,' *ana-δromí*$_{\mathrm{N}}$ 'recursion'

(8) *anti-*: *anti-ritiδikós*$_{\mathrm{A}}$ 'anti-wrinkle,' *anti-próeδros*$_{\mathrm{N}}$ 'vice-president,' *anti-paraθéto*$_{\mathrm{V}}$ 'to justapose, contrast'

(9) *apo-*: *apó-meros*$_{\mathrm{A}}$ 'remote, secluded,' *apo-prosanatolízo*$_{\mathrm{V}}$ ' to disorientate'

(10) *δia-*: *δia-valkanikós*$_{\mathrm{A}}$ 'inter-Balkan,' *δia-pernó*$_{\mathrm{V}}$ 'to penetrate,' *δia-γónios*$_{\mathrm{A/N}}$ 'diagonal'

(11) *δis-*: *δis-árestos*$_{\mathrm{A}}$ 'unpleasant,' *δis-évretos*$_{\mathrm{A}}$ 'difficult to find'

(12) *ef- /ev-*: *éf-kamptos*$_{\mathrm{A}}$ 'flexible,' *ev-íonos*$_{\mathrm{A}}$ 'auspicious'

(13) *ek-*: *ék-filos*$_{\mathrm{A}}$ 'licentious,' *ek-lojikévo*$_{\mathrm{V}}$ 'to rationalize,' *ek-próθesmos*$_{\mathrm{A}}$ 'overdue'

(14) *en-*: *em-fitévo*$_{\mathrm{V}}$ 'to implant,' *ém-psixos*$_{\mathrm{A}}$ 'animate'

(15) *epi-*: *epi-kalípto*$_{\mathrm{V}}$ 'to coat,' *epi-veveóno*$_{\mathrm{V}}$ 'to confirm,' *epi-táfios*$_{\mathrm{A/N}}$ 'epitaph'

(16) *kse-*: *kse-δéno*$_{\mathrm{V}}$ 'to untie,' *kse-kuf-éno*$_{\mathrm{V}}$ 'to make somebody completely deaf,' *kse-káθaros*$_{\mathrm{A}}$ 'unequivocal'

(17) *iper-*: *iper-ánθropos*$_{A/N}$ 'superhuman, superman,' *iper-oceánios*$_{A}$ 'transoceanic,' *iper-fortóno*$_{V}$ 'to overload'

(18) *ipo-*: *ip-éδafos*$_{N}$ 'subsoil, underground,' *ipó-γlikos*$_{A}$ 'sweetish,' *ipo-δieró*$_{V}$ 'to subdivide'

(19) *is-*: *is-pnéo*$_{V}$ 'to inhale,' *is-érxome*$_{V}$ 'to enter'

(20) *kata-*: *katá-mavros*$_{A}$ 'coal black,' *kata-céo*$_{V}$ 'to burn all over,' *kat-eδafízo*$_{V}$ 'to demolish'

(21) *para-*: *para-káno*$_{V}$ 'to overdo,' *para-θalásios*$_{A}$ 'littoral,' *par-erminévo*$_{V}$ 'to misinterpret'

(22) *peri-*: *perí-metros*$_{N}$ 'perimeter,' *peri-válo*$_{V}$ 'to surround,' *peri-xarís*$_{A}$ 'jubilant'

(23) *pro-*: *pro-jajá*$_{N}$ 'great grandmother,' *pro-δimosiefsi*$_{N}$ 'pre-publication,' *pro-istorikós*$_{A}$ 'prehistoric,' *pro-ástio*$_{N}$ 'suburb,' *pro-poló*$_{V}$ 'to sell in advance'

(24) *pros-*: *pros-arázo*$_{V}$ 'to run aground,' *pros-krúo*$_{V}$ 'to bump,' *pros-pernó*$_{V}$ 'to overtake,' *pros-céfalo*$_{N}$ 'cushion'

(25) *sin-*: *sin-oδiyós*$_{N}$ 'codriver,' *sin-ipárxo*$_{V}$ 'to coexist'

As can be seen in (6)–(25), all Greek prefixes can be used in verbal morphology with the exception of *a-*, *δis-*, and *ef-/ev-* (cf. also Table 4.2).[8] It is also interesting to note that all prefixes that participate in verb derivation are also used in parasynthetic verbs with the exception of *anti-*.

According to Ralli (2003), Modern Greek prefixes belong to two categories: those that are bound at all levels of analysis (e.g., *a-*, *δis-*, *kse-*) and those that derive from free morphemes. Among the latter, most linguistic analyses include the set of Ancient Greek prepositions (e.g., *ek-*, *kata-*, *pros-*, *para-*, etc.; see, among others, Philippaki-Warburton 1970, Malikouti-Drachman & Drachman 1992, Smirniotopoulos & Joseph 1998, Ralli 2004, 2005).

It is important to add, however, that the derivational status of several prefixes in not a clear-cut case (cf. among others Ralli 2004, 2005), and that in most traditional grammars (cf. Triandafyllidis 1991), the combination between an Ancient Greek preposition and a base is interpreted as part of a compounding process. For example, while *ksana* 'again' (e.g., *ksanaγrafo* "re-write") is treated as left-hand constituent of compounds by Ralli (2004, 2005), there are also analyses that refer to it as prefix (see for example Philippaki-Warburton 1970).

In what regards Modern Greek preverbation, Ralli (2004, 2005) argues that Modern Greek preverbs may be assigned to three classes: the first class includes prefixes like *kse-*, which are not attested as independent words. *Kse-* never appears

[8] For these prefixes, see also Klairis and Babiniotis (2009).

separately and bears no primary stress. As Mendez-Dosuna (1997) has shown, *kse-* is a formation of Medieval Greek, which developed from the combination of the Ancient Greek preverb *ek* (or *eks-* before vowels) with the verbal syllabic augment *e-* (i.e., *ek-* + *e-* > *eks* + *e* > *kse-*). (Note that, as seen in Table 4.2, the prefix *kse-* is by far the most frequent prefix in MGPVs; cf. also Efthymiou 2001, 2002, Ralli 2004).

The second class contains prefixes derived from Ancient Greek preverbs (e.g., *ana-*, *anti-*, *apo-*, *δia-*, *is-*, *en-*, *ek-*, *epi-*, *kata-*, *meta-*, *para-*, *peri-*, *pro-*, *pros-*, *sin-*, *iper-*, *ipo-*),[9] which can also be used as prepositions or conjunctions (cf. also Bortone 2010 for the diachrony of Greek prepositions). An interesting fact about the members of this class is that most still possess some of their original Ancient Greek properties but have also developed some new meanings or functions. For example, *para-* has developed a new meaning relating to the excessive realization of an event: for example, *káno* 'do' → *para-káno* 'over-do.'[10] Although Greek preverbs were free morphemes in Ancient Greek, most linguists agree that these items should be analyzed as affixes, synchronically (cf., for example, Philippaki-Warburton 1970, Malikouti-Drachman & Drachman 1992, Smirniotopoulos & Joseph 1998, Ralli 2004, 2005). The basic arguments for treating Modern Greek preverbs as prefixes are the following: (a) in most cases, the use of the similar prepositions is restricted to fixed or archaic expressions; (b) Modern Greek prepositions are always free (e.g., *ja* 'for,' *me* 'with'); and (c) the meaning of all prefixed verbs is not always compositional.

The members of the third class are adverbs that can appear either as phrasal elements or first constituents of compounds (e.g., *ksana-* 'again' and *poli-* 'much, a lot'; *ksana-γráfo* 'rewrite,' *poli-tróo* 'eat a lot'). According to Ralli (2004), the [adverb verb] formations are not derivatives but compounds. Interestingly, the members of this class are not used in parasynthetic verb formation, but can be added at the left-hand edge of the parasynthetic verbs: for example, *ksana-apocentróno* 'to re-decentralize' (see also Ralli 2004 for the co-occurrence and ordering of Modern Greek preverbs).[11]

It should be kept in mind, however, that there are cases where the distinction prefixed structure and a compound one is blurred to some extent, since some independent words may behave as prefixes (e.g., *ksana* 'again') and some preverbs

---

[9] The preverbs of Ancient Greek were the following: *αμφι* /amphi/, *ανα* /ana/, *αντι* /anti/, *απο* /apo/, *δια* /dia/, *εις* /eis/, *εν* /en/, *εκ* /ek/, *επι* /epi/, *κατα* /kata/, *μετα* /meta/, *παρα* /para/, *περι* /peri/, *προ* /pro/, *προς* /pros/, *συν* /syn/, *υπερ* /hyper/, *υπο* /ypo/ (see Humbert 1972). According to Ralli (2004), *αμφι* /amphi/ is not productive in Modern Greek.

[10] For more details on this see the discussion in Delveroudi & Vassilaki (1999), Efthymiou (2003), and Ralli (2004).

[11] In this paper I will not deal with the co-occurrence of the MGPVs' prefixes with other preverbs: e.g., *iper-apocentróno* 'to over-decentralize,' *ksana-apocentróno* 'to re-decentralize,' *para-apovlakóno* 'to make extremely stupid.' I will leave the investigation of the co-occurrence and ordering of preverbs in MGPVs for future research.

may display a rather loose relation with the base (e.g., the so called 'excessive' *para-*: cf. among others Ralli 2004: 299).

With respect to the semantic modifications they bring to the base, verbal prefixes can be distinguished into two types (Di Sciullo 1997, 2005). Internal prefixes modify the semantic core of the base, affecting its argument structure and Aktionsart, while external prefixes adverbially modify the base as a whole and do not affect its argument structure or Aktionsart. In accordance with Di Sciullo's definition, Ralli (2004) claims that *kse-*, *apo-*, and *para-* display a dual character, since they behave like internal prefixes when they affect the meaning of the base, and like external prefixes when they bring only external specifications to it. For example, *kse-* and *apo-* are internal when they function as intensives (e.g., *kuf-éno* 'to make somebody deaf' → *kse-kuf-éno* 'to make somebody completely deaf,' *trel-éno* 'to drive somebody mad' → *apo-treléno* 'to drive somebody completely mad') and external when found with a reversative meaning (e.g., *δéno* → 'to tie' *kse-δéno* 'to untie,' *sinδéo* 'to connect' → *apo-sinδéo* 'to disconnect'). *Para-* is internal when it expresses the basic idea of proximity or parallelism to the meaning of the base (e.g., *méno* 'stay, live' → *para-méno* 'stay on, remain') and external in its excessive meaning (*tróo* 'eat' → *para-tróo* 'over-eat'). Interestingly enough, privative, ablative, and causative–completive meanings (i.e., the internal variants of *apo-*, *kse-* and *para-*) appear mostly in MGPV parasynthetic formations. On the other hand, as will be shown, the external variants of *apo-*, *kse-*, and *para-* do not seem to participate in parasynthetic verb formation (see also section 5.2). Given the dual character of *kse-*, *apo-*, and *para-*, Ralli (2004) suggests that semantic modification is not a safe distinctive criterion for Modern Greek preverbs on its own. She instead proposes that other properties of these prefixes, such as phonological and combinatorial properties, are more relevant to their morphological status. For example, as stated by Ralli (2004), while the *para-* formations are subject to an optional vowel deletion when *para* is external (e.g., *para-éxo* 'to over-have'), an obligatory vowel deletion is attested when *para-* is internal (e.g., *par-érxome* 'elapse, pass by'). Interestingly enough, the external variant of *para-*, which does not participate in parasynthetic verb formation (cf. also section 5), can be added at the left-hand edge of the parasynthetic verbs: for example, *para-apovlakóno* 'to make extremely stupid.'

Moreover, it should be noted that the prefixes that participate in Modern Greek verb parasynthetic formation may cause phonological alternations, even though they seem to be stress-neutral. Most of them display vowel deletion or assimilation in the appropriate context: for example, *ks-eyn-jáz-o* 'to be free from care,' *em-psix-ón-o* 'to reanimate, to encourage,' *ep-arjir-ón-o* 'to cover with silver,' *eks-atomik-év-o* 'to individualize.' By contrast, Modern Greek verb-forming suffixes are always stress-shifting, since they bear primary stress: for example, *alat-íz-o* 'to salt' (*aláti* 'salt').

It should also be mentioned that although Modern Greek prefixes do not affect the stress of MGPVs, stress shift may occur in Greek deverbal formations:

for example, *fuskotós* 'inflated' versus *kse-fúskotos* 'uninflated.' As claimed by Nespor and Ralli (1996), this stress shift is generally due to the presence of prefixes (see also *γnósi* 'knowledge' versus *pró-γnosi* 'forecast, prognosis').

Finally, the fact that the Greek prefixes derive words of more than one grammatical categories (e.g., *kse-káθaros* 'unequivocal,' *pro-istoría*$_{N}$ 'prehistory,' *pro-klasikós*$_{Adj}$ 'pre-classical,' *katá-mavros*$_{Adj}$ 'coal black,' *ék-filos*$_{Adj}$ 'licentious'; cf. also the examples in 5–24) implies that in the case of MGPVs the element that is responsible for the verbal category of the derivative is the suffix and not the prefix. This means that in MGPVs the prefixes do not determine the output grammatical category,[12] and hence do not violate the Righthand Head Rule (RHR) (cf. Williams 1981; cf. Ralli 2004 for the RHR in Greek word formation).

5. Derivational Evidence for the Hierarchical Relation Between Prefixes and Suffixes

In this section the meanings of nonparasynthetic suffixed verbs will be compared to those of parasynthetic verbs. It will be shown that in the case of Modern Greek parasynthetic verb formation prefixes behave like internal prefixes and determine the core semantics of the parasynthetic verb.

5.1 THE SEMANTICS OF N/A-TO-V SUFFIXATION

Modern Greek has eight well-known verb-forming suffixes (cf. Ralli 2005, Efthymiou 2011, 2013, Efthymiou, Fragaki, & Markos 2012, among others). These are listed, along with examples, in (26):

(26) a. *-ízo*: *vurtsízo* 'to brush'
b. *-óno*: *vutiróno* 'to butter'
c. *-jázo*: *ritiδjázo* 'to wrinkle'
d. *-(i)ázo*: *δiplasiázo* 'to double'/*onomázo* 'to denominate'
e. *-évo*: *proeδrévo* 'to chair, to preside'
f. *-éno*: *kondéno* 'to shorten'
g. *-áro*: *stresáro* 'to stress'
h. *-íno*: *oksíno* 'to sharpen'

As shown by Charitonidis (2005, 2011) and Efthymiou (2011, 2013), Modern Greek verbal suffixes have a wide variety of meanings,[13] such as causative/resultative 'cause to become x/turn into x,' inchoative 'become x,' ornative 'provide with x,

---

[12] This assumption is in line with Ralli (2003: 43), who claims that *kse-* and *para-* do not bring any categorical change to the verbal base.

[13] All verbs are coded in terms of the semantic categories established by Plag (1999).

locative' 'put into/on x,' instrumental 'use x,' similative 'imitate x,' stative/essive 'be x,' and so on. Interestingly enough, some of the derivatives are ambiguous: for example *neróno* can mean 'add water into a liquid' or 'become like water' (*neró* 'water'; for discussion see Efthymiou 2011; see also Plag 1999: 127–133, Lieber 2004: 84 for similar remarks on English suffixed verbs).

The principal meanings of *-ízo* verbs can be described as '(cause) to become x,' 'imitate x,' 'perform/do/make x,' 'provide with x,' 'put in(to) x,' and 'use x.' Some examples of *-ízo* derivatives are given in (27):

(27)	a. *kaθarízo* 'to clean'	*kaθarós* 'clean'
	b. *piθicízo* 'to imitate ape's behavior'	*píθikos* 'ape'
	c. *alatízo* 'to salt'	*aláti* 'salt'
	d. *vurtzízo* 'to brush'	*vúrtsa* 'brush'

It must be pointed out, however, that, according to Anastassiadis-Symeonidis (1994), Charitonidis (2005), and Efthymiou (2011), the semisuffix *-pió*, which is more transparent and predictable in meaning and less sensitive to phonological, morphological, and lexical restrictions, has begun to replace *-ízo* in the domain of the causative/resultative meaning.

The meanings of *-óno* derivatives can be described as 'provide with x,' '(cause to) become x,' 'use x,' and 'put into x,' as seen in (28):

(28)	a. *vutiróno* 'to butter'	*vútiro* 'butter'
	b. *malakóno* 'to soften'	*malakós* 'soft'
	c. *karfóno* 'to nail'	*karfí* 'nail'

According to Efthymiou (2013), the forms *-jázo* and *-(i)ázo* are two different quasi-homonymous suffixes. The suffix *-jázo* is stylistically restricted and derives mainly [–learned] words with negative connotations. The most robust semantic pattern of *-jázo* derivatives is the meaning 'be provided with many and usually unwanted endogenous entities' (cf. Charitonidis 2011, Efthymiou 2011, 2013). In *-jázo*, [ia] is pronounced as one syllable, while the consonant of the base or [i] is palatalized. The other quasi-homonymous suffix has the [+/–learned] form *-(i)ázo* and is pronounced [*iázo*] or [*ázo*]. *-(I)ázo* verbs express causative, inchoative, ornative, and locative meanings. Examples of *-(i)ázo* and *-jázo* verbs are given in (29) and (30):

(29)	a. *ritiδjázo* 'to wrinkle'	*ritíδa* 'wrinkle'
	b. *komatjázo* 'to tear into pieces'	*komáti* 'piece'
	c. *dropjázo* 'to disgrace'	*dropí* 'disgrace'
(30)	a. *orimázo* 'to mature'	*órimos* 'mature'
	b. *δiplasiázo* 'to double'	*δiplásios* 'double'

Derivatives in *-évo* usually mean 'carry out the official activities of x,' 'make/become x,' 'do x,' 'provide with x,' 'put into x,' and 'use x.' Examples of these are given in (31).

(31)	a.	*proeδrévo* 'to chair/preside'	*próeδros* 'president'
	b.	*siγurévo* 'to ensure'	*síγuros* 'sure'
	c.	*xorévo* 'to dance'	*xorós* 'dance'
	d.	*δezmévo* 'to bind/tie'	*δezmós* 'bond'
	e.	*pajiδévo* 'to trap'	*pajíδa* 'trap'

According to Efthymiou (2011), *-évo* usually combines with both nominal [+animate] [+masculine] and adjectival bases. Note, however, that the vast majority of *-évo* verbs are intransitive and are formed on nominal bases.

The suffix *-áro* is of Italian etymology. The basic meanings of *-áro* derivatives can be described as 'provide with x,' 'do x,' 'produce x, 'put into x,' 'use x,' 'cause to become x' and 'act as/be x.' Some examples are given in (32).

(32)	a.	*stresáro* 'to stress'	*stres* 'stress'
	b.	*zumáro* 'to zoom'	*zum* 'zoom'
	c.	*pacetáro* 'to put into a packet, to pack'	*pacéto* 'packet'
	d.	*frenáro* 'to brake'	*fréno* 'brake'
	e.	*pilotáro* 'to pilot'	*pilótos* 'pilot'

Finally, *-éno* derivatives mean 'cause to become x,' 'become x,' and 'provide with x,' while *-íno* derivatives mean 'cause to become x' or 'provide with x.' Some examples are given in (33):

(33)	a.	*kondéno* 'to shorten'	*kondós* 'short'
	b.	*lipéno* 'to lubricate'	*lípos* 'fat, oil'
	c.	*vaθéno* 'to deepen'	*vaθís* 'deep'
	d.	*oksíno* 'to sharpen'	*oksís* 'sharp, acute'
	e.	*mejeθíno* 'to magnify'	*méjeθos* 'size, magnitude'

Interestingly, the last three suffixes, *-éno, -áro* and *-íno*, do not seem to participate in MGPV formation. The next section discusses the meaning of parasynthetic verbs.

5.2 THE SEMANTIC CATEGORIES OF MODERN GREEK PARASYNTHETIC VERBS

A curious fact about verb formation in Greek is that parasynthetic verbs can express privative and ablative meanings, whereas nonparasynthetic suffixed verbs cannot (cf. Efthymiou 2011). As examples (34)–(36) show, the vast majority of

Modern Greek parasynthetic verbs display privative, ablative, or causative–completive meanings (cf. also Table 4.2 and Tsakou 2010). On the other hand, as already mentioned (see section 4), reversative, intensive and excessive meanings seem to be absent from MGPVs:

(34) a. *kse-paraδ-jáz-o* 'to take/spend someone's money'
b. *kse-δond-jáz-o* 'to take one's teeth out'
c. *apo-cefal-íz-o* 'to decapitate'

(35) a. *apo-centr-ón-o* 'to decentralize'
b. *ek-θron-íz-o* 'to dethrone'
c. *kse-port-íz-o* 'to go away, to slip out'

(36) a. *eks-anθrop-íz-o* 'to humanize'
b. *apo-ksen-ón-o* 'to estrange, to alienate'
c. *apo-skelet-ón-o* 'to make someone very thin (like a skeleton)'
d. *eks-atomik-év-o* 'to individualize'

In some cases, Modern Greek parasynthetic verbs express ornative, locative, directional, or iterative–aspectual meanings, as in examples (37) and (38).

(37) a. *epi-nicel-ón-o* 'to nickel'
b. *ep-arjir-ón-o* 'to cover with silver'
c. *em-fial-ón-o* 'to bottle'

(38) a. *ipo-vaθm-íz-o* 'to downgrade'
b. *pros-eδaf-íz-o* 'to land'
c. *meta-top-íz-o* 'to displace'
d. *ana-ne-ón-o* 'to renew'
e. *pros-selin-ón-o* 'to land on the moon'

More specifically, the principal meanings of *-ízo* MGPVs can be described as causative, privative, ablative, locative, or iterative, as can be seen in Table 4.5. Similarly, *-óno* parasynthetic verbs tend to exhibit causative, privative, ablative, ornative, locative, or iterative meanings, as can be seen in Table 4.6. Finally, as shown in Table 4.7, the vast majority of *-jázo* verbs tend to display privative meanings, whereas *-évo* and *-(i)ázo* display mostly privative, ablative, or causative-completive meanings.

Given that some semantic categories (i.e., the privative and ablative meanings, which are displayed by the vast majority of MGPVs) cannot be expressed by nonparasynthetic suffixed verbs in Modern Greek (cf. also section 4 and Efthymiou 2011), it can be concluded that the meaning of MGPVs is mainly determined by the prefix, which has an internal character. Therefore, the ordering of the affixes that participate in MGPV formation reflects their information content; that is, their contribution to the construction of the meaning of MGPVs: the first affix (i.e., the prefix) affects the meaning of the derivative more strongly than the second affix (i.e., the suffix).

TABLE 4.5
MGPVs in *-ízo*

Parasynthetic verb	Base
eksanθropízo 'to humanize'	*ánθropos* 'man, human'
ekθronízo 'to dethrone'	*θrónos* 'throne'
enθronízo 'to enthrone'	*θrónos* 'throne'
kseﬂuδízo 'to peel, to flake'	*ﬂúδa* 'skin, flake, hull'
apocefalízo 'to decapitate'	*cefáli* 'head'
apolepízo 'to scale, to peel, to flake'	*lépi* 'scale'
kseportízo 'to go away, to slip out'	*pórta* 'door'
ipoγramízo 'to underline'	*γramí* 'line'
proseδafízo 'to land'	*éδafos* 'ground, land'
metatopízo 'to displace'	*tópos* 'place'
anaγramatízo 'to anagrammatize'	*γráma* 'letter'

TABLE 4.6
MGPVs in *-óno*

Parasynthetic verb	Base
apotoxinóno 'to detoxify'	*toxíni* 'toxin'
apocentróno 'to decentralize'	*céntro* 'centre'
apoksenóno 'to estrange, to alienate'	*ksénos* 'stranger, foreigner'
apoliθóno 'to petrify'	*líθos* 'stone'
empsixóno 'to reanimate, to encourage'	*psixí* 'soul, spirit'
apovlakóno 'to make stupid'	*vlákas* 'stupid'
eksaθlióno 'to impoverish, to make someone miserable'	*áθlios* 'miserable'
aposkeletóno 'to make someone very thin (like a skeleton)'	*skeletós* 'skeleton'
epinicelóno 'to nickel'	*nícel* 'nickel'
ananeóno 'to renew'	*néos* 'new'
proselinóno 'to land on the moon'	*selíni* 'moon'

TABLE 4.7
MGPVs in *-jázo*, *-évo*, and *-(i)ázo*

Parasynthetic verb	Base
kse-paraδ-jáz-o 'to take/spend someone's money'	*parás* 'money$_{informal}$' (pl.nom. *paráδes*)
kse-δond-jáz-o 'to to take one's teeth out'	*δóndi* 'tooth'
kse-cil-jáz-o 'to disembowel'	*cilja* 'belly'
eks-atomik-év-o 'to individualize'	*atomikós* 'individual'
ek-trox-iáz-o 'to derail'	*troxós* 'wheel'

TABLE 4.8
The Distribution of Transitive/ Intransitive in MGPVs

Transitive	295
Intransitive	13
Total	308

5.3 TELICITY AND TRANSITIVITY IN MODERN GREEK PARASYNTHETIC VERBS

Another interesting fact about Modern Greek verb formation is that all parasynthetic verbs are 'change-of -state' or 'change-of-place' verbs and denote telic events; that is, events that have a necessary endpoint (Comrie 1976): for example, *apotoxinóno* 'to detoxify,' *ekθronízo* 'to dethrone' (cf. also the examples in section 5.2). By contrast, nonparasynthetic suffixed verbs can express both telic and atelic meanings. For example, nonparasynthetic *-ízo* verbs which are derived from [+human] bases are intransitive and atelic, whereas parasynthetic *-ízo* verbs derived from similar bases are transitive and telic: for example, *amerikan-íz-o* 'to behave like an American' versus *ek-xristian-íz-o* 'to christianize.' Furthermore, as observed by Tsakou (2010), the vast majority of MGPVs verbs are transitive, as is confirmed by the figures in Table 4.8 for the MGPVs under discussion.

These findings suggest, in my view, that verbal prefixes in MGPVs behave like internal suffixes and provide the telic meaning to the parasynthetic verb.[14] This suggestion concurs with the well-known assumption that verbal prefixes can change the actional value (i.e., the Aktionsart) of verbs and express telicity and transitivity (cf. Brugmann 1885, Meillet 1948 Booij and Kemenade 2003, Ralli 2004, Di Sciullo 2005, Iacobini & Masini 2006, Nagano 2011, Tsangalidis & Valma 2011, among others). This finding is also in line with Iacobini and Masini (2006), who claim that the only means of actional marking in contemporary Italian seems to be parasynthetic verb formation. Finally, the telicizing (and transitivizing) character of the prefixes corroborates the assumption that the ordering of the affixes in MGPVs is related to their information content; that is, their contribution in the meaning and the argument structure of MGPVs.

6. Register Properties of Modern Greek Parasynthetic Verbs

In this section I investigate the register properties of MGPVs in order to show that prefixes play a major role in conditioning the register of the parasynthetic verb. Before analyzing the register properties of MGPVs, it is necessary to briefly explain the features [+learned], [+/– learned], and [–learned], which will be used below. Following Anastassiadis-Symeonidis and Fliatouras (2003), the feature [+learned]

[14] This finding accords with the previous observation (cf. sections 4 and 5.2) that the external variants of *apo-*, *kse-*, and *para-* do not seem to participate in parasynthetic verb formation.

is attributed to words that either (a) come from Ancient Greek; (b) constitute artificial formations of 'katharevousa' (i.e., the artificial, ancient-looking form of Greek developed by scholars in the nineteenth century); or (c) are only used in refined or written speech (cf. also Ralli 2004). Words characterized as [−learned] either have a popular origin or are used in informal or spoken (colloquial) speech. Finally, the feature [+/− learned] is ascribed to words that have an unmarked use or origin (i.e., they are neither learned nor nonlearned).

The investigation of MGPVs reveals that prefixes play a major role in determining the register of the derived verb. For example, all parasynthetic verbs prefixed by *ek-*, *en-*, and *epi-* can be characterized as [+learned], whereas parasynthetic verbs prefixed by *kse-* are [−learned] or [+/−learned], as is illustrated in Table 4.9. This is expected, since *kse-* is the only Modern Greek prefix of the list that was not present in Ancient Greek.

Furthermore, several other observations corroborate our suggestion about the register-conditioning role of prefixes and indicate that the linear ordering of the affixes reflects their information content. First, almost all MGPVs with *apo-*,

TABLE 4.9
The Register Conditioning Role of Prefixes

[+learned] *ek-*, *en-* and *epi-*	[−learned] *kse-*
eks-is-óno 'to equate'	*kse-port-íz-o* 'to go away, to slip out'
epi-nikel-óno 'to nickel'	*kse-cil-jáz-o* 'to disembowel'
ek-δimokrat-íz-o 'to democratize'	*kse-δond-jáz-o* 'to take one's teeth out'
eks-islam-íz-o 'to islamize'	*kse-paraδ-jáz-o* 'to take/spend someone's money'
en-θron-íz-o 'to enthrone'	*kse-brats-óno* 'to bare one's arms'

TABLE 4.10
Formations of Scientific Sublanguages/Terms Translated from French and English

Parasynthetic verb	English translation
apo-toxin-ón-o	'to detoxify'
em-fial-ón-o	'to bottle' (cf. also French *embouteiller*)
ek-viomixan-íz-o	'to industrialize' (cf. also French *industrialiser*)
eks-islam-íz-o	'to islamize' (cf. French *islamiser*)
ipo-vaθm-íz-o	'to downgrade'
ipo-γram-íz-o	'to underline' (cf. French *souligner*)
en-iδat-ón-o	'to hydrate' (cf. also French *hydrater*)
eks-ail-ón-o	'to annihilate' (cf. French *immaterialiser*)
δia-vaθm-íz-o	'to classify' (cf. French graduer, *classifier*)
ana-δas-ón-o	'to reforest' (cf. French *reboiser*)
en-armon-íz-o	'to harmonize' (cf. French *harmoniser*)
pros-anatol-íz-o	'to orient, to orientate' (cf. French *orienter*)
epi-nikel-ón-o	'to nickel' (cf. also French *nickeler*)

ana-, *δia-*, *sin-*, and *ipo-* can be characterized as [+learned] or [+/−learned]. As already observed by Efthymiou (2002) and Tsakou (2010), a considerable number of MGPVs with *apo-*, *ana-*, *ek-*, *epi-*, *en-*, *δia-*, *sin-*, *ipo-*, and *pros-* are extensively used in formations of scientific sublanguages or in terms translated mainly from French and English, as seen in Table 4.10.

In addition, as shown in Table 4.10, the cross-linguistic (i.e., English or French) equivalents of MGPVs are not always parasynthetic formations. This seems to indicate that some [+learned] MGPVs are not natural remnants of Ancient Greek but are constructions of 'katharevousa,' while some [+learned] prefixes have re-entered living usage and are more or less productive thanks to purism (cf. Bortone 2010 for similar remarks on Greek prepositions).

7. Affix Combinability in MGPVs and the Role of Frequency

In this section I will investigate the factors affecting prefix–suffix combinations in MGPVs.[15] It will also be argued that the preference for certain suffixes is mainly determined by frequency factors.

7.1 PREFIX–SUFFIX COMBINATIONS

The examination of the MGPVs extracted from *RDMG* reveals that all affix combinations are not possible, and that the combination of prefixes and suffixes is restricted by register, etymological, semantic, morphological, and frequency factors. In Table 4.11 frequent combinations (i.e., more than 40 verbs) are noted by three stars,[16] medium frequency combinations (i.e., 10–30 verbs) by two stars, and rare combinations (less than 4.10 verbs) by one star.

As can be seen in Table 4.11, the suffix *-ízo*, which is the most frequent suffix in the list, appears to combine with all prefixes that participate in MGPV formation. Similarly, the prefix *kse-*, which is by far the most frequent prefix in MGPVs, seems to appear in more combinations than the other prefixes. On the other hand, the [−learned] suffix *-jázo* combines only with the [−learned] prefix *kse-*. These findings indicate that prefix–suffix combinations in MGPVs are restricted by register or etymological factors and that *-ízo* seems to behave more or less like the default MGPV suffix pattern.

In addition, evidence for the fact that prefix–suffix combinations are influenced by semantic factors can be drawn from the MGPVs in *-jázo* and *-óno*. As

[15] The reason that I use the term combinability is that the structure of parasynthetic verbs involves the obligatory co-occurrence of two affixes (i.e., a prefix and a suffix).

[16] It should be noted that the most frequent prefix–suffix combinations listed in the *RDMD* are the following: (1) *kse+N/A+ízo* (49 verbs), and (2) *ek+N/A+ízo* (41 verbs).

TABLE 4.11

Prefix-Suffix Combinations in Modern Greek Parasynthetic Verbs

	-ízo	*-óno*	*-jázo*	*-iázo*	*-évo*
kse-	***	**	**	-	
ek-	***	**	-	*	*
apo-	**	**	-	-	*
epi-	*	**	-	-	-
en-	*	*	-	*	-
δia	*	*	-	-	*
kata-	*	*	-	-	-
pros-	*	*	-	-	-
ana-	*	*	-	-	*
sin-	*	*	-	-	-
meta-	*	*	-	-	*
ipo-	*	*	-	-	-
para-	*	-	-	-	-
peri-	*	-	-	-	-
pro-	*	-	-	-	-
iper	*	-	-	-	-

the examples in (39) and (40) illustrate, both parasynthetic and nonparasynthetic verbs in *-jázo* tend to have negative or ironic connotations and convey pejorative meanings (cf. Efthymiou 2013).

(39) a. *kse-paraδ-jáz-o* 'to take/spend someone's money'
b. *kse-δond-jáz-o* 'to take one's teeth out'

(40) a. *ritiδ-jáz-o* 'to wrinkle, to become wizened'
b. *spir-jáz-o* 'to be covered with pimples'

Furthermore, as observed by Tsakou (2010), the [+learned] prefix *epi-* 'on, over' combines almost exclusively with the suffix *-óno*, which denotes mostly ornative and causative meanings (e.g., *epi-nicel-ón-o* 'to nickel'). On the other hand, no ornative meanings are attested for parasynthetic *-ízo* verbs. Given that in MGPVs with *-óno* both derivational affixes can exhibit ornative meanings, one may wonder whether the addition of the prefix serves to disambiguate the meaning of the derivative, to reinforce the derivative's semantic transparency, or to signal its register (cf. also section 8).

It must also be pointed out that *kse-* prefers combining with *-ízo* and *-jázo*, whereas *apo-* prefers combining with *-óno* (cf. also Efthymiou 2001, 2002, Tsakou 2010). As already mentioned, *apo-* does not combine with *-jázo*, most probably due to register, etymological, or semantic reasons, which need to be further examined. This observation accords with Ralli (2005), who claims that the fact that *apo-* has undergone fewer changes than other Greek preverbs explains why it does not attach to new forms of ancient verbs: for example, *δíno* 'to give,' *kse-δíno* 'to relax' versus **apo-δíno*.

Finally, affix combinations in MGPVs seem to be also conditioned by the phonological restrictions of the suffixes. As observed by Tsakou (2010), the phonological behavior of suffixes in MGPVs follows the same tendencies observed in Efthymiou (2011). For example, both *-ízo* and *-jázo* attach primarily to consonant-final bases and avoid attaching to bases ending in /z/. Instead, *-óno* attaches both to consonant-final and vowel-final bases (apart from bases in /a/).

7.2 THE ROLE OF FREQUENCY AND PRODUCTIVITY IN SUFFIX PREFERENCE

As already indicated (see Table 4.1), *-ízo* and *-óno* are the most frequent prefixes of MGPVs. The same high frequency for these suffixes has also been found in three other studies, namely a study of lemmas in the *Reverse Dictionary of Modern Greek*, a study of tokens and types in a corpus of Modern Greek schoolbooks (Efthymiou 2011, 2013), and a corpus study of 4,143,583 words (Efthymiou, Fragaki, & Markos 2012). At the same time, it is interesting to note that suffixes like *-éno*, and *-íno*, which have been found to be the least productive in these three studies, do not participate in MGPVs. It can thus be concluded that the occurrence of derivational suffixes in MGPVs correlates to their frequency and productivity in nonparasynthetic derived verbs.

It should be noted, however, that the picture with *-évo* is a bit more complicated: although it is the third most productive suffix in the data of Efthymiou (2011) and Efthymiou, Fragaki, and Markos (2012), it is not preferred in MGPV formation. In my view, the low frequency of *-évo* in MGPVs can be related to its noncombinability with the [–learned] *kse-* and its semantic preferences. A similar remark can be made about *-áro*, which, although it is the fourth most productive suffix in the above-mentioned studies, seems to be excluded from MGPVs. It can be suggested that the exclusion of *-áro* from MGPVs is due to its etymology, as well as its recent history and preference for nonnative bases.

8. The Compulsory Co-occurrence of Prefixes and Suffixes and the Parameters of Morphological Naturalness

As regards the compulsory presence of the suffix, it was already suggested in section 4 that this is due to the fact that prefixes derive words of multiple categories and do not seem to function as category-determining affixes. In this section I will offer a brief overview of the theory of Natural Morphology (henceforth NM) and propose that the co-occurrence of prefixes and suffixes in MGPVs can also be explained with reference to a number of parameters of morphological naturalness, namely optimal shape, iconicity, transparency, and biuniqueness.

8.1 THE PARAMETERS OF MORPHOLOGICAL NATURALNESS

Natural Morphology is a cognitively oriented theory of morphological organization. 'Natural' is a central notion in NM and means cognitively simple, less marked, and therefore universally preferred (for a detailed presentation see Dressler et al. 1987, Dressler 2005, Manova 2011, among others). The main assumptions of NM are formulated in three subtheories: (1) the theory of universal naturalness (i.e., system-independent naturalness), (2) the theory of typological naturalness, and (3) the theory of language-specific naturalness. According to Dressler (2005: 267), these subtheories interact in a hierarchical fashion and function as subsequent filters on possible and probable words of a language: what is allowed or preferred universally may be disallowed or dispreferred by typological naturalness, and then by language-specific system naturalness.

The subtheory of system-independent naturalness operates with a set of naturalness parameters: iconicity, indexicality, transparency, biuniqueness, binarity, and optimal shape of a morphological word.

The best-known parameter is *iconicity*, based on Peirce's (1965) concept of icons with its subtypes; that is, images, diagrams, and metaphors. According to Dressler (2005: 269), the most important instance of morphological iconicity occurs in the subparameter of *constructional iconicity* (also called constructional diagrammaticity; cf. also Dressler et al. 1987); that is, addition of semantics implies addition of form. On this parameter, affixation is more natural than conversion, because of the analogy of addition in meaning in form. Conversion appears to be noniconic, because addition of meaning is not paralleled by change in form.

The parameter of *indexicality* is also derived from Peircean semiotics. On this parameter, affixation of a marker to an immediately adjacent base is preferred to having a meaningless interfix inserted between an affix and its base.

Biuiniqueness holds if one and the same form always has the same meaning and vice versa. Ambiguity and uniqueness are less natural.

On the parameter of *morphosemantic transparency*, full transparency means fully compositional meaning. According to this parameter, compounds are more transparent than derivations, because they are more descriptive (cf. Seiler 1971).

On the parameter of *morphotactic transparency*, the most natural forms are those where there is no opacifying obstruction to ease the perception. On this parameter, suffixation and prefixation is preferred over infixation (discontinuous base) or circumfixation (discontinuous affix).

The parameter of *binarity* is based on the binary nature of the neurological information transmittance. According to this parameter, binary grammatical relations are the most natural grammatical relations.

Finally, according to the parameter of the *optimal shape*, the most natural word shape is bisyllabic (cf. Dressler 1985: 276).

In the next section I will propose that the co-occurrence of prefixes and suffixes in MGPVs can be explained with reference to the parameters of optimal shape, constructional iconicity, transparency, and biuniqueness.

8.2 THE COMPULSORY CO-OCCURRENCE OF PREFIXES AND SUFFIXES

As observed in Anastassiadis-Symeonidis and Masoura (2009), the most frequent shape for Modern Greek verbal endings is the nonaccented inflectional suffix *-o*, whereas verbs with the accented inflectional suffix *-ó* appear to be much less frequent in contemporary Greek. In their view, disyllabic suffixes like *-ízo*, *-óno*, and so on represent the most natural Modern Greek verb shape, and this is why they serve as ideal endings for MGPVs. It is interesting to note that this observation accords with our previous finding (see section 7) that the occurrence of derivational suffixes in MGPVs correlates to their frequency and productivity in nonparasynthetic derived verbs.

The structure of some MGPVs can also be explained with reference to other parameters, such as constructional iconicity, biuniqueness, and morphosemantic transparency. Evidence for the claim that the addition of the prefix serves to disambiguate the meaning of MGPVs or to reinforce their semantic transparency can be drawn from MGPVs that express ornative or causative–completive meanings (cf. Zwanenburg 1997 for similar remarks about the Dutch prefixes). For example, as already observed, the [+learned] prefix *epi-* 'on, over' combines almost exclusively with the suffix *-óno*, which is ambiguous between ornative, causative, and inchoative meanings (e.g., *epi-nicel-ón-o* 'nickel'). Given that in these verbs both derivational affixes can exhibit ornative meanings, it can be argued that the addition of the prefix serves to disambiguate the meaning of the parasynthetic verb. This assumption is also corroborated by the fact that parasynthetic verbs with *epi-* are preferred in the terminological jargon, a domain that enhances biuniqueness and favors [+learned] elements.

In what regards MGPVs with causative meanings, we have already seen that nonparasynthetic *-ízo* verbs do not seem to derive transitive telic verbs from [+human] [+national] bases in contemporary Greek and that new *-ízo* verbs derived from this type of bases are more likely to be interpreted as similative (e.g., *amerikan-íz-o* 'to behave like an American'). Given that all verbal suffixes are ambiguous and that the causative semantic domain no longer seems to be a frequent pattern for *-ízo* (cf. Efthymiou 2011, Efthymiou et al. 2012), it can be argued that the addition of the prefix appears to be paralleled by change in meaning (e.g., *ek-xristian-íz-o* 'to christianize'). In my view, the occurrence of prefixes in the structure of parasynthetic *-ízo* verbs serves to produce verbs that appear to be more iconic and morphosemantically more transparent.[17] Interestingly enough, this argument accords with the previous claim about the major role of prefixes in the structure of MGPVs and indicates that the linear ordering of the affixes reflects their information content.

9. Conclusion

In this paper 308 Modern Greek parasynthetic verbs were examined, namely parasynthetic verbs in *-ízo*, *-jázo*, *-(i)ázo*, *-óno*, and *-évo*. On the basis of their semantic interpretation and their formal characteristics it was suggested that the linear ordering of the affixes reflects their information content and their contribution to the morphological structure of the parasynthetic verb. More specifically, it was claimed that the prefixes of these verbs behave like internal prefixes and determine the core semantics of the parasynthetic verb. It was also argued that affixation is determined by various factors, such as register and frequency. Finally, the appearance of multiple derivational affixes in the structure of MGPVs was explained in terms of morphological biuniqueness, constructional iconicity, optimal shape, and semantic transparency.

Abbreviations

Modern Greek parasynthetic verbs (MGPVs)
Natural Morphology (NM)
Reverse Dictionary of Modern Greek (RDMG)
Righthand Head Rule (RHR)

Acknowledgments

I am grateful to Stela Manova, Matthias Gerner, and an anonymous reviewer for their invaluable comments and suggestions. I would also like to thank M. Camilleri and G. Fragaki for discussing earlier drafts of this paper.

References

Anastassiadis-Symeonidis, Anna. 1994. *Neological Borrowing in Modern Greek*. Thessaloniki: Filosofiki Scholi [in Greek].

---

[17]Similar remarks can be made about MGPVs in *-óno* (e.g., *apovlakóno* 'to make someone stupid'). Given that nonparasynthetic *-óno* verbs prefer attaching to [+material] bases (e.g., *kokalóno* 'turn into a bone/ begin to have properties of bone, be stunned,' *zaxaróno* 'to crystallize, to sugar'; cf. also Charitonidis 2011: 38 for similar remarks), the addition of the prefix in parasynthesis appears to be paralleled by addition of meaning.

Anastassiadis-Symeonidis, Anna. 2002. *The Reverse Dictionary of Modern Greek*. Thessaloniki: Institouto Neoellinikon Spoudon [in Greek].

Anastassiadis-Symeonidis, Anna, and Asimakis Fliatouras. 2003. "The features [learned] and [-learned] in Greek: Definition and classification." In *The Proceedings of the 6th International Conference on Greek Linguistics*, edited by Georgia Katsimali, Alexis Kalokairinos, Elena Anagnostopoulou, and Ioanna Kappa. University of Crete. www.philology.uoc.gr/conferences/6thICGL/gr.htm. [in Greek].

Anastassiadis-Symeonidis, Anna, and Elvira Masoura. 2009. "Ending-part and memory: A theoretical account." In *The Proceedings of the 8th International Conference on Greek Linguistics*, edited by Georgios Giannakis, Mary Baltazani, Georgios Xydopoulos, and Anastasios Tsangalidis, 616–634. Ioannina: University of Ioannina. [in Greek].

Anastassiadis-Symeonidis, Anna, and Elvira Masoura. 2012. "Word ending-part and phonological memory: A theoretical approach." In *Irregularity in Morphology (and beyond)*, edited by Thomas Stolz, Hitomi Otsuka, Aina Urdze, and Johan van der Auwera, 127–140. Berlin: Akademie Verlag,

Aronoff, Mark. 1976. *Word Formation in Generative Grammar*. Cambridge, MA: The MIT Press.

Baayen, Harald. 2008. "Corpus linguistics in morphology: Morphological productivity." In *Corpus Linguistics. An International Handbook*, edited by Anke Lüdeling, and Merja Kytö, 899–919. Berlin: Mouton de Gruyter.

Bell, Melanie, and Ingo Plag. 2012. "Informativeness as a determinant of compound stress in English." *Journal of Linguistics* 48(3): 485–520.

Booij Geert. 2010. *Construction Morphology*. Oxford: Oxford University Press.

Booij, Geert, and Ans Van Kemenade. 2003. "Preverbs: An introduction." In *The Yearbook of Morphology* 2003, edited by Geert Booij, and Jaap van Marle, 1–22. Dordrecht: Kluwer.

Bortone, Pietro. 2010. *Greek Prepositions. From Antiquity to the Present*. Oxford: Oxford University Press.

Brugmann, Karl. 1885. *Griechische Grammatik*. Munich: Beck.

Charitonidis, Chariton. 2005. *Verb Derivation in Modern Greek*. Frankfurt: Peter Lang.

Charitonidis, Chariton. 2011. *Making Verbs Happen: Interviews on Greek Verb Endings*. München: Lincom Europa.

Comrie, Bernard. 1976. *Aspect*. Cambridge: Cambridge University Press.

Corbin, Danielle. 1987. *Morphologie dérivationelle et structuration du lexique*. 2 Vols. Tübingen: Niemeyer.

Crocco Galèas, Grazia, and Claudio Iacobini. 1993. "The Italian parasynthetic verbs: A particular kind of circumfix." In *Natural Morphology. Perspectives for the Nineties*, edited by Livia Tonelli, and Wolfgang U. Dressler, 127–142. Padova: Unipress,.

Delveroudi, Rea, and Sophia Vassilaki. 1999. "Préfixes d'intensiteé en Grec Moderne: *para-*, *kata-*, *poly-* et *olo-*." In *Les Opérations de Détermination: Quantification/Qualification*, edited by Alain Deschamps and Jacqueline Guillemin-Flescher, 149–167. Paris: Ophrys.

Di Sciullo, Anna Maria. 1997. "Prefixed-verbs and adjunct identification. " In *Projections and Interface Conditions*, edited by Anna Maria Di Sciullo, 52–73. New York: Oxford University Press.

Di Sciullo, Anna Maria. 2005. *Asymmetry in Morphology*. Cambridge, MA: The MIT Press.

Dressler, Wolfgang U., Mayerthaler Willi, Panagl Oswald, and Wolfgang. U. Wurzel. 1987. *Leitmotifs in Natural Morphology*. Amsterdam: Benjamins.

Dressler, Wolfgang U. 2005. "Word formation in natural morphology." In *Handbook of Word-Formation*, edited by Pavol Stekauer, and Rochelle Lieber, 267–284. Dordrecht: Springer.

Efthymiou, Angeliki. 2001. "The Modern Greek prefix *kse-*. The concepts of move away and change of state." *Studies in Greek Linguistics* 2000: 202–213 [in Greek].

Efthymiou, Angeliki. 2002. "Semantic considerations for the Modern Greek prefixes *kse-*, *ek-* and *apo-*." *Studies in Greek Linguistics* 2001: 199–209 [in Greek].

Efthymiou, Angeliki. 2003. "Prefixes or first elements denoting intensification in Modern Greek." *Studies in Greek Linguistics* 2002: 519–528 [in Greek].

Efthymiou, Angeliki. 2011. "The semantics of verb forming suffixes in Modern Greek." In *Proceedings of the 19th International Symposium of Theoretical and Applied Linguistics, 3–5 April 2009*, edited by Eliza Kitis, Nicolaos Lavidas, Nina Topinzi, and Tasos Tsangalidis, 175–184. School of English: Aristotle University of Thessaloniki.

Efthymiou, Angeliki. 2013. "How many factors influence the meaning of denominal and deadjectival verbs? The case of Modern Greek verbs in -(i)azo." In *The Semantics of Word Formation and Lexicalization*, edited by Pius Ten Hacken, and Claire Thomas, 225–246. Edinburgh: Edinburgh University Press.

Efthymiou, Angeliki, Fragaki Georgia, and Angelos Markos. 2012. "Productivity of verb forming suffixes in Modern Greek: A corpus-based study." *Morphology* 22(4): 515–543.

Giannakis, Georgios. 1997. *Studies in the Syntax and Semantics of the Reduplicated Presents of Homeric Greek and Indo-European*. Innsbruck: Innsbrucker Beiträge zur Sprachwissenschaft.

Humbert, Jean. 1972. *Syntaxe Grecque*. 3rd ed. Paris: Kincksieck.

Iacobini, Claudio. 2010. "Les verbes parasynthétiques: de l'expression de l'espace à l'expression de l'action. "*De Lingua Latina* 3. www.paris-sorbonne.fr/IMG/pdf/Iacobini_parasynthetiques.pdf.

Iacobini, Claudio, and Francesca Masini. 2006. "The emergence of verb-particle constructions in Italian: locative and actional meanings." *Morphology* 16(2): 155–188.

Klairis, Christos, and George Baniniotis 2009. *Modern Greek Grammar*. 11th ed. Athens: Ellinika Grammata [in Greek].

Lieber, Rochelle. 2004. *Morphology and Lexical Semantics*. Cambridge: Cambridge University Press.

Lieber, Rochelle. 2010. *Introducing Morphology*. Cambridge: Cambridge University Press.

Manova, Stela. 2011. *Understanding Morphological Rules: With Special Emphasis on Conversion and Subtraction in Bulgarian, Russian and Serbo-Croatian*. Dordrecht: Springer.

Malikouti-Drachman, Angeliki, and Gaberell Drachman. 1992. "Greek clitics and lexical phonology." In *Phonologica* 1988: 197–206.

Meillet, Antoine. 1948. *Aperçu d'histoire de la langue grec*. 6th ed. Paris: Hachette.

Mendez-Dosuna, Julian. 1997. "Fusion, fission and relevance in language change: De-univerbation in Greek verb morphology." *Studies in Language* 21(3): 577–612.

Nagano, Akiko. 2011. "The right-headedness of morphology and the status and development of category-determining prefixes in English." *English Language and Linguistics* 15(1): 61–83.

Nespor, Marina, and Angeliki Ralli. 1996. "Morphology-phonology interface. Phonological domains in Greek compounds." *The Linguistic Review* 13(3–4): 357–382.

Plag, Ingo. 1999. *Morphological Productivity. Structural Constraints in English Derivation*. Berlin Mouton de Gruyter.

Plag, Ingo. 2003. *Word-Formation in English*. Cambridge: Cambridge University Press.

Plag, Ingo. 2006. Productivity. In *Handbook of English Linguistics*, edited by Bas Aarts and April McMahon, 537–556. Oxford: Blackwell.

Papanastasiou, Georgios. 2011. "The preverb *apo-* in Ancient Greek." In *Proceedings of the 9th International Conference on Greek Linguistics*, edited by Katerina Chatzopoulou, Alexandra Ioannidou, and Suwon Yoon, 97–11. Chicago, Illinois.

http://dl.dropbox.com/u/12639124/icgl2009/proceedings/10_papanastassiou_edited_97.pdf.

Peirce, Charles. 1965. *Collected Papers*. Cambridge, MA: Harvard University Press.

Philippaki-Warburton, Irene. 1970. *On the Verb in Modern Greek*. Bloomington: Indiana University Press

Ralli, Angeliki. 2003. "Preverbs in Greek: the case of *ksana, kse-, para-*." In *Asymmetry in Grammar: Morphology, Phonology and Language Acquisition*, edited by Anna Maria Di Sciullo, 37–65. Amsterdam John Benjamins Publishing Company.

Ralli, Angeliki. 2004. "Stem-based versus word-based morphological configurations: The case of Modern Greek preverbs." *Lingue e Linguaggio* 2004(2): 241–275.

Ralli, Angeliki. 2005. *Morphology*. Athens: Patakis [in Greek].

Rainer, Franz. 1989. *I nomi di qualitá nell' italiano contemporaneo*. Wien: Braumüller.

Rice, Keren. 2000. *Morpheme Order and Semantic Scope*. Cambridge: Cambridge University Press.

Scalise, Sergio. 1984. *Generative Morphology*. Dordrecht: Foris.

Seiler, Hansjakob. 1975. "Die Prinzipien der deskriptiven und der etikettierenden Benennung. In *Linguistic Workshop* III, edited by Hansjakob Seiler, 2–57. München: Wilhelm Fink.

Smirniotopoulos, Jane, and Brian Joseph. 1998. "Syntax versus the lexicon: Incorporation and compounding in Modern Greek." *Journal of Linguistics* 34(2): 447–488.

Thomadaki, Evangelia. 1996. "Derivatives in Modern Greek." In *Issues of Modern Greek Language. A Didactic Approach*, edited by Georgia Katsimali and Fotis Kavoukopoulos, 69–96. Rethymnon: University of Crete [in Greek].

Triandafyllidis Manolis. (1991) [1941]. *Modern Greek Grammar*. 3rd revised edition. Thessaloniki: Manolis Triandafyllidis Foundation [in Greek].

Tsakou, Eleni. 2010. "Derivative verbs of Modern Greek without correspondence between form and meaning. The case of prefixed verbs ending in *–ízo, –iázo* and *–óno*." Master's thesis, Democritus University of Thrace [in Greek].

Tsangalidis, Anastasios, and Eleni Valma. 2011. "Aspectual marking and the resolution of ambiguity: Prepositions in Modern Greek." Paper presented at the conference 'Prépositions et aspectualité,' University of Provence, France, 3–4 June.

Williams, Edwin. 1981. "On the notions 'lexically related and 'head' of word." *Linguistic Inquiry* 12: 245–274.

Zwanenburg, Wiecher. 1997. "Dutch prefixes and prepositions in complex verbs." In *Advances in Morphology*, edited by Wolfgang U. Dressler, Martin Prinzhorn, and John R. Rennison, 63–77. Berlin: Mouton de Gruyter.

PART II:

Phonological and Morphological Ordering

5

Mobile Affixation Within a Modular Approach to the Morphology–Phonology Interface

Yuni Kim

1. Introduction

The phonological conditioning of mobile affix placement in Huave, as analyzed in Noyer (1994) and Kim (2010), appears to be problematic for a modular feed-forward architecture of grammar in which all morphological operations precede all phonological operations. Following McCarthy and Prince's (1993) approach to phonologically conditioned morphology, the analyses implement a "P >> M" strategy within Optimality Theory where phonological (P) and morphological (M) constraints are evaluated in parallel. Mobile affix positioning then results from a global optimization of phonological structure, at the expense of morphological preferences and defaults. In other words, the outcome of phonological processes such as epenthesis and syllabification must be evaluated in order to determine whether the mobile affix surfaces as a prefix or a suffix, and this type of analysis is incompatible with an ordered separation of the two modules.

On the other hand, Paster (2009) argues that true cases of phonological affix order do not exist. In an extensive survey of putative cases of phonological affix order, she finds that nearly all of them are better analyzed as cases of purely phonological operations taking place after affixation, of phonologically conditioned suppletive allomorphy, or of phonological subcategorization (where phonological underlying forms are visible to the morphology, but not touched by it; cf. Bermúdez-Otero's 2012 "insect trapped in amber" analogy), all of which are compatible with a modular feed-forward architecture. Huave is the only case that is not reanalyzed (although a potential direction for reanalysis is suggested), and so the question remains of whether any strictly modular analysis is possible. If so, this is consistent with Paster (2006a, 2009) and Yu (2007)'s arguments against the P >> M model of phonologically conditioned morphology. If not, we are left in the uncomfortable position of having very few counterexamples to the claim that phonologically driven affix ordering does not exist. Aside from Huave, these include Athapaskan (Rice 2011: 183) and Moro (Jenks & Rose, to appear).

In this chapter I make three points. The first is that Paster's (2009: 34) tentative reanalysis of Huave mobile affix placement, as a phonologically predictable procedure of associating floating features to skeletal positions (cf. Rose 1995), is not compatible with the data. The second is that Huave mobile affixation *is* compatible with a modular feed-forward architecture and the subcategorization-based approach, if we pursue an alternative generalization that allows us to follow Paster's (2009: 35) reanalysis of Afar mobile affixes and view the Huave case as a type of phonologically conditioned suppletive allomorphy. The P >> M architecture is thus no longer strictly needed. The third point, however, is that the suppletion analysis is potentially powerful enough to generate the type of data predicted by P >> M models—that is, permitting this type of analysis undermines the empirical basis on which phonological affix order has previously been defined. Both the empirical issue and the theoretical debate are consequently rendered less compelling.

It is then an open issue whether there is a more theory-neutral way of understanding and classifying phonological influences on the position of affixes within a word, regardless of where one draws the boundary between what is and is not considered to be phonological affix order. An issue with this term is that it can refer ambiguously to a structurally diverse variety of phonologically conditioned affix orders (i.e., those making reference to phonological structure in any part of the morpheme–linearization algorithm), as noted by Kim (2010: 158) and Rice (2011: 178). For example, it can be deployed in a way that highlights a modular architecture of grammar, referring specifically to the putative subset of affix linearizations that are determined within the phonological component, or otherwise in parallel with phonology, in opposition to phonologically conditioned processes taking place within the morphosyntactic component (see Paster 2009: 23–24).

Here I follow Manova and Aronoff (2010: 115) in taking phonological conditioning of any kind as the primary defining criterion for phonological affix order, mainly for heuristic reasons. With the caveat that inclusion or exclusion of specific cases will unavoidably be analysis-dependent, since an analysis is inherent to any generalization, this criterion seems to circumscribe a maximal set of cases that can then be examined to determine the most relevant parameters of pattern description, any unexpected typological gaps, and so on; I suggest a classification in section 5. The main thrust of my reanalysis of Huave mobile affixation, then, is just to propose an improved generalization, one that confirms its status as a case of phonologically conditioned affix ordering. At the same time, it moves the phenomenon to a slightly different place in the typology, triggering a need to refine our theories—which, as this case throws into relief, are only as good as the insightfulness of their predictions. It appears that within a modular architecture we currently lack data on which to base sharper analytical distinctions between morphologically conditioned phonological processes, locally determined suppletion, and global optimization. In the future, more definitive evidence should be sought in interactions of phonologically conditioned affix ordering with other phonological processes, of a kind which to my knowledge are not found in Huave.

2. Huave Affix Mobility as Epenthesis Avoidance?

Kim (2010) characterizes Huave verb structure as basically hierarchical, with affixes surfacing at a morphologically specified distance from the stem relative to other affixes. The point of interest is that Huave has not only fixed prefixes and fixed suffixes, but also mobile affixes that can surface on either side of the root. Regardless of whether a mobile affix is a prefix or a suffix, however, it still appears at its fixed hierarchical distance from the stem. Kim (2010: 135) models the Huave verb template with symmetrical layers numbered L1 through L4 expanding from either side of the stem, as in (1); Layer 0 is unproductive and does not figure further in the analysis.

(1) [L4 [L3 [L2 [L1 [[Stem] Lo] L1] L2] L3] L4]

Layers 1 and 3 contain mobile affixes. In (2a) in the word *ʃ-i-n-a-mut* 'I will write (it),' we see the Layer 1 mobile affix *n-* surfacing in prefixal position adjacent to the stem, while the Layer 3 the first-person mobile affix *s-* (allophonically realized as *ʃ-*) occurs farther outside. The mirror-image case is the word *ndil-i-t-ej-as* 'I turned around; I came back' in (2b), where the Layer 1 mobile affix *-t* surfaces in suffixal position next to the stem, while the first-person *s* is realized suffixally but again in an outside position. Note the presence of a preceding vowel, which is epenthetic.

(2)	[L4 [L3	[L2	[L1	[[Stem]	L1]	L2]	L3]	L4]
a.	ʃ-	i-	n-	a-mut				
	1	FUT	1SUB	TV-write				
b.				ndil-i	-t	-ej	-as	
				turn-V	CPL	RFL	1	

The generalization in Kim (2010) is that mobile affixes surface by default as suffixes but are placed in prefixal position wherever this would obviate the need for epenthesis (since Huave does not allow consonant clusters). In (2b), epenthesis is tolerated in the first-person suffix *-s*, but since the base begins with a consonant, prefixation would equally have required epenthesis; so the default preference emerges. However, since the base in (2a) is vowel-initial (but ends with a consonant), moving the first-person affix to prefixal position is the option that will avoid epenthesis. The formal analysis is implemented in Optimality Theory, with phonological epenthesis-avoidance constraints outranking morphological constraints that require mobile affixes to be suffixal. Since phonological factors determine the linear placement of affixes with respect to other affixes in this description and analysis of the phenomenon, it appears to be a clear case of phonologically driven affix ordering.

I will expand on Paster's (2009: 34) suggestion for an alternative analysis in section 3, before proposing a new generalization and analysis in section 4. The new generalization is argued to be preferable, in that it also accounts for some exceptions to the epenthesis-avoidance story.

3. Floating-Feature Analysis

Paster (2009: 34) proposes that a different analysis of Huave affix placement could be attempted by assuming that a mobile affix consists of two objects: a skeletal slot (or slots, such as CV), plus unassociated segmental material (cf. Rose 1995 on Chaha). After affixation has taken place in the morphology, the phonological component associates floating segments left to right (or by another standard phonological association convention) to empty slots in the template. A principled separation between morphology and phonology could thus be maintained.

For example, (3) shows the derivation of *t-a-mut-us* 'I wrote (it),' which contains the stem *a-mut* and the mobile affixes *t* (completive; Layer 1) and *s* (1st person; Layer 3). The floating segments associate left to right to empty slots around the stem; in the absence of another deciding factor, association happens in the morphological order of attachment: first *t*, then *s*. An epenthetic vowel, indicated here in parentheses, would then be inserted to break up the final consonant cluster. The illustration in (3) is schematic; for expository purposes I show association to a slot of the corresponding Layer number, wherever there are several empty ones that would produce the same linear affix order. I do not resolve issues such as which of whether the L2 slots would be present in the absence of L2 affixation, or the precise shape of the skeletal components of the affixes. In any case, the result of *t*-prefixation and *s*-suffixation is correctly obtained.

(3) Derivation of *t-a-mut-(u)s* 'I wrote (it)'

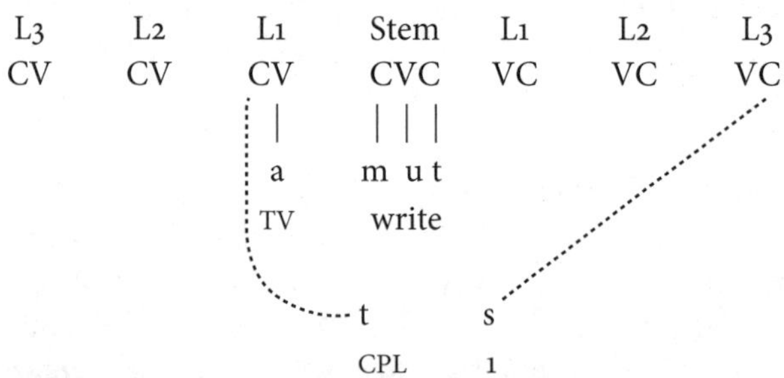

To get the basic pattern of mobile-affix suffixation, it is necessary to stipulate that a consonant can associate to the C slot of a CV morphological unit if and only if the V is already filled. Otherwise, it must look for the next available empty slot. An example of this type is shown in (4); again, epenthetic segments are indicated by parentheses. Here, both affixes surface as suffixes because the lack of a base-initial vowel prevents their association in prefixal positions.

(4) Derivation of *ndil-i-t-ej-(a)s* 'I turned around'

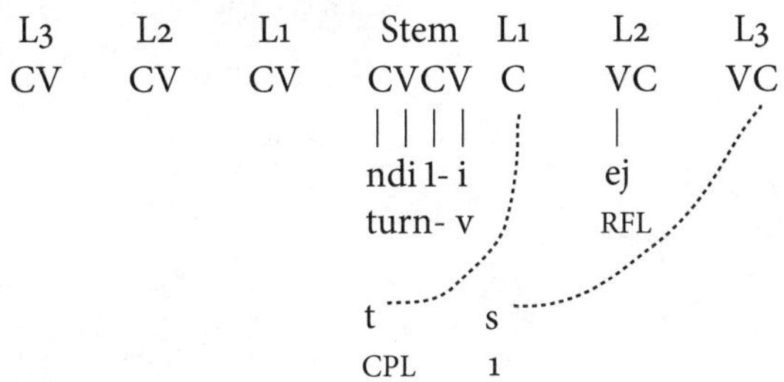

Fell-swoop association of all floating affixes at once, such as in (3) and (4), can only account for cases where the left-to-right order of mobile affixes mirrors their order of attachment. As can be seen from the symmetrical order of layers on either side of the stem, this is not the case in words where two mobile affixes both surface as prefixes, like in (5).

(5) Structure of *ʃ-i-n-a-mut* 'I will write (it)'

L3	L2	L1	Stem	L1	L2	L3
s	i	n	a-mut			
1	FUT	1SUB	TV-write			

Left-to-right attachment, illustrated in (6), incorrectly predicts **n-i-s-a-mut*. Because both linear orders are possible, a postmorphological association procedure fails.

(6) Incorrect derivation of *ʃ-i-n-a-mut* 'I will write (it)'

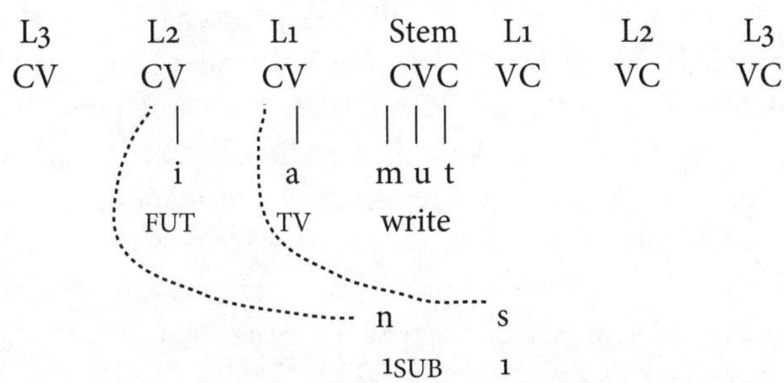

An obvious solution is to interleave morphology and phonology (Kiparsky 1982) such that the 1SUB Layer 1 affix *n* is already attached by the time Layer 3 is added. The first-person *s*, being the only floating segment at this stage, will associate to the first available slot, which will be some consonantal position preceding the vowel *i* of the future prefix.

(7) Cyclic derivation of *ʃ-i-n-a-mut* 'I will write (it)'

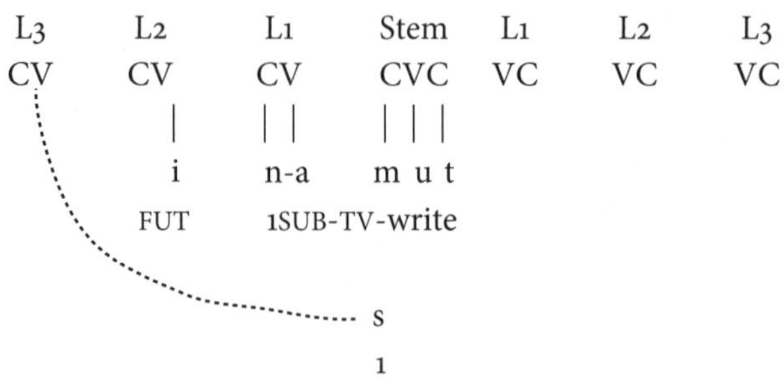

Complications arise when the morphological identity of the preassociated segments must be known in order to determine a floating segment's eligibility for attachment to a preceding position. The condition in (4) that association to a C slot must be licensed by a following V is stated purely phonologically and is thus compatible with strict modularity. However, stems of the shape VCV behave differently depending on which of the vowels is a root vowel and which is a stem-formative. The examples in (8) illustrate the differential placement of the subordinate affix *m*, as a prefix in (8a) but a suffix in (8b).

(8) a. m-a-la
SUB-TV-gobble
'that s/he gobbles (it)'

b. uy-u-m
circle-V-SUB
'that it spins'

Since a main advantage of the floating-feature analysis was that its phonological nature and lack of reference to morphological conditioning made it compatible with a modular architecture, the data in (8) represent a challenge; it is not clear how the more restrictive theory can model them. In the P >> M analysis in Kim (2010), the nonmodular nature of evaluation and operations enables constraints to refer simultaneously to morphological and phonological information. This type of mechanism was used to argue that the constraints responsible for the basic phonological pattern were sometimes outranked by a higher, morphologically sensitive constraint (i.e., AFFIXSYLL, a constraint penalizing monosyllabic root-plus-affix combinations).

4. A Suppletive Allomorphy Analysis

A second type of alternative analysis is that prefixal and suffixal variants are not in fact the same affix, but rather suppletive allomorphs. The issue then shifts from determining the algorithm for placement of a mobile affix to determining which of the allomorphs will surface in any given form. For example, the completive *t* would have the prefixal allomorph *t-* and the suffixal allomorph *-t*. This analysis has the theoretical advantage of being compatible with the more restrictive, modular feed-forward architecture. Allomorph selection is done entirely within the morphology, with conditioning either from morphosyntactic features, or from the phonological environment in lexically underlying forms; Paster's (2006a) Phonological Subcategorization theory of phonologically conditioned suppletive allomorphy is an extended treatment of the latter case.

The original justification for a phonological analysis of Huave mobile affixes is that the morphosyntactic contexts for prefixal and suffixal realizations are not coherent. Taking into account the lexical–semantic distinction between verbs with a prefixal theme vowel (the majority; labeled "prefixing" in Kim 2010) and verbs without one ("suffixing"; argued by Kim 2009 to be the set of externally caused change-of-state verbs, and abbreviated here as ExtCoS), one possible characterization of the morphosyntactic contexts for the prefixal and suffixal allomorphs of completive marker *t* and first-person marker *s* is listed in (9).

(9) Suppletive allomorphy analysis: morphosyntactic subcategorization

a. Completive *-t*	ExtCoS non-second person
b. Completive *t-*	Everywhere else
c. 1st person *-s*	Completive; ExtCoS (all TAM)
d. 1st person *s-*	Everywhere else

For both, the prefixal allomorph (9bd) can be seen as a default. The suffixal allomorph of the completive marker only appears in a special subset of intransitive verbs in the first (both inclusive and exclusive) and third persons. The suffixal allomorph of the first-person marker appears in the completive of all verbs, and in all TAM categories for the special set of intransitives.

These sets of morphosyntactic environments are arguably not complicated to the point of unlearnability, but they do reduce to a more streamlined list if reconfigured in morphophonological terms, as in (10). Seen this way, the suffixal allomorph is the default: the prefixal allomorph has priority and appears wherever the base-initial segment is a nonroot vowel. The suffixal allomorph only chosen if the base is consonant-initial or begins with a vowel belonging to the root, as in (8b). This analysis is similar to Paster's (2009: 35) reanalysis of phonologically conditioned affix mobility in Afar, although the relevant environment in Afar is stateable purely in phonological terms (rather than incorporating morphological information as is done here).

(10) Suppletive allomorphy analysis: morphophonological subcategorization

a. Completive *t-*	[$V_{[\text{-root}]}$
b. Completive *-t*	Everywhere else
c. 1st person *s-*	[$V_{[\text{-root}]}$
d. 1st person *-s*	Everywhere else

The morphophonological subcategorization analysis captures the entire pattern of Huave affix mobility, including the problematic cases in (8), in a simple way. Because it is applicable to all mobile affixes (beyond the two shown here), it preserves the generalization in the Kim (2010) analysis that all mobile affixes are placed according to a unified algorithm, with the added advantage that further add-ons such as the AFFIXSYLL constraint are not needed. Allomorph selection is done without look-ahead to phonological processes such as vowel epenthesis, and without a global optimization procedure. Therefore, we can conclude that the P >> M model is not needed for the analysis of Huave mobile affix placement, and that this particular pattern does not provide conclusive evidence against the modular feed-forward architecture of grammar.

Is Huave then not a robust case of phonologically conditioned affix ordering? Before drawing this conclusion, it is necessary to reconsider the definition of the phenomenon. Paster (2009: 23) characterizes it as follows: "Phonology can produce morpheme orderings that disobey other principles." Identification of phonologically conditioned affix order is thus dependent on an analysis of what the default or basic principles are. In many cases, sustained research will converge on a generally accepted best analysis of what the principles are, however they may be formalized. Nevertheless, where these principles are defined in terms of phonology, as in the phonological subcategorization approach, the lines may become blurred.

Huave is a case in point. If we accept that phonology has some effect on whether completive marking manifests itself as a prefix or a suffix, the analytical problem is how to analyze this conditioning. The essence of the suppletive–allomorphy analysis is that each distinct phonological environment is considered as evidence for a separate allomorph, despite all allomorphs being homophonous. Presumably, then, any case of phonologically influenced affix positioning could be modeled in this way, and any case of deviance from an otherwise robust principle would be considered a separate allomorph. Without further constraints on what can be considered a suppletive allomorph, claims about the presence or absence of phonologically driven affix order risk becoming unfalsifiable.

I illustrate the problem with the hypothetical language in (11), which illustrates a pattern of apparent global optimization of affix order of the kind predicted not to exist in Paster (2006b). This language has three prefixes, *b*, *l*, and *a*, which configure themselves in order to avoid vowel hiatus (11ac) and coda consonants (11bd). From (11e) we assume that *b* is structurally an inner affix whereas *l* is an outer one, since *l-a-bag* and **b-a-lag* would have been phonotactically legal. In

(11fgh), though, we see that they can be reordered to avoid vowel epenthesis wherever possible (given that *bl* is a legal cluster), so that we get *b-l-ag* rather than **lə-b-ag*.

(11)

	V-initial root		C-initial root
a.	a-l-ag	b.	l-a-ga
c.	a-b-ad	d.	b-a-ga
e.	l-a-b-ag	f.	b-l-a-ga
g.	b-l-ag	h.	b-lə-ga

Assuming for the sake of illustration that the prefix *a* is attached first with no phonological conditions, followed by *b* and then *l*, one possible suppletive–allomorphy analysis of this toy data set is given in (12). The allomorphs are listed in order of priority. Here I assume that the order follows automatically from degree of specificity, since the environments form a nest of proper subsets, but an analysis drawing on the formal Priority concept of Bonet, Lloret, and Mascaró (2007) and Mascaró (2007) would also be conceivable. First is the allomorph subcategorized to appear intervocalically, but if no suitable environment is present, a prevocalic version will surface. Failing that, the affix attaches to the left edge of any other base (i.e., a consonant-initial one) and phonology supplies an epenthetic vowel. Epenthesis feeds cyclic allomorph selection such that (11h) goes from *b-ga* to *bə-ga* upon prefixation of *b*, providing a site for Allomorph 2 of the *l* prefix to attach.

(12) Suppletion analysis of phonologically driven, globally optimizing affix order

	b	*l*
a. Allomorph 1	V] _ [V	V] _ [V
b. Allomorph 2	_ [V	_ [V
c. Allomorph 3	_ [	_ [

One possible objection is that the subcategorization frame in (12a) violates the normal assumption that affixes subcategorize for either the left-hand and right-hand environment, but not both (Anderson 1992: 210, Yu 2007: 49). However, alternative nonoptimization analyses are available. The examples in (11fgh), where *l* and *b* are rearranged to create a well-formed onset cluster *bl*, could be analyzed as purely phonological (albeit possibly morphologically specific) metathesis, along the lines of Paster's (2009: 31ff.) reanalyses of Doyayo (Wiering & Wiering 1994), Witsuwit'en (Hargus & Tuttle 1997), and Hamer (Zoll 1998). Or, one or more of the affixes could be analyzed as floating, being associated to a phonotactically appropriate slot later on in a phonological part of the derivation (assuming that it is truly floating; that is, that its position in the morphological hierarchy does not impose strict linear precedence relations with respect to other morphemes). In short, the available mechanisms seem powerful enough to derive a large array of patterns that are also compatible with a "P >> M" phonological optimization analysis. The more fruitful challenge may be to pursue a different theoretical approach that still incorporates the morphophonological conditions in (10). As Manova and Aronoff (2010: 116) observe, "of course, it is also possible that affixation uses phonological and morphological information at the same time."

5. Conclusion

I have argued, contra Kim (2010), that epenthesis avoidance may not be the driving principle behind Huave mobile affix ordering, even though it may still be considered under the umbrella of phonological affix ordering. This leaves only a few remaining challenges to Paster's (2009) hypothesis that there is no language where "phonology can produce morpheme orderings that disobey other principles." Jenks and Rose (to appear) argue that in the Kordofanian language Moro, mobile affix placement depends on the tonal composition of the base, and is driven by tonal well-formedness factors. In several Athapaskan languages, Rice (2011: 183) argues that there is a set of morphemes that is ordered from prosodically shortest to prosodically longest.

Generally, analyses of phonological affix ordering have differed in whether the relevant phonological conditions are attributed to the lexical entries of individual affixes, to the phonology of the language in general, or to a certain morphological domain, context, or construction (Paster 2009: 23–24; see also Kim 2010: 158). The revised analysis presented here changes Huave's place in the typology: it is now a variant of the type where affix-specific subcategorization frames require an affix to be placed in specific phonological environments, for example Chintang (Bickel et al. 2007) and Choguita Rarámuri (Caballero 2010); in Huave, the relevant environment is defined both morphologically and phonologically. Conversely, in Moro (Jenks & Rose, to appear) and the Kim (2010) analysis of Huave, the placement of an individual affix is driven by phonological requirements independent of the affix itself; and in Athapaskan (Rice 2011: 183), a set of multiple affixes is ordered along a single phonologically defined scale. Paster (2006b) argues convincingly that Pulaar is precisely *not* a case of this last type.

It is not yet clear whether there is an objective and discrete empirical break between affix ordering based on phonological principles, and affix placement based on phonological subcategorization. As more languages are studied (and the same languages are studied more), the typology of phonological influences on affix placement is continuously refined, and it is from the porous membrane between empirical generalizations and corresponding theoretical analyses that changes in our concept of the landscape arise. For phonologically conditioned allomorphy in general, however, Paster (2006a), Embick (2010), and Bermúdez-Otero (2013) point out one crucial difference in prediction: phonological subcategorization should be sensitive to underlying forms, while output-driven optimization must be conditioned by derived phonological properties; see also Deal and Wolf (2013) on Nez Perce. We have not been able to test this prediction for Huave, since known phonological processes in Huave do not affect the conditions for mobile affixation. If more languages with mobile affixation are found, interactions between phonological processes and affix placement would provide key data.

It is worth noting that the subcategorization analysis in (12) of the prototypically optimizing pattern, while possible, is impressionistically somewhat ugly, and this may not be unconnected to the fact that such clear cases of this type are not really attested (Paster 2006b). Meanwhile, the subcategorization analysis in (10) of Huave, which is attested, arguably achieves more generality with less theoretical apparatus than does the nonmodular P >> M approach. Overall, previous literature (including Yu 2007 on infixation) provides a strong case for the existence of phonologically conditioned alternations in surface morpheme order, but deeper analysis of a wider variety of languages will be needed in order to establish a solid typology of the grammatical sources of these effects.

Abbreviations

1	first person
1SUB	first-person subordinate
CPL	completive
FUT	future
RFL	reflexive
SUB	subordinate
TV	theme vowel
V	stem vowel

Acknowledgments

I am grateful to Ricardo Bermúdez-Otero and Mary Paster for helpful comments on earlier versions of this work. Needless to say, all errors are my own.

References

Anderson, Stephen. 1992. *A-Morphous Morphology*. Cambridge: Cambridge University Press.

Bermúdez-Otero, Ricardo. 2012. "The architecture of grammar and the division of labour in exponence." In *The Morphology and Phonology of Exponence*, edited by Jochen Trommer, 8–83. Oxford Studies in Theoretical Linguistics 41. Oxford: Oxford University Press.

Bermúdez-Otero, Ricardo. 2013. "The Spanish lexicon stores stems with theme vowels, not roots with inflectional class features." *Probus* 25:3–103.

Bickel, Balthasar, Goma Banjade, Martin Gaenszle, Elena Lieven, Netra Prasad Paudyal, Ichchha Purna Rai, Manoj Rai, Novel Kishore Rai, and Sabine Stoll. 2007. "Free prefix ordering in Chintang." *Language* 83: 43–73.

Bonet, Eulàlia, Maria-Rosa Lloret, and Joan Mascaró. 2007. "Allomorph selection and lexical preferences: two case studies." *Lingua* 117:903–927.

Caballero, Gabriela. 2010. "Scope, phonology and morphology in an agglutinating language: Choguita Rarámuri (Tarahumara) variable suffix ordering." *Morphology* 20:165–204.

Deal, Amy Rose, and Matthew Wolf. 2013. "Outwards-sensitive phonologically conditioned allomorphy in Nez Perce." http://ling.auf.net/lingbuzz/001791.

Embick, David. 2010. *Localism versus Globalism in Morphology and Phonology*. Linguistic Inquiry Monographs 60. Cambridge, MA: MIT Press.

Hargus, Sharon, and Siri Tuttle. 1997. "Augmentation as affixation in Athabaskan languages." *Phonology* 14:177–220.

Jenks, Peter and Sharon Rose. To appear. "Mobile object markers in Moro: the role of tone." *Language*. Pre-publication version in PDF last consulted 30 July 2014 at http://linguistics.berkeley.edu/~jenks/Research_files/ObjectorderingMarch2013_Distrib.pdf.

Kim, Yuni. 2009. "Alternancias causativas y estructura de eventos en el huave de San Francisco del Mar." *Proceedings of CILLA IV*. www.ailla.utexas.org/site/cilla4/Kim_CILLA_IV.pdf.

Kim, Yuni. 2010. "Phonological and morphological conditions on affix order in Huave." *Morphology* 20:133–163.

Kiparsky, Paul. 1982. "Lexical morphology and phonology." In *Linguistics in the Morning Calm: Selected Papers from SICOL-1981 (vol. 1)*, edited by I.-S. Yang, 3–91. Seoul: Hanshin.

Manova, Stela, and Mark Aronoff. 2010. "Modeling affix order." *Morphology* 20:109–131.

Mascaró, Joan. 2007. "External allomorphy and lexical representation." *Linguistic Inquiry* 38:715–735.

McCarthy, John, and Alan Prince. 1993. "Generalized alignment." In *Yearbook of Morphology* 1993, edited by Geert Booij and Jaap van Marle, 79–153. Dordrecht: Kluwer.

Noyer, Rolf. 1994. "Mobile affixes in Huave: optimality and morphological well-formedness." In *Proceedings of the Twelfth West Coast Conference on Formal Linguistics*, edited by E. Duncan, D. Farkas, and P. Spaelti, 67–82. Stanford: Center for the Study of Language and Information.

Paster, Mary. 2006a. "Phonological Conditions on Affixation." PhD diss., UC Berkeley.

Paster, Mary. 2006b. "A survey of phonological affix order with special attention to Pulaar." In *Proceedings of the 35th Annual Meeting of the North Eastern Linguistic Society: Volume 2*, edited by L. Bateman and C. Ussery, 491–506. Amherst: University of Massachusetts Graduate Linguistic Student Association.

Paster, Mary. 2009. "Explaining phonological conditions on affixation: evidence from suppletive allomorphy and affix ordering." *Word Structure* 2:18–47.

Rice, Keren. 2011. "Principles of affix ordering: An overview." *Word Structure* 4:169–200.

Rose, Sharon. 1995. "Ethio-Semitic inflectional affix order: a phonological solution." *Langue Orientales Anciennes—Philologie et Linguistique* 5–6:259–291.

Wiering, Elisabeth, and Marinus Wiering. 1994. *The Doyayo Language: Selected Studies.* Arlington, TX: Summer Institute of Linguistics and University of Texas at Arlington.

Yu, Alan. 2007. *A Natural History of Infixation.* Oxford: Oxford University Press.

Zoll, Cheryl. 1998. *Parsing Below the Segment in a Constraint-Based Framework.* Stanford: Center for the Study of Language and Information.

6

Hierarchy-Governed Affix Order in Eastern Kiranti

Eva Zimmermann

1. The Phenomenon: Hierarchy-Governed Affix Order

It is a well-known phenomenon that agreement markers on a transitive verb can be restricted to a specific order regardless of whether they realize agreement features of the agent or the patient (='template morphology,' cf. for example Stump 1996 or Stump 2006 for an overview).[1] In (1), this situation is illustrated with examples from Athpare, a Southeastern Kiranti language. The verb stem 'to beat' is followed by different suffixes marking person and/or number agreement with one or both of the arguments and the past-tense marker *-e* in (1). All Athpare data in the following are taken from the grammar by Ebert (1997b).[2]

(1) Affix order in Athpare (Ebert 1997b: 182)
   a. a-lemsa-tsi-ŋa-e
      2-BEAT-NSG-1-PST
      'You beat us two (excl.)' (2s→1de)
   b. a-lemsa-tsi-ŋa-e
      2-BEAT-NSG-1-PST
      'You two beat me' (2d→1s)
   c. lemsa-u-ŋ-tsi-e
      BEAT-3P-1SGA-NSG-PST
      'I beat them (two)' (1s→3d)

The agreement and tense suffixes following the stem always occur in a specific order irrespective of whether they realize features of the agent or the patient. This is most apparent in the contrast between (1-a) and (1-b). The suffix string is identical in both

[1] For valuable comments and discussion I am grateful to Daniela Henze, Stela Manova, Sebastian Bank, Jochen Trommer, and the audience at the 3rd Vienna Workshop on Affix Order at the University of Vienna (January 15–16, 2011). This research was supported by a grant by the Deutsche Forschungsgemeinschaft to the project *Micro- and Macro-variation: Hierarchy Effects in Kiranti and Broader Algic* (TR 521-3).

[2] See the appendix for a list of the glosses and abbreviations I use throughout.

forms, but the markers differ in whether they mark agreement with the agent or the patient argument. In (1-a), *-tsi* (dual) and *-ŋa* (1st person) both mark agreement with number and person of the patient argument whereas in (1-b), *-tsi* marks the dual number of the agent argument and *-ŋa* first-person features of the patient argument. In (1-c) on the other hand, *-u* and *-tsi* mark agreement with the third-person dual patient argument and both are separated by *–ŋ*, indicating first person of the agent argument. These different orders of affixes marking agreement with the agent or patient argument can be summarized with the abstract labels A(gent) and P(atient) as PP (1-a), AP (1-b), and PAP (1-c). One class of theoretical approaches to affix order simply assigns exponents to specific position slots that reflect such a fixed linear order of affixes, illustrated for Athpare in (2). An example for such an approach is the assumption of word formation rules that are ordered in blocks (Anderson 1992).

(2) Suffix slots in Athpare

Σ	-u	-m	-tsi	-ŋa	-e
		-ŋ	-i		
			-na		

Another line of research derives the order of affixes inside Optimality Theory from a number of morpheme-specific constraints (Paster 2006; Ryan & Schuh 2010; Ryan 2010). In contrast, I present an optimality-theoretic analysis for the Kiranti facts that is based on Alignment constraints (McCarthy & Prince 1993; Trommer 2003c) about morphosyntactic features, not morphemes (Trommer 2001, 2003a, 2003c). The first argument for such an analysis is therefore its economy, and the fact that it uses general constraints rather than morpheme-specific mechanisms.

The second main argument for my optimality-theoretic proposal is the fact that there are clearly defined exceptions to the expected hierarchy-governed order in some Kiranti languages. These exceptional reorderings between some morphemes pose severe problems for a templatic account, and can only be derived through additional arbitrary mechanisms like the suspension of the assumption that rule blocks apply sequentially and the introduction of specific portmanteau rule blocks (Stump 2001). I argue for the existence of a markedness constraint demanding that arguments must be marked prominently. Evidence for such a constraint type can be found in a cross-language comparison of different Kiranti languages where different repair strategies can be found to ensure a morphologically prominent argument. Such a situation where a marked structure is repaired through different operations in different contexts is a straightforward prediction of an optimality-theoretic system as the one I propose here.

I begin with an exemplifying case study of the affix order in Athpare in section 2. First, the inflectional affixes and their meaning are introduced in subsection 2.1 before the relevant generalizations about their order are discussed in subsection 2.2. In section 3, my optimality-theoretic analysis is presented that is based on the assumption of Alignment constraints predicting a hierarchy-governed system of affix order (3.1) and a markedness constraint demanding departures from the expected ordering patterns (3.2). Further predictions and cross-language evidence for the proposed analysis are discussed in subsection 3.2.3 where I present facts from the closely related Kiranti language Camling, which differs from Athpare in an interesting way. In section 4, I briefly discuss alternative Optimality Theory (OT) analyses for affix order, especially the accounts in Trommer (2003c), Hyman (2003), and Caballero (2010). I conclude in section 5.

2 Case Study: Affix Order in Athpare

2.1 THE INFLECTIONAL SUFFIXES OF ATHPARE

2.1.1 Theoretical Background: Realizational Morphology

I argue that the affix order in Athpare reflects a hierarchy of morphosyntactic feature categories. Such a hierarchy demands that affixes realizing a certain feature must precede or follow another affix realizing another kind of feature. The meaning (i.e., the set of morphosyntactic features) of the inflectional markers is therefore crucial in accounting for their linear order. Since the assignment of meaning to a morpheme is always the result of abstract analysis, I briefly present the principles that underlie my analysis of agreement affixes in Kiranti. The overall background to which many of the following assumptions refer is a realizational approach to morphology where abstract morphosyntactic feature bundles the syntax provides are realized by the insertion of markers (Distributed morphology (=DM); Halle & Marantz 1993).

For all Kiranti languages that are discussed in the following, the segmentation into morphemes as well as their meaning is roughly equivalent to the morpheme lists in the descriptive grammars. Nevertheless, some formal criteria must justify the segmentation and especially the assignment of morphosyntactic features to morphemes as well. The three criteria in (3) describe a morpheme segmentation that is plausible from the viewpoint of learnability and economy at the same time (Bierwisch 2006; Pertsova 2007; Bank 2010; Trommer 2011). The principle (3-c) is what Müller (2006b, c) calls the Syncretism Principle, stating that identity of form implies identity of function, unless there is good evidence for the contrary.

(3) Preferences for the segmentation into morphemes and the assignment of meaning
   a. Only those segment strings that occur as free forms (for suffixes: occur after the stem on their own) are possible morphemes.
   b. A morpheme is assigned to a feature specification that is necessary and sufficient to describe all its occurrences in the paradigm.
   c. Homophonous morphemes are avoided.

These criteria are obviously not inviolable principles, but preferences governing a learning algorithm that results in an economic lexicon. Such a lexicon avoids redundant information but is learnable at the same time. Related to the demand to be nonredundant is the avoidance of co-occurring markers expressing one and the same feature. In a realizational framework like DM, this follows from the notion of feature discharge stating that whenever a marker is inserted, the features it realizes are unavailable for further insertion of another marker specified for the same feature (Noyer 1992; McGinnis 1996; Müller 2006a). A final important assumption is the distinction into substantial and context features. The former are necessary for the insertion of a marker and cannot be expressed twice (=they are discharged), and the latter are necessary for the presence of a marker but stay available for insertion of another marker (Carstairs-McCarthy 1987; Noyer 1992; Halle & Marantz 1993). Context features are notated as '__[X]' in the following. This distinction into substantial and context features is briefly illustrated with the abstract example in Table 6.1. The markers *-A* and *-B* surface in quite similar contexts, and it seems at first glance reasonable to specify *-A* as a specific [+α,+β] marker and *-B* as more general [+β] marker. However, given the assumption of feature discharge, it is predicted that both markers can never co-occur since they are specified for the same substantial feature [+β]. Specifying *-A* for the substantial feature [+α] and for the context feature [+β] is more in line with the principles I discussed above (cf. the list of morphemes in (4)). From this specification it follows that *-A* can only be inserted if [+β] is present but the feature is not discharged and remains visible for further insertion.

(4) List of morphemes: abstract example
-A ↔ [+α] / __ [+β]
-B ↔ [+β]
-C ↔ [-β]

2.1.2 Morphological Analysis for Inflectional Suffixes in Athpare

In Tables 6.2 and 6.3, a segmented paradigm for transitive and intransitive agreement suffixes in Athpare is given. In the following, I concentrate only on the agreement suffixes of the Kiranti languages and ignore agreement prefixes for reasons of convenience. Only few prefixes can be found in the verbal paradigms of Kiranti, and their analysis adds nothing new to an investigation of affix order. In addition, I

TABLE 6.1
Abstract Example

		α	
		+	–
β	+	-A-B	-B
	–	-C	-C

TABLE 6.2
Suffix Paradigm for Athpare, 1/2 Object (Past)

		Agent							
		1s	1de	1pe	1di	1pi	2s	2d	2p
Patient	1s						-na-e	-na-tsi-e	-na-ni-e
	1de						-na-e	-na-tsi-e	-na-ni-e
	1pe						-na-e	-na-tsi-e	-na-ni-e
	1di								
	1pi								
	2s	-ŋa-e	-tsi-ŋa-e	-i-ŋa-e					
	2d	-tsi-ŋa-e	-tsi-ŋa-e	-i-ŋa-e					
	2p	-i-ŋa-e	-tsi-ŋa-e	-i-ŋa-e					
	3s	-ŋa-e	-tsi-ŋa-e	-i-ŋa-e	-tsi-e	-e	-e	-tsi-e	-i-e
	3d	-tsi-ŋa-e	-tsi-ŋa-e	-i-ŋa-e	-tsi-e	-tsi-e	-e	-tsi-e	-i-e
	3p	-ŋa-e	-tsi-ŋa-e	-i-ŋa-e	-tsi-e	-e	-e	-tsi-e	-i-e
	intr	-ŋa-e	-tsi-ŋa-e	-i-ŋa-e	-tsi-e	-ie	-e	-tsi-e	-i-e

TABLE 6.3
Suffix Paradigm for Athpare, 3 Object (Past)

		Agent		
		3s	3d	3p
Patient	1s	-u-ŋ-e	-u-ŋ-tsi-e	
	1de	-tsi-u-ŋa-e	-tsi-u-ŋa-e	
	1pe	-u-m-ŋa-e	-u-m-tsi-ŋa-e	
	1di	-tsi-u-e	-tsi-u-e	
	1pi	-u-m-e	-u-m-tsi-e	
	2s	-u-e	-u-tsi-e	
	2d	-tsi-u-e	-tsi-u-e	
	2p	-u-m-e	-u-m-tsi-e	
	3s	-u-e	-u-tsi-e	
	3d	-tsi-u-e	-tsi-u-e	
	3p	-u-e	-u-tsi-e	
	intr	-e	-tsi-e	-e

only list underlying abstract forms for all morphemes and abstract away from any phonological processes obscuring the surface forms.[3]

The 'meaning' for an inflectional suffix is taken to be a set of morphosyntactic features, described with binary features. Table 6.4 lists the morphosyntactic categories and their decomposition into binary features that are relevant for the analysis of Athpare. Kiranti languages distinguish the three numbers 'singular,' 'dual,' and 'plural'

[3] In Athpare, the relevant phonological processes that result in predictable surface alternation of morphemes are nasal place assimilation and hiatus-avoiding vowel deletion. Apart from these more or less standard phonological rules, there is an interesting morpheme-specific copying process in Athpare: an affix-nasal preceding *-tsi* is copied after these morphemes, resulting in, e.g., [umtsimma] from the underlying affix string *-u-m-tsi-ŋa*. This copying is actually a common process in Kiranti in general; cf. Zimmermann (2012). For all languages, the surface paradigms can be found in the Appendix.

TABLE 6.4

Agreement Categories and Their Decomposition in Kiranti

Category	Binary features					
	1	2	3	sg	pl	
1s	+	–	–	+	–	
1de	+	–	–	–	–	(dual exclusive)
1di	+	+	–	–	–	(dual inclusive)
1pe	+	–	–	–	+	(plural exclusive)
1pi	+	+	–	–	+	(plural inclusive)
2s	–	+	–	+	–	
2d	–	+	–	–	–	
2p	–	+	–	–	+	
3s	–	–	+	+	–	
3d	–	–	+	–	–	
3p	–	–	+	–	+	

and have a contrast between a first-person nonsingular exclusive and a first-person nonsingular inclusive. In addition to the binary features in Table 6.4, I assume the two privative features 'A' and 'P' abbreviating the thematic roles of transitive agent and transitive patient arguments. I refer to those as 'case' features in the following. With using these abstract labels and avoiding the terms 'accusative/nominative' or 'ergative/absolutive,' I remain agnostic about the thematic alignment in Kiranti.

The assignment of morphosyntactic feature sets to the Athpare suffixes is given in (5). The morphemes are listed in the order in which they appear in longer suffix combinations, reflecting the affix slots in (2). This order is most apparent in one of the combinations of four suffixes *-u-m-tsi-e* (e.g., 1pe→3Ns). I briefly go through these marker specifications below.

(5) Morphemes in Athpare

a.	-u	↔ [P,−1,−2,+3]		**P**
b.	-ŋ	↔ [A,+sg,−pl,+1,−2,−3] / __[+3]		
c.	-m	↔ [A,−sg,+pl,−3] / __[+3]		**A**
d.	-na	↔ [A,+1,−2] / __[+2]		
e.	-i	↔ [−sg,+pl,−3]		
f.	-ni	↔ [−sg,+pl,−3]		**N**
g.	-tsi	↔ [−sg,+pl,+2] / __[A,+1]		
h.	-ŋa	↔ [+1,−2]		**Ps**
i.	-e	↔ [+past]		**Pst**

The assignment of a morphosyntactic feature set that is a necessary and sufficient condition (cf. (3-b)) describing all occurrences of a marker in the segmented paradigms in Tables 6.2 and 6.3 is trivially possible for *-e*. The suffix occurs in the whole paradigm but is absent in the nonpast paradigms. It is therefore arguably the past-tense marker.[4] Such an assignment of an unambiguous meaning is possible as well for *-u*, *-na*, *-ŋ*,[5] and *-m*. The suffix *-u* always marks a third-person patient argument, and most importantly it occurs in all contexts with a third-person patient. Its feature specification (5-a) therefore includes not only the case feature P but all (binary) features explicitly specifying third person. The morphemes *-m* and *-ŋ* occur only in third-person patient contexts as well. However, their distribution is more restricted since they are bound to contexts of first-person singular agents (for *–ŋ*) and to contexts of first- and second-person plural agents (for *–m*). Their resulting feature specifications (5-b+c) mirror that. Finally, the suffix *-na* only occurs in the context of a first-person agent acting upon a second-person patient (1→2). It is therefore taken as an agent marker realizing first person in the context of a second-person patient (5-d)[6]—a feature specification that is once again necessary and sufficient for describing all occurrences of the marker. The suffix *-ŋa* generally marks agreement with a first-person argument. In contrast to the marker specifications of *-e*, *-u*, *-m*, and *-na*, the morphosyntactic context of first person is not sufficient for the occurrence of *-ŋa* since it does not surface in 1→2 and 1s→3 contexts, although its feature specification is met. This simply follows from the assumption of feature discharge and the avoidance of multiple marking for features. The suffixes *-na* and *-ŋ* realize first-person agent case features and are consequently more specific than *-ŋa* only realizing first-person features.[7] After *-na* and *-ŋ* are inserted, subsequent insertion of *-ŋa* is impossible since the features [+1,–2] are already realized and discharged.

A similar blocking relation holds between *-i* and *-m*. The suffix *-i* can only be found when one argument is second-person plural or first-person exclusive plural. The feature set that is common to all these occurrences is [–sg,+pl,–3]. This is a subset of the feature specification of *-m* ↔ [A,–sg,+pl,–3] / __[+3] that is consequently more specific than *-i*. As soon as *-m* is inserted, the features [–sg,+pl,–3] are discharged that are necessary for the insertion of *-i*—the marker is blocked by a more specific marker.[8] A third number marker in Athpare is *–tsi,* which generally

---

[4] Athpare distinguishes only between past and nonpast tense as is common in Kiranti.

[5] That *-ŋ* and *-ŋa* are different morphemes is obscured at the surface in the past paradigm: *-ŋa* + *-e* are realized as [ŋe], which is perfectly predictable given the general phonology of the language. In the nonpast paradigm where the final *-e* is absent, however, it is clear that there are two suffixes, *-ŋ* and *-ŋa*.

[6] One could actually specify it the other way around, that is, as a second-person patient marker in a first-person agent context. In any case, it is a marker that must be specified for case, that is, i.e., for A or P and nothing hinges on the choice for one or the other specification in the following discussion of affix order.

[7] For the concept of specificity deciding the competition of markers about insertion in DM cf., for example, Halle and Marantz (1993, 1994), Halle (1997), Noyer (1998), or Harley and Noyer (1999).

[8] Given the feature specification [–sg,+pl,+3], the marker *-i* is actually expected in more contexts: in 1piP contexts and in the context of 2p→1de. The standard account in DM for such an absence of an otherwise expected marker is the concept of impoverishment rules that manipulate the morphosyntactic context prior to insertion of morphological markers (Halle and Marantz 1993; Halle 1997; Noyer 1998; Frampton 2003; Müller 2006a). In Athpare, an impoverishment rule deletes the number feature of the second-person agent in the context of [+1,–sg,–pl] and the number features for every 1pi patient argument. An alternative is the assumption of more specific homophonous markers ($-i_1$↔ [+pl,+2] /__ [+sg,+1], $-i_2$ ↔ [P,+pl,+1,–2], and $-i_3$ ↔ [P,+pl,+2]). However, for the present discussion of affix order, the choice between one or the other analysis is irrelevant.

TABLE: 6.5

Two Alternative Analyses for -tsi

'1tsi-Analysis'	'2tsi's-Analysis'
tsi ↔ [–sg]	tsi ↔ [–sg,–pl]
	tsi ↔ [–sg] / [_+3]

marks dual number for agents and first- and second-person patients. In the context of a third-person patient, *-tsi* is not only restricted to dual contexts but generally marks third-person nonsingular patient arguments. This distribution of *-tsi* can be observed in many Central-Eastern Kiranti languages,[9] and opens up two alternative analyses. One can take *-tsi* as a general [–sg] marker ('1tsi-Analysis' in Table 6.5) or assume two homophonous markers *-tsi*: one being a dual marker and the other being a nonsingular marker restricted to the context of third-person patients ('2tsi's-Analysis').

Under the '1tsi-Analysis,' the marker is expected in all nonsingular contexts and several nonappearances of the marker especially in first- and second-person plural contexts remain mysterious. The '2tsi's-Analysis' analysis avoids some of these mispredictions through restricting one *-tsi* to dual contexts and the other one to nonsingular third-person patient contexts.[10] However, it introduces homophonous lexical entries with identical forms but different meanings. As was already discussed in section 2.1.1, this is a dispreferred option according to the Syncretism Principle (Müller 2006b, c).

The choice between the two analyses is in fact irrelevant for the discussion of affix order, since the hierarchy-governed order approach predicts the same ordering properties all markers realizing number features. As became clear in the list in (5), I adopt the '1tsi-Analysis' of one general [–sg] *-tsi*, which is in accordance with the Syncretism Principle (Müller 2006b, c). I return to this discussion in section 2.2 where another argument for assuming a single *-tsi* is discussed: in some contexts, two occurrences of *-tsi* are expected (=one for each argument) but only one surfaces. This blocking phenomenon receives a more straightforward account in an analysis with one *-tsi*.

2.2 ORDER OF SUFFIXES IN ATHPARE

With the exception of *-tsi* and *-e*, all Athpare suffixes in (5) are specified for more than one morphosyntactic feature category. For example, *-i* ↔ [–sg,+pl,–3] is

---

[9] At least for Athpare (Ebert 1997b), Bantawa (Doornenbal 2009), Belhare (Bickel 1998; 2003), Camling (Ebert 1997a), Chintang (Bickel, Banjade, Gaenszle, Lieven, Paudyal, Rai, Rai, Rai, and Stoll 2007), Limbu (van Driem 1987), Lohorung (van Driem 1992), Mewahang (Gaenszle 1995), Puma (Bickel, Gaenszle, Rai, Rai, Rai, Rai, and Sharma 2007), and Yakkha (Schackow 2014).

[10] Some nonappearances of *-tsi* remain unexplained even under the '2tsi's-Analysis.' This is the case in the context of dual agents acting upon 1pe or 2p. Cf. footnote 8 for a discussion of possible solutions for such neutralizations inside DM.

specified for number and person features. However, one can still establish a clear connection between the features a marker realizes and its position in a sequence of markers: every suffix that precedes another suffix realizes an additional type of morphosyntactic feature, marked in boldface in (5). This generalization about feature specification and precedence allows the establishment of a simple hierarchy of morphosyntactic features that governs the order of affixes. This hierarchy given in (6) does not only hold in Athpare, but is very similar for many other Eastern Kiranti languages.[11]

(6) Feature hierarchy for the affix order in Athpare
P >> A >> N >> Ps

This hierarchy simply states that a marker realizing patient features always precedes a marker realizing agent features, and the latter always precedes a marker realizing number features, and so on. If a marker is specified for more than one feature type, it is always the feature higher on the hierarchy that determines its position with respect to another morpheme. For example, *-u* and *-ŋa* are both specified for person but the former always precedes the latter, since *-u* realizes the case feature P as well and P is higher on the hierarchy than person. Some examples for the order in longer affix strings are given in (7). The form in (7-a) illustrates that affixes specified for number (*-i* ↔ [−sg,+pl,−3]) precede affixes specified only for person (*-ŋa* ↔ [+1,−2]). In (7-b), the suffix *-u* specified for person and the case feature P precedes the suffix *-m* specified for number, person and the case feature A. And both are followed by *-tsi* realizing only number features.

(7) a. lemsa-tsi-ŋa-e
BEAT-NSG-1-PST
'He beats us (excl)' (3s→1pe)

b. lemsa-u-m-tsi-e
BEAT-3P-1/2PLA-NSG-PST
'We (incl) two beat them' (1di→3p)

The hierarchy in (6) excludes the category tense. The position of tense markers is relatively unstable in the Kiranti languages, which is in sharp contrast to all other inflectional categories. In the Athpare paradigms I gave above, the tense marker *-e* is always the last suffix in the suffix string. In Limbu, on the other hand, another Kiranti language that is discussed in later sections, the past marker *-ε* always occurs directly after the stem and before all other agreement suffixes. In my OT account, such differences easily fall out with simply reranking of the ALIGNMENT

[11] Some more examples from Eastern Kiranti are discussed below. For a classification of the Kiranti languages cf., for example, Michailovsky (1994), van Driem (2001), Opgenort (2005), and Bickel (2008).

constraints for tense categories. However, I restrict myself to the agreement features in the following and remain agnostic about the question whether tense categories are integrated in the hierarchy of morphosyntactic features.[12]

A close look at the paradigms in Tables 6.2 and 6.3 actually reveals that the hierarchy in (6) is able to predict most affix ordering relations correctly (74 paradigm cells out of 86), but fails to capture the facts in 12 contexts, namely in the contexts of a dual agent acting upon a third-person patient argument. As can be seen in the paradigm extract in Table 6.6, the number marker *-tsi* precedes *-u* in those contexts. This is unexpected since *-u* is a case marker specified for P and *-tsi* is only specified for number.

This departure from a hierarchy-governed affix order is actually quite common in Kiranti languages and can be found in, for example, Limbu as well, another language of Eastern Nepal that is described in van Driem (1987).[13] A portion of the transitive nonpast paradigm of Limbu is given in Table 6.7. The inflectional markers differ slightly in shape and distribution from the suffixes in Athpare, as can be seen in the list in (8).[14] For example, a suffix *-ge* generally marks the first person, and two instances of *-si* (=Athpare *-tsi*) surface in third-person nonsingular patient contexts—one marking number of the agent, the other number features of the patient. The general order of affixes in Limbu, however, follows the same hierarchy P >> A >> N >> Ps, and the order *-si-u* in 1/2d–3 contexts is again unexpected.

TABLE 6.6
Third-Person Patients in Athpare

		Agent	
		3s	3d/p
Patient	1s	-u-ŋ	-u-ŋ-tsi
	1de	**-tsi-u-ŋa**	**-tsi-u-ŋa**
	1pe	-u-m-ŋa	-u-m-tsi-ŋa
	1di	**-tsi-u**	**-tsi-u**
	1pi	-u-m	-u-m-tsi
	2s	-u	-u-tsi
	2d	**-tsi-u**	**-tsi-u**
	2p	-u-m	-u-m-tsi
	3s	-u	-u-tsi
	3d	**-tsi-u**	**-tsi-u**
	3p	-u	-u-tsi

[12] This is consistent with the analysis in Trommer (2001) and Trommer (2003c), where only the order of agreement markers follows from ALIGNMENT constraints. Tense as a lexical head is already present in the syntax and has another status.

[13] Other languages exhibiting this reordering of *-tsi* (and its cognates) are, e.g., Bantawa (Doornenbal 2009), Belhare (Bickel 1998; 2003), and Yakkha (Schackow 2014).

[14] Abstract and surface paradigms for all the exemplifying languages are given in the Appendix. Cf. B.3 for Limbu.

(8) List of morphemes: Limbu, nonpast

-u	↔ [P,+3]
-ŋ	↔ [A,+1,−2,+sg]
-m	↔ [A,−3,+pl] /__[+3]
-si	↔ [−sg]
-ge	↔ [+1,−2,−sg]

As I already mentioned, these well-defined exceptions to a hierarchy-governed order are an important argument for the optimality-theoretic approach I argue for in section 3. What is the crucial generalization about these two different positions of *-tsi*? If one compares the two suffix strings *-u-tsi* (e.g., 3s→3d/p) and *-tsi-u* (e.g., 3d→3s) in Athpare, the systematic difference between the two orderings consists in the fact that *-tsi* marks the number of the patient in the former but the number of the agent in the latter case. The two positions of *-tsi* are therefore crucially bound to whether number features of the patient or the agent argument are marked. As became clear in the very first example in (1), it is impossible to generally predict the affix order between two affixes in Kiranti simply from taking into account the question whether they realize agent or patient features. This became apparent in the different orders PP, AP, and PAP we observed in the data in (1) (recall especially the order *-tsi-ŋa* that is identical in contexts in which *-ŋa* realizes agent features and patient features). The possible assumption that all markers realizing features of the patient head are inserted first and the other argument's features are realized afterward can therefore only predict the *-u* ↔ *-tsi* minimal pair, but is doomed to fail in the rest of the paradigm.

Another possible explanation is apparently the assumptions of two different instances of *-tsi*, with different marker specifications that predict different positions in the suffix string. This alternative assumption of two homophonous morphemes *-tsi* was already discussed in section 2.1.2. One *-tsi* could be specified for dual, the other for the nonsingularity of a third-person patient argument, and both appear in different positions: the dual *-tsi* before the patient case marker *-u* and the third-person nonsingular patient *-tsi* after all agent case markers, specifically after

TABLE 6.7

Limbu, Nonpast (van Driem, 1987: 368–374)

		Agent	
		3s	3d/p
Patient	1s	-u-ŋ	-u-ŋ-si
	1de	**-si-u-ge**	**-si-u-si-ge**
	1pe	-u-m-ge	-u-m-si-ge
	1di	**-si-u**	**-si-u-si**
	1pi	-u-m	-u-m-si
	2s/3s	-u	-u-si
	2d/3d	**-si-u**	**-si-u-si**

-u, *-m*, *-na*, and *-ŋ*. There are three problems with such an analysis. The first one, as already discussed, is simply the dispreference for homophonous marker entries, especially if the two homophonous markers have such a strikingly similar feature specifications. The second problem is theory-internal—such a solution is impossible given the assumptions I took for granted in my hierarchy-governed analysis. If it is truly the feature specification of a morpheme that demands its position inside the string, it is impossible to find a general hierarchy of feature categories which makes the following predictions:

1. the first *-tsi* realizing number is ordered before *-u* realizing person and P,
2. *-u* (Ps,P) is ordered before *-m*, *-ŋ* and *-na* that realize A and person (and number), and
3. the second *-tsi* realizing number is ordered after all these markers.

There is an obvious ranking paradox in this list: N >> P >> N. And a third problematic point concerns the marker distribution in d→3Ns forms. Whether we assume one general morpheme *-tsi* or two homophonous markers in different positions, we expect two occurrences of *-tsi* in these contexts, one for the dual agent and one for the nonsingular patient. However, only one *-tsi* surfaces. If one morpheme specified for [-sg] is assumed, the absence of the second number marker is due to haplology and avoidance of multiple occurrence of the same morpheme in the same suffix string (Menn & McWhinney 1984; Yip 1998; Plag 1998; de Lacy 1999; Nevins 2012). Since both instances of *-tsi* are not expected to occur phonologically adjacent, the alternative analysis with two markers is not as straightforward. If it is only their phonological representation that is accidentally homophonous, phonological adjacency seems necessary to exclude multiple insertion. The complementary distribution of agent and patient *-tsi* would therefore remain a coincidence in the alternative assuming two homophonous markers.

The analysis I present in section 3 therefore assumes that one number marker *-tsi* exists in Athpare that is generally predicted to appear after *-u* according to the morphosyntactic feature hierarchy (6), but can exceptionally be ordered before it if this is the only possibility to mark the agent prominently.

3. Analysis of Affix Order in OT

3.1. HIERARCHY EFFECTS THROUGH ALIGNMENT CONSTRAINTS

I assume a realizational morphological system where insertion of morphemes and their order is evaluated in a parallel fashion. This is the model of Distributed Optimality (Trommer 2003a), where the optimal morpheme sequence is evaluated for a given set of morphosyntactic features through the ranking of violable constraints. The constraints determining affix order are ALIGNMENT constraints (McCarthy & Prince 1993) that demand that a certain edge (left or right) of a phonological or morphological category must be aligned with a certain edge of another category.

This concept of ranked ALIGNMENT constraints has been proposed in the domain of affix order by various researchers where it actually implements (phonological or morphological) subcategorization frames for morphemes (cf., for example, Caballero 2010; Kim 2010). A more generalized version that avoids reference to specific morphemes assumes ALIGNMENT constraints for morphological categories rather than for morphemes (Hargus & Tuttle 1997; Trommer 2001, 2003a, 2003c). The abstract form of these constraints is given in (9).

(9) ⇐X
Assign a violation mark for every morpheme that intervenes between a marker realizing a morphological feature of class X and the right edge of the stem.

Since we are dealing with suffixes in Kiranti, all ALIGNMENT constraints in the following refer only to the right edge of the stem.[15] The notion 'feature class' in (9) refers to morphosyntactic categories of agreement features: 'person' and 'number,' or the two case features I introduced above (A and P). In section 2 where the Athpare suffixes and their meaning were discussed, I concluded that the hierarchy in (6), repeated in (10), governs the order of affixes. A morpheme that is specified for a morphosyntactic feature that is ranked higher on the hierarchy in (10) precedes all morphemes that only realize lower-ranked features.

(10) Feature hierarchy for the affix order in Athpare
P >> A >> N >> Ps

This hierarchy is straightforwardly implemented in an OT system with ALIGNMENT constraints through the constraint ranking in (11).

(11) Constraints predicting the affix order in Athpare (preliminary)
⇐P >> ⇐A >> ⇐N >> ⇐Ps

The effect of the constraint ranking is illustrated in Table 6.8. The tableau optimizes the suffix string in the context 1de→3s that surfaces as *-u-m-tsi*. It takes the feature specification for both heads as its input and evaluates full strings of suffixes calculating the optimal ordering between these suffixes. Below every morpheme, the feature categories it realizes are given. This abbreviates their full feature specification (cf. the list in (5)) to the information that is relevant for the ALIGNMENT constraints. The subscribed indices indicate whether the marker is inserted to realize features of the agent or patient head.

[15] In Trommer (2003a, b) the constraint type is actually proposed to capture the typological generalization that person affixes have a strong tendency to precede number affixes (for a discussion of such typological tendencies cf. also Siewierska 2004). This is captured through assuming left-ALIGNMENT for person and right-ALIGNMENT for number. Kiranti languages are a clear counterexample to this typological claim, since number markers systematically precede genuine person markers that always occur in final position (only followed by a potential tense suffix). Cf. section 4 for discussion.

TABLE 6.8

Athpare Affix Order

[A,–sg,–pl,+1,–2,–3] [P,+sg,–pl,–1,–2,+3]	⇐P	⇐A	⇐N	⇐Ps
a. –m –u –tsi [A,N,Ps]$_A$ [P, Ps]$_P$ [N]$_A$	*!		**	*
☞ b. –u –m –tsi [P, Ps]$_P$ [A,N,Ps]$_A$ [N]$_A$		*	*,**	*
c. –u –tsi –m [P, Ps]$_P$ [N]$_A$ [A,N,Ps]$_A$		**!	*,**	**

The tableau nicely illustrates that it is impossible to fulfill all ALIGNMENT constraints if more than one suffix is attached to a stem. This follows since only one morpheme can be perfectly aligned with the right edge of the stem. For the moment, it is taken for granted that noninsertion of an expected affix is no possible repair to avoid violations of ALIGNMENT constraints. Consequently, all suffixes that are the most specific affixes whose feature specification is met by the input are inserted.

The highest-ranked and therefore most important constraint is ⇐P, demanding that a case marker for the patient must directly follow the stem. Since only one affix specified for P is inserted (*-u*), perfect satisfaction of this ALIGNMENT constraint is possible and the winning candidate must order *-u* before all other suffixes. This can be seen in the contrast between candidates a. and b.+c. in Table 6.8, where the former orders *-u* after *-m*. The second constraint demands that a marker specified for A must be aligned with the stem. Again, only one marker is specified for this case feature, namely *-m*. At least one violation for ⇐A must be tolerated, since it is more important that the patient case marker *-u* is adjacent to the stem. Any other marker that intervenes between *-u* and *-m* (as in candidate c.) causes an additional violation of ⇐A. The constraint ⇐N now demands adjacency to the stem for more than one marker in this optimization, namely for *-m* and *-tsi*, which are both specified for number (violations for different markers are separated by a comma). Because the case-marker *-m* is also forced to be aligned with the stem from higher-ranked ⇐A, the candidate b. with *-m* before *-tsi* has a better constraint profile than candidate c. with the reversed order. This is a good example for the generalization that was drawn earlier: only the highest-ranked feature category for which a marker is specified is decisive in a hierarchy-governed affix order system.

3.2 MORPHOLOGICAL PROMINENCE FOR AGENT ARGUMENTS

3.2.1 Morphological Prominence in OT

In this subsection, the departures from the hierarchy-governed affix order in Athpare (discussed in subsection 2.2) are further investigated. I propose a new constraint type that accounts for these reorderings: a markedness constraint that demands the agent argument to be marked prominently. A close look at the cases of unexpected orderings and suffix distributions in Kiranti reveals that the exceptions of the hierarchy-governed order are not morpheme-specific and arbitrary, but can be summarized under the concept of 'morphological prominence.' This concept unites the two independently motivated concepts of prominence by position and prominence through case marking. The unexpected reorderings in Athpare are always bound to the first position in the affix string—an arguably perceptually prominent position. It is not only the first position in the affix string (cf. the concept of prominence by position, Steriade 1995; Beckman 1998; Zoll 1998; Nelson 2003), but also a position adjacent to the stem (McCarthy & Prince 1995; Urbanczyk 2001; Alderete 2001). All affixes in this position are therefore taken to be morphologically prominent. This concept of morphological prominence through position in an affix string extends to argument heads quite easily: whenever features of an argument are realized by an affix that is realized in a prominent position, the head is marked morphologically prominently.

A second important factor for morphological prominence of arguments is the presence or absence of case-markers. If agreement affixes specified for the case features of an argument are inserted, the argument is marked prominently as well. This 'prominence through case' refers to, for example, the concept 'discourse prominence' of a case bearing argument (de Hoop & Malchukov 2008). In summary, morphological prominence for arguments is defined as in (12), unifying the two concepts of morphological prominence by position and the concept of prominence through case marking.

(12) Argument prominence through affixation
An argument α is marked prominently if:

a. An affix is present that is specified for the case feature A or P and realizes features of α, or
b. An affix realizing morphosyntactic features of α is in a prominent position adjacent to the stem.

The preference for the agent argument to be marked prominently is ensured by the constraint in (13).

(13) PROMAGENT! (=PAG!)
Assign a violation mark for every verbal affix string where the agent is not marked prominently.

This preference for marking agent agreement through either a case marker or an agreement affix in a prominent position is reminiscent of the typological correlation between word order and case marking (Comrie 1981; Haspelmath 2000; Müller 2002). It has been argued that whenever a language has free word order, it also has morphological case. The functional motivation for this correlation is the fact that there is a preference for unambiguous identification of the arguments of a grammatical function. Example (13) demands an unambiguous marking for agent arguments on the level of affixation: the agent argument must be marked through an affix specified for case or through an affix in a specific prominent position.

3.2.2 Reordering in Athpare

In the examples from Athpare and Limbu, a departure from the expected hierarchy-governed order was observed whenever a dual agent acts upon a third-person patient. The expected order-*u*-*tsi* that follows from the hierarchy P >> N is obscured, and -*tsi*-*u* surfaces. The important generalization discussed in 2.2 was that the number marker unexpectedly occurs next to the stem and before -*u* when it marks the number of an agent in the absence of an agent case-marker. I argue now that -*tsi* is forced to appear in the morphologically prominent position directly after the stem, because no agent-case marker ensures agent prominence that is demanded by the high-ranked constraint PROMAGENT! The following tableaux illustrate the point with examples from Athpare. In Table 6.9, the tableau for the affix combination in the context 1pi→3s is evaluated. Given the feature specification of the input, the inflectional affixes -*u* and -*m* are inserted, the former realizing features of the patient and the latter features of the agent. Since -*m* is marked for case, the agent is marked prominently in this affix string and PROMAGENT! is satisfied. The default ordering -*u*-*m* that is expected from the ranking ALIGNMENT constraints surfaces in this context.

Table 6.10 now shows the tableau that derives a situation where a departure from the hierarchy-governed order is predicted. In the context 2d→3sg, the suffixes -*u* and -*tsi* are expected to occur and no inflectional affix that is specified for A is available for insertion. The highest-ranked PROMAGENT! nevertheless demands that the agent is marked prominently, and the only chance to satisfy the constraint is to order a suffix that realizes agent features in the prominent position right after the stem. The nonsingular marker -*tsi* is the only affix that is coindexed with the agent head in this context. Realizing -*tsi* in the position after the case marker -*u* as is expected from the hierarchy ensured by ALIGNMENT constraints is therefore impossible (candidate a.). Rather, the candidate b. with the prominently marked agent is the optimal output.

However, although PROMAGENT! is the highest-ranked constraint in Athpare, there is a situation in which a violation of PROMAGENT! is tolerated for the optimal candidate. This is the case if simply no marker realizing agent features is available.

TABLE 6.9

Prominent Agent Through Case Marker

[A,–sg,+pl,+1,+2,–3] [P,+sg,–pl,–1,–2,+3]	PAG!	⇐P	⇐A	⇐N	⇐Ps
☞ a. –u [P, Ps]$_P$ –m [A,N,Ps]$_A$			*	*	*
b. –m [A,N,Ps]$_A$ –u [P, Ps]$_P$		*!			*

TABLE 6.10

Athpare Affix Order: -tsi-u

[A,–sg,–pl,–1,+2,–3] [P,+sg,–pl,–1,–2,+3]	PAG!	⇐P	⇐A	⇐N	⇐Ps
a. –u –tsi [P, Ps]$_P$ [N]$_A$	*!			*	
☞ b. –tsi –u [N]$_A$ [P, Ps]$_P$		*			*

TABLE 6.11

Unprominent Agent in Athpare

	3s	3d/p
2s	**-u**	**-u-tsi**
2d	-tsi-u	-tsi-u
2p	-u-m	-u-m-tsi
3s	**-u**	**-u-tsi**
3d	-tsi-u	-tsi-u
3p	**-u**	**-u-tsi**

There are only agent case markers specified for first person (*-m*, *-ŋ*, or *-na*) in Athpare, and no agreement suffixes specified for singular or third-person plural. Consequently, there are some contexts where no inflectional suffix that realizes agent features can be inserted. Table 6.11 gives some examples of verbal forms, with an unprominent agent marked in boldface.

In the Table 6.12, the tableau for an unprominent agent is given. In the exemplifying context 2s→3d, no marker for the second-person singular head is available and PROMAGENT! is violated by all candidates. Now the lower-ranked ALIGNMENT constraints determine a hierarchy-governed order, and since ⇐P is higher-ranked than ⇐N, the nonsingular marker *-tsi* must follow the patient case marker *-u*.

The ranking of constraints that was illustrated in the tableaux 6.9, 6.10, and 6.12 now derives the order of all agreement suffixes in Athpare (cf. the paradigms in Tables 6.2 and 6.3). General ALIGNMENT constraints derive a hierarchy-governed order that crucially hinges on the features the markers realize. And in some contexts, the demand to mark the agent head prominently results in a situation where a number marker for a dual agent occurs in a prominent position right after the stem.

TABLE 6.12

Athpare Affix Order: -u-tsi

[A,+sg,–pl,–1,+2,–3] [P,–sg,–pl,–1,–2,+3]	PAG!	⇐P	⇐A	⇐N	⇐Ps
☞ a. –u –tsi [P, Ps]$_P$ [N]$_P$	*			*	
b. –tsi –u [N]$_P$ [P, Ps]$_P$	*	*!			*

3.2.3 A Prediction: Nonrealization of a Marker

In this subsection, I discuss some further predictions about attested patterns in the languages of the world that follow from the introduction of PROMAGENT! into the inventory of OT constraints. It is taken for granted in Distributed Optimality (=DO, Trommer 2003a) that GEN only generates candidates with morphemes whose features are part of the input. For example, in the tableau in 6.12 a candidate like *-u-m-tsi* ($[P,Ps]_P$ $[A,N,Ps]_A$ $[N]_A$) is impossible, since the marker *-m* realizes features that are not present in the input (+pl). The fact that morphemes are inserted in the first place and realize as much input features as possible follows in DO from PARSE constraints (14), demanding that a feature in the input must be realized through insertion of a morpheme that is specified for this feature in the output.

(14) PARSE FS (Trommer 2003a: 106)
Assign a violation mark for each feature structure FS' in the input that is subsumed by FS and not realized by a feature structure in the output that parses FS in FS.'

A prediction of OT is that all possible rankings of constraints should result in an existing grammar of a language (the 'factorial typology' of a proposed constraint set; Prince & Smolensky 2002 [1993]). Given the existence of the constraint (14), it is predicted that not only reordering (violations of ALIGNMENT) but also the absence of expected suffixes (violations of PARSE) is a possible repair to avoid violations of PROMAGENT! This prediction is borne out in Kiranti. There are languages where all markers that are expected to intervene between the number marker *-tsi* and the stem are absent in the contexts where reordering is observed in Athpare. An example is Camling, which is very similar to Athpare with respect to its marker inventory. Some suffix forms for the third-person patient contexts are given in Table 6.13; the relevant markers and their specifications are listed in (15). The order of suffixes in Camling follows the same hierarchy as the suffixes in Athpare and Limbu, namely P >> A >> N >> Ps.

(15) Camling: list of morphemes
-u ↔ [P,–1,–2,+3]
-m ↔ [A,–sg,+pl,–3] /__[+3]
-tsi ↔ [–sg]
-ka ↔ [–sg,+1, –2]

If the ordering of suffixes in Camling were due to the same constraint-ranking as in Athpare;—that is, if PROMAGENT! dominated all ALIGNMENT constraints—we would expect the same reordering between *-u* and *-tsi* in d→3 contexts. As can be seen in the paradigm in Table 6.13, Camling shows a different pattern in all these contexts where no agent case marker and no marker realizing agent features is expected to appear adjacent to the stem: the suffix *-u* is absent.

TABLE 6.13

Avoidance of a Nonprominent Agent in Camling, Past (Ebert 1997a: 70–75)

	3s	*Expected given the marker specification*
1de	-tsi-ka	*-u-tsi-ka
1pe	-u-m-ka	
1di	-tsi	*-u-tsi
1pi	-u-m	
2d	-tsi	*-u-tsi
2p	-u-m	

I argue that Camling -*u* has the same feature specification [P,+3] as in Athpare, and that its absence in certain contexts follows from my assumptions about the demand to mark agents prominently in Kiranti. More concretely, the blocking of the expected -*u* is simply another strategy to satisfy high-ranked PROMAGENT! The hierarchy of ALIGNMENT constraints predicts that the patient marker -*u* occurs closer to the stem than -*tsi*—an ordering that would result in a situation where the agent head is not marked prominently. If the -*u* is simply absent, the number marker -*tsi* is in a prominent position without contradicting the expectations of a hierarchy-governed order. This repair strategy violates PARSE, since morphosyntactic features from the input remain unrealized. Exactly such blocking is predicted in DO, where realization and order of morphemes is calculated at the same time. In Camling, the constraint demanding the realization of all patient case features (PARSE -P) is ranked below ⇐P, and therefore the suffix -*u* is rather not inserted than realized in a position that is not adjacent to the stem.

Camling is therefore very similar to Athpare: it strives to mark the agent prominently. It differs from Athpare in that the constraints ensuring the hierarchy-governed affix order are higher-ranked than PROMAGENT!, and reordering is impossible to achieve agent prominence. The derivation for this state of affairs in Camling is illustrated in the tableaux given in Table 6.14. In the first competition, I., for the context 2p→3d, the dual suffix -*tsi* realizes patient features and appears after -*u*, as is predicted by the hierarchy of ALIGNMENT constraints. High-ranked PROMAGENT! is irrelevant, since a marker specified for agent case, -*m*, is present. In optimization II, however, no case marker for the agent argument can be inserted. The agent can nevertheless be marked prominently since -*tsi* realizing agent number features is present. If this number marker now appears in the prominent position right after the stem, PROMAGENT! will be satisfied. One strategy to realize -*tsi* in this position is reordering between -*u* and -*tsi* as in candidate c. This is the strategy that becomes optimal in Athpare (cf. Table 6.10) or Limbu (cf. Table 6.7), but is excluded in Camling. Another possible repair, namely noninsertion of -*u* in candidate b., fares better since it only violates lower-ranked PARSE-P rather than higher-ranked ⇐P.

TABLE 6.14

Order and Nonrealization of -u in Camling

		PAG!	⇐P	PARSE-P	⇐A	⇐N	⇐Ps
I.	[A,–sg,+pl–1,+2,–3,]$_A$ [P,–sg,–pl–1,–2,+3,]$_P$						
a.	-tsi -u -m [N]$_P$ [P, Ps]$_P$ [A,N,Ps]$_A$		*!		**	**	*,**
b.	-tsi -m [N]$_P$ [A,Ps,N]$_A$			*!	*	*	*
☞ c.	-u -m -tsi [P, Ps]$_P$ [A,Ps,N]$_A$ [N]$_P$				*	*,**	*

		PAG!	⇐P	PARSE-P	⇐A	⇐N	⇐Ps
II.	[A,–sg,–pl–1,+2,–3,] [P,+sg,–pl–1,–2,+3,]						
a.	-tsi -u [N]$_A$ [P, Ps]$_P$		*!				*
☞ b.	-tsi [N]$_A$			*			
c.	-u -tsi [P, Ps]$_P$ [N]$_A$	*!				*	

TABLE 6.15

Different Repair Strategies to Ensure Agent Prominence

	E.g. 1di–3s	Excluded from PAG!	Repair
Athpare	Σ-tsi-u	*Σ-u-tsi	➢ reordering of *-u* and *-tsi*
Camling	Σ-tsi	*Σ-u-tsi	➢ absence of *-u*

The different strategies of reordering and nonrealization of an otherwise expected marker to ensure that the agent argument is marked prominently are summarized in Table 6.15.

The comparison with a closely related Kiranti language is striking evidence for the optimality-theoretic approach I adopt here. The markedness constraint PROMAGENT! is active in different languages and due to a slightly different ranking of general constraints, different repair strategies to achieve agent prominence are predicted: noninsertion of a marker or reordering. This is a straightforward prediction of an OT system where all possible rankings of existing constraints should yield existing grammars ('factorial typology').

4. Alternative Accounts

In this section, I compare several alternative OT accounts for linear order with my own proposal and conclude that an approach based on ALIGNMENT and PROMAGENT! is not only empirically adequate for the Kiranti facts, but also more economic from a theoretical perspective. In Trommer (2003c), the ordering properties of agent agreement markers for person and number features are the focus of the research. A sample of 58 languages is investigated that consists of languages where "one agreement affix marks only one category C1 (person and number) while the other affix marks the other one, C2, and possibly also C1" (Trommer 2003c: 288). A clear tendency is found that Ps >> N is the dominant order. This typological finding is hard-wired into the proposed framework through the assumption of the two ALIGNMENT constraints in (16).

(16) Person-number asymmetry in Trommer (2003c)
  a. L ← [+PER] (Person-Agreement is at the left edge)
  b. [+NUM] → R (Number-Agreement is at the right edge)

Person strives always to be aligned with the left edge of the stem, whereas number strives to be aligned with the right edge of the stem. In contrast, my analysis of the Kiranti affix order is based on ALIGNMENT constraints referring only to the right edge of the stem that correctly predict the hierarchy-based affix order P >> A >> N >> Ps. The constraint system in (16) does not allow us to predict a hierarchical affix order system where person affixes generally follow number affixes. Trommer (2003c) argues that all the apparent counterexamples to this left-right asymmetry are reanalyzable and adduces them to one of the following three instances:

1. the morphemes in question are rather incorporated quantifiers than affixes, or
2. the morphemes in question realize additional features as well, or
3. the morphemes in question have additional independently motivated context-restriction to their position.

The Kiranti facts cannot be reanalyzed in any of these ways. There is no evidence at all that the number markers that precede person markers in Kiranti are quantifiers. It is also highly implausible that the markers realize additional features (as in Trommer's reanalysis of Isthmus Zapotec facts, where person markers are fused with gender marking) if one takes a look at their distribution throughout the Kiranti paradigms. An additional context restriction that, for example, certain markers must remain adjacent to the stem or another affix cannot be motivated either: there is simply no such subcategorization requirement that would apply consistently to all number or person affixes. The fact that number marking follows person marking is argued to be a statistical finding in Trommer (2003c), and it is not surprising at all to find exceptions to typological tendencies. However, the Kiranti data are true counterevidence for the optimality-theoretic analysis on affix order that Trommer (2003c) proposes on the basis of his typological findings.[16]

Other optimality-theoretic approaches to affix order might be very well able to predict the Kiranti ordering facts, but are less general and consequently

[16]Another example for an apparent violation of the universal tendency for person affixes to precede number affixes can be found in the verbal paradigm of Chamacoco, as is discussed in Bertinetto (2011).

suffer from a severe lack of predictive power. This holds for morpheme-specific approaches like the ones in Paster (2006) or Ryan and Schuh (under preparation), where constraints explicitly referring to specific morphemes are assumed. The same critique applies to the approaches I summarized under the heading of 'templatic approaches' in section 1; that is, approaches that assign exponents to specific position classes (Anderson 1992; Stump 2001). Although the hierarchy of morphosyntactic features looks apparently similar to a list of affix slots, the crucial difference is that the former is specified for morphosyntactic categories and the latter for specific morphemes, and consequently only active in a single language.

My approach for the affix order in Kiranti is at first glance very similar to the analysis for affix order in Bantu languages argued for in Hyman (2003). Morpheme order in Bantu "represents a language-specific resolution of a basic tension between two competing pressures: the pressure for affix order to be compositional versus the pressure for affix order to be fixed (invariant)" (Hyman 2003: 246). The former pressure is ensured by a constraint MIRROR and the latter by the constraint TEMPLATE demanding that a language-specific hierarchy of morphosyntactic features is obeyed in the affix order. The formal implementation is therefore different from the one I propose: Only a single constraint ensures the hierarchy-governed order. For one, this assumption makes the approach less general, since language-specific hierarchies must be stored in the grammar in addition to a constraint ranking of TEMPLATE and MIRROR. On the other hand, this approach seems to make the prediction that as soon as the hierarchy-governed order cannot be obtained between two morphemes, it is absent between all other morphemes of this string as well. This follows since TEMPLATE is apparently not a gradient constraint. It is defined as a constraint demanding that a "morphosyntactic input {CAUS, APP} is realized according to CARP" (Hyman 2003, 249).[17] As soon as the template is violated once, further departures from the templatic order in the affix string do not worsen the constraint profile.

This prediction is not borne out—at least not in Kiranti, where the affix order always remains as close as possible to the hierarchy of morphosyntactic feature categories, even if a markedness constraint makes this expected order impossible for a single morpheme. Although the unexpected order *-tsi-u* surfaces in some contexts, all suffixes following these two markers are nevertheless ordered according to the hierarchy: for example *-tsi-u-ŋa-e* (1de→3).

Another OT-analysis on affix order can be found in Caballero (2010), where the variable suffix order in Choguita Rarámuri is analyzed. Her system is based on the interaction of scope, morphotactic constraints, and phonological subcategorization. First, semantic compositionality conditions affix order. Second, local morphotactic constraints of the form A >> B ('morphosyntactic category A must precede morphosyntactic category B') are crucial. In addition, morphemes can be marked for a phonological subcategorization requirement that is ensured through

[17]CARP = the default suffix ordering in most Bantu languages: Causative-Applicative-Reciprocal-Passive (Hyman 2003).

morpheme-specific Alignment constraints (e.g., Align_{EV} "the left edge of the evidential marker is aligned to the right edge of the foot" (Caballero 2010: 193). Such a system relying on three different constraint types that govern the order of affixes is not only quite complex, it is also quite unrestrictive and allows innumerable different interactions between scope, morphology, and phonological subcategorization—and it remains to be shown that more of those patterns are borne out in the languages of the world. In addition, Caballero's system is inherently language-specific: the morphotactic constraints and the phonological subcategorization constraints are formulated for specific morphemes in a specific language.

The OT system in Hyman (2003) is similar to my own approach in assuming the general architecture that there are at least two interacting constraint types: one predicting a default ordering and one overwriting their effect on affix order in certain contexts (cf. also Stiebels 2003). The crucial difference between his and my own approach consists in the fact that semantic scope is crucial in the affix order phenomena he analyzes. I remained completely agnostic about possible interpretations of the Kiranti affix order in terms of scope, and presented an analysis that only referred to the morphological information of the suffix string. As I already argued above, a second difference is the fact that an approach where the nonscopal order is predicted from Alignment constraints is the most general approach with fewer and more general constraints as in, for example, the approach of Caballero (2010), where a lot of morpheme-specific constraints interact. In my approach, there is also no need to store a language-specific hierarchy independently as in the approach proposed in Hyman (2003).

5. Conclusion

I presented an optimality-theoretic analysis for affix order patterns in Kiranti languages that crucially refers to Alignment constraints on feature classes. Such an analysis avoids any morpheme-specificity but directly implements a hierarchy of morphosyntactic features. Since constraints are violable in OT, the analysis allows and predicts departures from this hierarchy-governed order in contexts where higher-ranked markedness constraints intervene. I argued that reordering patterns found in Kiranti all follow from the concept of morphological prominence demanding that the agent argument must be marked prominently and unambiguously: either through the presence of a case marker or through an affix-realizing agent feature that is in a prominent position in the affix string. A comparison with a related language provides evidence for this kind of analysis, where a slightly different pattern can be observed that results from another repair strategy employed to ensure morphological prominence for an argument.

This instance of hierarchy-governed affix order is particularly interesting from a typological perspective as well, since it orders number features above person features—contra to the typological tendency on affix order found in Trommer (2003a, c).

Appendix A: Abbreviations

GLOSSES

A	agent-like argument of canonical transitive verb
NSg	nonsingular
P	patient-like argument of canonical transitive verb
Pst	past

ABBREVIATIONS USED IN THE TEXT

A	agent-like argument of canonical transitive verb
d	dual
de	dual exclusive
di	dual inclusive
DM	Distributed morphology
N	Number
OT	Optimality Theory
P	patient-like argument of canonical transitive verb
p	plural
Ps	person
Pst	past
pi	plural inclusive
s	singular

Appendix B: Paradigms

B.1 ATHPARE

(Ebert 1997b)

TABLE 6.16

Surface Forms for Athpare Past (Suffixes, C-final Stem)

	1s	1de	1pe	1di	1pi	2s	2d	2p	3s	3d	3p
1s						-ne	-natse	-nane	-uŋe	-uŋtsiŋe	
1de						-ne	-natse	-nane	-tsuŋe	-tsuŋe	
1pe						-ne	-natse	-nane	-umme	-umtsimme	
1di									-tsue	-tsue	
1pi									-ume	-umtsime	
2s	-ŋe	-tsiŋe	-iŋe						-ue	-utse	
2d	-tsiŋe	-tsiŋe	-iŋe						-tsue	-tsue	
2p	-iŋe	-tsiŋe	-iŋe						-ume	-umtsime	
3s	-ŋe	-tsiŋe	-iŋe	-tse	-e	-e	-tse	-ie	-ue	-utse	
3d	-tsiŋe	-tsiŋe	-iŋe	-tse	-tse	-e	-tse	-ie	-tsue	-tsue	
3p	-ŋe	-tsiŋe	-iŋe	-tse	-e	-e	-tse	-ie	-ue	-utse	
intr	-ŋe	-tsiŋe	-iŋe	-tse	-ie	-e	-tse	-ie	-e	-tse	-e

B.2 CAMLING

(Ebert 1997a; Rai 2003)

TABLE 6.17

Morphemes: Nonpast Paradigm for Camling (Suffixes)

	1s	1de	1pe	1di	1pi	2s	2d	2p	3s	3d	3p
1s						-na-e	-na-tsi-e	-na-ni-e	-ə̃i	-ə̃i-tsi	
1de						-na-e	-na-tsi-e	-na-ni-e	-tsi-ka-e	-tsi-ka-e	
1pe						-na-e	-na-tsi-e	-na-ni-e	-u-m-ka-e	-u-m-tsi-ka-e	
1di									-tsi-e	-tsi-e	
1pi									-u-m-e	-u-m-tsi-e	
2s	-ə̃i	-tsi-ka-e	-i-ka-e						-jo	-jo-tsi	
2d	-ə̃i	-tsi-ka-e	-i-ka-e						-tsi-e	-tsi-e	
2p	-ə̃i	-tsi-ka-e	-i-ka-e						-u-m-e	-u-m-tsi-e	
3s	-ə̃i	-tsi-ka-e	-i-ka-e	-tsi-e	-i-e	-e	-tsi-e	-i-e	-jo	-jo-tsi	
3d	-ə̃i	-tsi-ka-e	-i-ka-e	-tsi-e	-i-e	-e	-tsi-e	-i-e	-tsi-e	-tsi-e	
3p	-ə̃i	-tsi-ka-e	-i-ka-e	-tsi-e	-i-e	-e	-tsi-e	-i-e	-e	-e	
intr	-ə̃i	-tsi-ka-e	-i-ka-e	-tsi-e	-i-e	-e	-tsi-e	-i-e	-e	-tsi-e	-e

TABLE 6.18

Surface Forms for Camling Nonpast (Suffixes, V-final Stem)

	1s	1de	1pe	1di	1pi	2s	2d	2p	3s	3d	3p
1s						-ne	-natse	-nane	-ə̃i	-ə̃itsə̃i	
1de						-ne	-natse	-nane	-tske	-tske	
1pe						-ne	-natse	-nane	-umke	-umtsumke	
1di									-tse	-tse	
1pi									-ume	-umtsumne	
2s	-ə̃i	-tske	-imke						-jo	-jotsjo	
2d	-ə̃i	-tske	-imke						-tse	-tse	
2p	-ə̃i	-tske	-imke						-umne	-umtsumne	
3s	-ə̃i	-tske	-imke	-tse	-ine	-e	-tse	-ine	-jo	-jotsjo	
3d	-ə̃i	-tske	-imke	-tse	-ine	-e	-tse	-ine	-tse	-tse	
3p	-ə̃i	-tske	-imke	-tse	-ine	-e	-tse	-ine	-e	-e	
intr	-ə̃i	-tske	-imke	-tse	-ine	-e	-tse	-ine	-e	-tse	-e

B.3 LIMBU

(van Driem 1987)

TABLE 6.19

Morphemes: Nonpast Paradigm for Limbu (Suffixes)

	1s	1de	1pe	1di	1pi	2s	2d	2p	3s	3d	3p
1s						-nɛ	-nɛ-si-ŋ	-nɛ-i-ŋ	-u-ŋ	-u-ŋ-si	
1de						-nɛ-si-ge	-nɛ-si-ge	-nɛ-si-ge	-si-u-ge	-si-u-si-ge	
1pe						-nɛ-si-ge	-nɛ-si-ge	-nɛ-si-ge	-u-m-ge	-u-m-si-ge	
1di									-si-u	-si-u-si	
1pi									-u-m	-u-m-si	
2s	-ʔe								-u	-u-si	
2d									-si-u	-si-u-si	
2p									-u-m	-u-m-si	
3s	-ʔe	-si-ge	-i-ge	-si			-si	-i	-u	-u-si	
3d	-ʔe	-si-ge	-i-ge	-si			-si	-i	-si-u	-si-u-si	
3p	-ʔe	-si-ge	-i-ge	-si			-si	-i	-u	-u-si	
intr	-ʔe	-si-ge	-i-ge	-si			-si	-i		-si	

TABLE 6.20

Surface Forms for Limbu Nonpast (Suffixes, C-final Stem)

	1s	1de	1pe	1di	1pi	2s	2d	2p	3s	3d	3p
1s						-nɛ	-nɛtsiŋ	-niŋ	-uŋ	-uŋsiŋ	
1de						-nɛtsige	-nɛtsige	-nɛtsige	-suge	-susige	
1pe						-nɛtsige	-nɛtsige	-nɛtsige	-umbe	-umsimbe	
1di									-su	-susi	
1pi									-um	-umsim	
2s	-ʔe								-u	-usi	
2d									-su	-susi	
2p									-um	-umsim	
3s	-ʔe	-sige	-ige	-si			-si	-i	-u	-usi	
3d	-ʔe	-sige	-ige	-si			-si	-i	-su	-susi	
3p	-ʔe	-sige	-ige	-si			-si	-i	-u	-usi	
intr	-ʔe	-sige	-ige	-si			-si	-i		-si	

References

Ackerman, Farrell. 2009. "Affix ordering and the morphosyntax of object marking in Moro." Lexical Function Grammar 09, Cambridge University, July 2009.

Alderete, John. 2001. "Dominance effects as transderivational anti-faithfulness." *Phonology* 18:201–253.

Anderson, Stephen R. 1992. *A-Morphous Morphology*. Cambridge: Cambridge University Press.

Bank, Sebastian. 2010. "Automatic segmentation of transitive paradigms." Talk given at the Workshop of Fine Structure of Grammatical Relations, Leipzig. www.uni-leipzig.de/~sbank/bank-fsgr.pdf.

Beckman, Jill. 1998. "Positional Faithfulness." PhD diss., University of Massachusetts at Amherst.

Bertinetto, Pier Marco. 2011. "How the Zamuco languages dealt with verb affixes." In *Affixes and Bases*, edited by S. Manova, 215–230. Edinburgh: Edinburgh University Press.

Bickel, Balthasar. 1998. "Rhythm and feet in Belhare morphology." Ms., University of California, Berkeley. ROA-287, Rutgers Optimality Archive, http://roa.rutgers.edu/.

Bickel, Balthasar. 2003. "Belhare." In *The Sino-Tibetan languages*, edited by G. Thurgood and R. J. LaPolla, 546–70. London: Routledge.

Bickel, Balthasar. 2008. "Aspects of Kiranti syntax: grammatical relations." Paper presented at the Central Dept. of Linguistics. Tribhuvan University, Kirtipur, August 2008.

Bickel, Balthasar, G. Banjade, Martin Gaenszle, Elena Lieven, Netra Paudyal, I. Rai, M. Rai, N. Rai, and Sabine Stoll. 2007. "Free prefix ordering in Chintang." *Language* 83:43–73.

Bickel, Balthasar, Martin Gaenszle, Arjun Rai, Prem Dhoj Rai, Shree Kumar Rai, Vishnu Singh Rai, and Narayan P. (Gautam) Sharma. 2007. "Two ways of suspending object agreement in Puma: Between incorporation, antipassivization, and optional agreement." *Himalayan Linguistics* 7: 1–18.

Bierwisch, Manfred. 2006. "Luxury in natural language." www.zas.gwz-berlin.de/fileadmin/material/40-60-puzzles-for-krifka/pdf/bierwisch.pdf.

Caballero, Gabriela. 2010. "Scope, phonology and morphology in an agglutinating language: Choguita Rarámuri (Tarahumara) variable suffix ordering." *Morphology* 20:165–204.

Carstairs-McCarthy, Andrew. 1987. *Allomorphy in Inflexion*. London: Croom Helm.

Comrie, Bernard. 1981. *Language Universals and Linguistic Typology*. Chicago: Blackwell.

de Hoop, Helen, and Andrej L. Malchukov. 2008. "Case-marking strategies." *Linguistic Inquiry* 39:565–587.

de Lacy, Paul. 1999. "Morphological haplology and correspondence." In *University of Massachusetts Occasional Papers in Linguistics 25: Papers from the 25th Anniversary*, edited by P. de Lacy and A. Nowak, 51–88. Amherst, MA: Graduate Linguistics Student Association.

Doornenbal, Marius. 2009. *A Grammar of Bantawa*. Utrecht: Landelijke Onderzoekschool Taalwetenschap Publications.

Ebert Karen, H. 1997a. *Camling*. München: Lincom Europa.

Ebert Karen, H. 1997b. *A Grammar of Athpare*. München: Lincom Europa.

Frampton, John. 2003. "Syncretism, impoverishment, and the structure of person features." In *Papers from the 2002 Chicago Linguistic Society Meeting*, edited by M. Andronis, E. Debenport, A. Pycha, and K. Yoshimura, 207–22. Chicago: Chicago Linguistic Society.

Gaenszle, Martin. 1995. "Aspects of Mewahang verbal morphology." Unpublished.

Halle, Morris. 1997. "Distributed morphology: Impoverishment and fission." In *Papers at the Interface*, edited by Y. K. Benjamin Bruening and M. McGinnis, 425–449. Cambridge, MA: MIT Working Papers in Linguistics

Halle, Morris, and Alec Marantz. 1993. "Distributed Morphology and the pieces of inflection." In *The View from Building 20*, edited by K. Hale and S. J. Keyser, 111–176. Cambridge, MA: MIT Press.

Halle, Morris, and Alec Marantz. 1994. "Some key features of Distributed Morphology." In *Papers on Phonology and Morphology*, edited by A. Carnie and H. Harley, 275–288. Cambridge, MA: MIT Working Papers in Linguistics.

Hargus, Sharon, and Siri Tuttle. 1997. "Augmentation as affixation in Athabaskan languages." *Phonology* 14:177–220.

Harley, Heidi, and Rolf Noyer. 1999. "Distributed morphology." *Glot International* 4(4), wwwlot.let.uu.nl/zs2001/papersMarantz/harley&noyer.pdf.

Haspelmath, Martin. 2000. "Optimality and diachronic adaption." *Zeitschrift für Sprachwissenschaft* 18:180–205.

Hyman, Larry. 2003. "Suffix ordering in Bantu: A morphocentric account." In *Yearbook of Morphology 2002*, edited by G. Booij and J. van der Marle, 245–281. Dordrecht: Kluwer.

Kim, Yuni. 2010. "Phonological and morphological conditions on affix order." *Morphology* 20:133–163.

McCarthy, John, and Alan Prince. 1993. "Generalized ALIGNMENT." *Yearbook of Morphology* 1993:79–153.

McCarthy, John, and Alan Prince. 1995. "Faithfulness and reduplicative identity." In *University of Massachusetts Occasional Papers in Linguistics*, edited by J. Beckman, L. Dickey, and S. Urbanczyk, 249–384. Amherst, MA: Graduate Linguistics Student Association

McGinnis, Martha. 1996. "Two kinds of blocking." In *Morphosyntax in Generative Grammar: Proceedings of the 1996 Seoul International Conference on Generative Grammar*, edited by H.-D. Ahn, M.-Y. Kang, Y.-S. Kim, and S. Lee, 357–368. Seoul: Hankuk Publishing.

Menn, Lise, and Brian McWhinney. 1984. "The repeated morph constraint." *Language* 60: 519–541.

Michailovsky, Boyd. 1994. "Manner vs. place of articulation in the Kiranti initial stops." In *Current Issues in Sino-Tibetan Linguistics*, edited by H. T. Nishida and Y. Nagano, 59–78. Osaka: National Museum of Ethnology.

Müller, Gereon. 2002. "Free word order, morphological case, and sympathy theory." In *Resolving Conflicts in Grammars*, edited by G. Fanselow and C. Féry, 9–48. Hamburg: Buske.

Müller, Gereon. 2006a. "Global impoverishment in Sierra Popoluca." In *Linguistische Arbeits Berichte Leipzig*, edited by G. Müller and J. Trommer, 23–42. Leipzig: Institut für Linguistik.

Müller, Gereon. 2006b. "Notes on paradigm economy." In *Linguistische Arbeits Berichte Leipzig*, edited by G. Müller and J. Trommer, 161–195. Leipzig: Institut für Linguistik.

Müller, Gereon. 2006c. "Subanalyse verbaler Flexionsmarker." In *Grammatische Untersuchungen*, edited by E. Breindl, L. Gunkel, and B. Strecker, 183–203. Tübingen: Narr.

Nelson, Nicole Alice. 2003. "Asymmetric Anchoring." PhD diss., Rutgers University.

Nevins, Andrew. 2012. "Haplological dissimilation at distinct stages of exponence." In *The Morphology and Phonology of Exponence: The State of the Art*, edited by Jochen Trommer, 84–116. Oxford: Oxford University Press.

Noyer, Robert. 1992. "Features, Positions and Affixes in Autonomous Morphological Structure." PhD diss., MIT.

Noyer, Robert. 1998. "Impoverishment theory and morphosyntactic markedness." In *Morphology and its Relation to Phonology and Syntax*, edited by S. G. Lapointe, D. K. Brentano, and P. M. Farrell, 264–286. Palo Alto, CA: Center for the Study of

Language and Information Publications Opgenort, Jean. 2005. *A Grammar of Jero.* Leiden: Brill.

Paster, Mary. 2006. "Pulaar verbal extensions and phonologically driven affix order." *Yearbook of Morphology* 2006:155–199.

Pertsova, Katya. 2007. "Learning Form-Meaning Mappings in Presence of Homonymy." PhD diss., UCLA.

Plag, Ingo. 1998. "Morphological haplology in a constraint-based morpho-phonology." In *Phonology and Morphology of the Germanic Languages*, edited by W. Kehrein and R. Wiese, 199–215. Tübingen: Niemeyer.

Prince, Alan, and Paul Smolensky. 2002 [1993]. "Optimality theory: Constraint interaction in generative grammar." Technical Reports of the Rutgers University Center of Cognitive Science, ROA-5370802, Rutgers Optimality Archive, http://roa.rutgers.edu/.

Rai, Vishnu. 2003. "Descriptive Grammar of Chamling." PhD diss., Leiden University.

Ryan, Kevin. 2010. "Variable affix order: grammar and learning." *Language* 86:758–791.

Ryan, Kevin, and Russell Schuh 2010. "Suffix doubling and suffix deletion in Bole." http://www.linguistics.ucla.edu/people/hayes/205/Readings/ryan_bole_handout.pdf.

Schackow, Diana. 2014. "Aspects of Yakkha Grammar." PhD diss., Zürich University.

Siewierska, Anna. 2004. *Person.* Cambridge: Cambridge University Press.

Steriade, Donca. 1995. "Positional neutralization." Ms. University of California, Los Angeles.

Stiebels, Barbara. 2003. "Transparent, restricted and opaque affix orders." In *Syntactic Structures and Morphological Information*, edited by U. Junghanns and L. Szucsich, 283–315. Berlin: Mouton de Gruyter.

Stump, Gregory. 1996. "Template morphology and inflectional morphology." In *Yearbook of Morphology 1996*, edited by G. Booij and J. van der Marle, 217–241. Dordrecht: Kluwer.

Stump, Gregory. 2001. *Inflectional Morphology.* Cambridge: Cambridge University Press.

Stump, Gregory. 2006. "Template morphology." In *Encyclopedia of Language and Linguistics*, edited by K. Brown, 559–563. Oxford: Elsevier.

Trommer, Jochen. 2001. "A Hybrid Account of Affix Order." In *Papers from the 37th Meeting of the Chicago Linguistic Society*, edited by M. Andronis, C. Ball, H. Elston, and S. Neuvel, 469–480. Chicago: Chicago Linguistic Society.

Trommer, Jochen. 2003a. "Distributed Optimality." PhD diss., University of Potsdam.

Trommer, Jochen. 2003b. Hungarian has no portmanteau agreement." In *Approaches to Hungarian*, edited by P. Siptár and C. Pinón, 283–302. Budapest: Akadémiai Kiadó.

Trommer, Jochen. 2003c. "The interaction of morphology and syntax in affix order." In *Year-book of Morphology 2002*, 283–324. Dordrecht: Kluwer.

Trommer, Jochen. 2011. "Paradigmatic generalization of morphemes." *Linguistische Arbeits Berichte Leipzig 88*, edited by S. Bank, D. Georgi, and J. Trommer, 227–246. Leipzig: Institut für Linguistik.

Urbanczyk, Suzanne. 2001. *Patterns of reduplication in Lushootseed.* New York: Garland.

van Driem, George. 1987. *A Grammar of Limbu.* Berlin: Mouton de Gruyter.

van Driem, George. 1992. "Le Proto-Kiranti revisité, morphologie verbale du Lohorung." *Acta Linguistica Hafniensia* 24:33–75.

van Driem, George. 2001. *Languages of the Himalayas: An Ethnolinguistic Handbook of the Greater Himalayan Region.* Leiden: Brill.

Yip, Moira. 1998. "Identity avoidance in phonology and morphology." In *Morphology and its relation to phonology and syntax, edited by* S. G. Lapointe, D. K. Brentari, and

P. M. Farrell, 216–247. StanfordPalo Alto, CA: Center for the Study of Language and Information Publications.

Zimmermann, Eva. 2012. "Affix copying in Kiranti." In *Proceedings of ConSOLE XIX*, edited by E. Boone, K. Linke, and M. Schulpen. http://www.hum.leiden.edu/lucl/research/sole/proceedings/console14.html

Zoll, Cheryl. 1998. "Positional asymmetries and licensing." Ms. ROA-282-0998.

7

Negation in Kurmanji

Songül Gündoğdu

1. Introduction

Kurmanji is a dialect of Kurdish, which is a member of the Western Iranian branch of Indo-European languages. Kurdish has three main subgroups; (i) Northern, (ii) Central and (iii) Southern Kurdish. In terms of the number of speakers, Kurmanji, which is also called the Northern Group, is the largest group of Kurdish dialects. Kurmanji (also written as Kurmancî or Kurmanc) is mostly spoken in Turkey but also has speakers living in Northern Iraq, parts of Syria, Iran, and the former Soviet Union (Bedirxan & Lescot 2000).

The data for this study are elicited from native speakers of Kurmanji living in Muş, a city located in the eastern part of Turkey.[1] Moreover, the study includes examples taken from Thackston (2006) and Bedirxan and Lescot (2000).

The current study aims to analyze affix ordering and negation in Kurmanji via a templatic morphology approach. It takes a close look at the internal structure of the verbs (verbal categories) in Kurmanji and focuses on affixes on verb stems, particularly on the prefix position, which can be occupied by negation and certain tense-aspect-mood (TAM) morphemes. The proposal of this study is that affix ordering has templatic organization that is stipulated by the morphology of the language in question. In addition, some affixes occupy the same position, and the emergence of one affix blocks the appearance of the other due to the fact that they cannot co-occur. Lastly, in negative sentences of Kurmanji, the negation prefix blocks any TAM prefixes from occupying the same slot.

The article has the following structure: section 2 provides a brief summary of the internal structure of the verbs and verbal categories in Kurmanji. Section 3 presents the arguments for a template morphology analysis. In section 4, adopting

---

[1]I would like to thank my informants, my parents, and my friends for their help and useful comments on the data. My heartfelt thanks to Aslı Göksel for stimulating discussion we had on verbal inflection in Kurmanji Kurdish. I am indebted to Stela Manova for her invaluable guidance and very detailed feedback on this study. I am also thankful to the anonymous reviewers and the audiences at the 3rd Vienna Workshop on Affix Ordering in February 2011 for their extensive comments and suggestions on this study.

the main premises of template morphology, affix ordering, and negation in Kurmanji are analyzed. Finally, concluding remarks and points requiring further attention are presented in the last section.

2. Basic Verbal Structure and Verbal Categories in Kurmanji

Internal structure and affix ordering are hitherto unanalyzed aspects of Kurmanji verbs, and the whole internal structure of the verb and the complete explanation of the order of affixes are issues beyond the scope of this study, which focuses on the affix ordering in negative verb forms. Nevertheless, in the following two subsections I introduce not only negation but the whole verb paradigm, as it is essential for the proper understanding of the analysis. It should also be noted that it is considerably difficult to decide what the verb root is; therefore, this study uses the term "verb stem" instead of "verb root" when segmenting the data.

Kurmanji displays interesting agreement and inflection (TAM) patterns that I describe with the help of the following template:

(1) *[prefix 2] + [prefix 1] + [verb stem] + [suffix 1] + [suffix 2] + [suffix 3]*
negation + TAM + lexical stem + TAM + agreement + TAM TAM

As will be illustrated in the following subsections, there are two different verb stems in Kurmanji. Based on the tense information they encode, we differentiate between past and present stems. TAM is expressed in Kurmanji through prefixes and suffixes, but agreement uses only suffixes. I will show that negation prefix and the two TAM prefixes *di-* and *bi-* are mutually exclusive, in the sense that they cannot co-occur in the same slot. Thus, I will propose that negation and TAM affixes occupy the same prefix slot, namely the Prefix 2 slot in (1). I will write Ø in a slot if and only if Ø competes with other affixes in that slot.

2.1 AGREEMENT IN KURMANJI

Kurmanji displays three different agreement patterns determined by the tense and the transitivity of the verb, namely subject-verb agreement, object-verb agreement, and zero agreement (Gündoğdu 2011a, 2011b). In all intransitive (2a/b) and present tense transitive (2c) constructions, the verb agrees with the subject of the sentence in person and number. The singular subject agrees with the verb both in person and number, whereas the plural subject agrees with the verb only in number (2).[2]

[2]The Kurmanji data used in this study is written with the letters used in the Modern Kurmanji orthography. In this orthography, "i" stands for the back high unrounded vowel while "î" stands for the front high unrounded vowel. Similarly, "û" is the back high rounded vowel whereas "u" is a sound between the back high unrounded vowel and the back high rounded vowel.

(2) a. *Tu ket-î*
2SG fall.PST-2SG
'You fell down.'

b. *Hûn di-kev-in*
2PL PROG-fall.PRS -PL
'You (all) fell down.'

c. *Ez cil-an di-şû-m*
1SG cloth-PL PROG-wash.PRS-1SG
'I am washing clothes.'

In past-tense transitive sentences the verb agrees with the object, hence the person-number marker on the verb reflects the properties of the object but not the subject (3a). However, in the Muş dialect of Kurmanji, in past-tense transitive sentences the verb agrees with neither the subject nor the object of the sentence but it is found in the default third-person singular form (3b) (Gündoğdu 2011a, 2011b).[3]

(3) a. *Wan ez dît-im*
3PL-SBJ 1SG-OBJ see.PST-1SG
'They saw me.'

b. *Wan min dît-Ø*
3PL-SBJ 1SG-OBJ see.PST-3SG
'They saw me.'

There are also two sets of pronouns and agreement markers that change according to the tense of the verb as shown in Table 7.1 and 7.2. The agreement markers appear in verb final position. Person and number features are fused into a single marker and found as a suffix on the verb. There is only one morpheme for all plural subjects denoting number, whereas singular subjects have different markers on the verb, as shown in Table 7.1. The form of person–number suffixes alternates according to whether the verb stem ends in a vowel or a consonant, as in *te-m* 'I am coming' versus *diç-im* 'I am going' (Gündoğdu 2011a). Furthermore, as will be shown in the section 2.2, agreement markers change depending on the TAM markers the verb bears; for instance, the third-person singular agreement marker is *-(y)e* in the present/progressive, present-past subjunctive, and present perfect, while it is null (-Ø) in the past, past progressive, and past perfect (see Table 7.6).

---

[3]Kurmanji is known to be an ergative language. Sentences like (3a) are called *canonical ergative construction*, whereas the ones like (3b) are called *double oblique constructions* in Haig (2004). For detailed information on the ergative nature of Kurmanji see Haig (2004), Gündoğdu (2011a, 2011b). Muş is a city located in the eastern part of Turkey. This phenomenon is not restricted to Kurmanji spoken only in Muş but in the area called the Serhed Region, which mainly comprises some cities in the eastern part of Turkey such as Muş, Erzurum, and Bingöl.

TABLE 7.1

Person–Number Agreement Markers in Kurmanji: Subjects in Intransitives and in the Present Tense Transitives in all Dialects Objects in the Past Tense Transitives in Standard Kurmanji

Person	Marker (person + number)
Ez (1SG)	-(i)m
Tu (2SG)	-(y)î/-y
Ew (3SG)	-(y)e/ Ø
Em /Hûn /Ew (1/2/3PL)	-(i)n / -(i)ne

TABLE 7.2

Subjects and Objects in the Past Tense Transitives in the Muş Dialect of Kurmanji

Person	Marker (person+number)
Ez (1SG)	-(y)e/ Ø
Tu (2SG)	-(y)e/ Ø
Ew (3SG)	-(y)e/ Ø
Em /Hûn /Ew (1/2/3PL)	-(y)e/ Ø

2.2 TENSE-ASPECT-MOOD IN KURMANJI

The person–number inflection is more or less regular; in fact, it is much more regular than the tense inflection. The TAM markers appear either as prefixes or suffixes on the verb, depending on its tense. To illustrate, the whole inflection paradigm for the verb *çûn* 'to go' is presented in Table 7.3. Note that the present/progressive, present subjunctive, and past progressive are reflected on the verb in prefix forms as *di-*, *bi-*, and *di-*, respectively, while the present perfect and past perfect are marked by the suffixes *-e* and *-bû*, respectively.[4] Also, the past subjunctive is formed via circumfixation; that is, the verb gets inflected by the subjunctive prefix *bi-* and the present subjunctive form of *bûn* (past form of *to be*) as a suffix.[5] The person-number agreement markers are always in suffix form; they precede the present perfect suffix *-e* but follow the past-perfect and past subjunctive suffixes *-b* and *-bû*. The simple past tense is expressed in terms of the absence of an overt form, and there are only person-number suffixes attached to the past verb stem.[6]

Prefix forms change depending on whether there is a vowel or a consonant following them, and similarly, suffix forms are altered according to whether there is a vowel or a consonant preceding them. For instance, if the stem begins with a

---

[4]The term "tense" is used here in the traditional sense; otherwise, the prefix *di-* is obviously not a tense marker but an aspectual morpheme standing for Progressive Aspect.

[5]For more examples of past subjunctive forms, see Thackston (2006).

[6]In the Muş dialect of Kurmanji, the present perfect suffix has the form of *-e* for all persons except for the second-person singular, which is expressed by the suffix *-î*. However, in Standard Kurmanji, the suffix is *-e* for all persons.

TABLE 7.3
Inflection Paradigm of the Verb *çûn* 'go'

çûn "go"	Present/prog.	Present subjunctive	Simple past	Past prog.	Past subjunctive	Present perfect	Past perfect
Ez (1SG)	diçim	biçim	çûm	diçûm	biçûbim	çûme	çûbûm
Tu (2SG)	diçî	biçî	çûy	diçûy	biçûbî	çûye/çûyî	çûbûyî
Ew (3SG)	diçe	biçe	çûØ	diçûØ	biçûbe	çûye	çûbûØ
Em /Hûn /Ew (1/2/3PL)	diçin	biçin	çûn	diçûn	biçûbin	çûne	çûbûn

vowel, the prefix for the past progressive is *d-* as in *davêže* 'he is throwing,' yet if it begins with a consonant the prefix takes the form of *di-* as in *diçe* 'he is going.' In the same way, if a verb stem ends in a vowel, the suffix for the past perfect is *-bû* as in *çûbûm* 'I had gone,' but if it ends in a consonant, the suffix for this tense is *-ibû* as in *gotibûm* 'I had said/told' (Gündoğdu 2011a).

The future tense in Kurmanji is different from other tenses in terms of affixation. As shown in Table 7.4, a present subjunctive prefix *bi-* appears on the verb and the marker *-(y)ê* follows the subject. When the subject is pronominal, it receives the clitic *-(y)ê*, whereas when it is a full noun, it is followed by the particle *dê/wê*.

The imperative is formed through the use of the subjunctive prefix *bi-* and person-number agreement markers. The second-person singular is reflected on the verb via the third-person singular agreement marker *-e* while the second-person plural has its standard agreement marker *-(i)n* as shown in (4).

(4) Imperative (verb *çûn* "go")

2 SG ***bi-ç-e***
SBJV-go.PST-3SG
'You (sg), go!'

2 PL ***bi-ç-in***
SBJV-go.PST-3PL
'You (pl), go!'

(Bedirxan & Lescot 2000)

There are two varieties of irrealis mood in Kurmanji Kurdish; past conditional and past perfect conditional. The past conditional is made by circumfixation: subjunctive ***bi-*** is prefixed and the following endings are added to the past stem

TABLE 7.4
Future Tense in Kurmanji

Pronoun+-*(y)ê*		Proper noun *wê/dê*	
Ezê ***biçim.***	'I will go.'	*Ruken* **wê** ***biçe.***	'Ruken will go.'
Tuyê ***biçî.***	'You will go.'		
Ewê ***biçe.***	'S/he will go.'	*Zana* **dê** ***biçe.***	'Zana will go.'
Emê/Hunê/Ewê ***biçin.***	'We/You/They will go.'		

(Thackston 2006). These endings can be considered as person-number agreement markers plus conditional suffix *-a* since *-m*, *-y*, *-Ø*, and *-n* are already person-number agreement markers in Kurmanji.

CONSONANT-FINAL STEM		VOWEL-FINAL STEM	
1 SG	bi-stem-ama	1 SG	bi-stem-ma
2 SG	bi-stem-ayî	2 SG	bi-stem-ya
3 SG	bi-stem-a	3 SG	bi-stem-ya
1/2/3 PL	bi-stem-ana	1/2/3 PL	bi-stem-na

(5) The past conditional form of the verbs *hatin* 'to come' and *çûn* 'to go' for all persons:

hatin 'to come'		*çûn* 'to go'	
ez	*bihatama*	*ez*	*biçûma*
tu	*bihatayî*	*tu*	*biçûya*
ew	*bihata*	*ew*	*biçûya*
em/hun/ew	*bihatana*	*em/hun/ew*	*biçûna*

The past perfect conditional is formed again via circumfixation: prefixing the subjunctive prefix ***bi***- and suffixing the conditional suffix *-(y)a* to the past perfect form of the verb (Thackston 2006). Note that the suffix *-(y)a* follows the person-number agreement markers (6).

(6) The past perfect conditional form of the verbs *hatin* 'to come' and *çûn* 'to go' for all persons: (person-number markers are in italics; past perfect conditional prefix and suffix are in bold)

hatin 'to come'		*çûn* 'to go'	
ez	*bihatibûma*	*ez*	*biçûbûma*
tu	*bihatibûya*	*tu*	*biçûbûya*
ew	*bihatibûya*	*ew*	*biçûbûya*
em/hun/ew	*bihatibûna*	*em/hun/ew*	*biçûbûna*

Ability mood is expressed with the verb *karin* 'to be able to'; *karin* is regularly conjugated in the present and it is followed by a subjunctive complement. The subjunctive complement contains a present subjunctive verb bearing person-number agreement markers (7).

(7) (Conjugated in present tense for all persons)

Ez	*di-kar-im*	*bi-ç-im*
1SG-SBJ	PROG-be able.PRS -1SG	SBJV-go.PRS-1SG

'I am able to go.'

Tu	*di-kar-î*	*bi-ç-î*
2SG-SBJ	PROG-be able.PRS -2SG	SBJV-go.PRS-2SG

'You (sg) are able to go.'

Ew	*di-kar-e*	*bi-ç-e*
3SG-SBJ	PROG-be able.PRS -3SG	SBJV-go.PRS-3SG

'S/he is able to go.'

Em/Hun/Ew	*di-kar-in*	*bi-ç-in*
1/2/3 PL-SBJ	PROG-be able.PRS -PL	SBJV-go.PRS-PL

'We/You/They are able to go.'

As is evident in the data presented so far, the verb stem has two different shapes in the present and past tenses; one is used in the present/progressive, present subjunctive, and future tenses (first group), whereas the other is used in the present perfect and in all past tenses including past subjunctive (second group). For instance, in the first group the roots for the verbs 'go' and 'say' are -ç- and -bej-, respectively, whereas they are -çû- and -got- in the second group, implying that the verb stem already embodies tense information. Such verbs in fact constitute the most striking irregularities in the verb paradigm: suppletion and apophony. What is significant about these irregular forms is that although suppletion and apophony apply to the verb root, the verb stem as a whole is also inflected by the TAM and person-number affixes (Gündoğdu 2010). The irregular forms of the verbs and their inflection are shown in Table 7.5 (the first two verbs are suppletive, while the last two verbs undergo apophony). Note that third-person agreement suffix is zero morpheme in the past, past perfect, and past progressive tenses.

In addition to suppletion and apophony, there are some further cases where the verb stem undergoes only one sound change. This may be analyzed as apophony as well. What happens in these cases is that sometimes a consonant in the verb stem becomes a different consonant or drops, or a new consonant is inserted. These sound changes in the verb stem are again related to the tense as they reveal tense information: present versus past tense. The relevant examples for these cases are given in Table 7.6 below.

Finally, although the nouns in this language have a gender distinction (feminine vs. masculine), this distinction is not reflected on the verb.

TABLE 7.5

Irregular Forms of Some Verbs and Their Inflection in Kurmanji –1

Ew (3SG)	Present/prog.	Present subjunctive	Simple past	Past prog.	Past subjunctive	Present perfect	Past perfect
hatin "come"	tê	bê/were	hat	dihat	bihatibe	hatîye	hatibû
gotin "say"	dibêje	bibêje	got	digot	bigotibe	gotîye	gotibû
man "stay"	dimîne	bimîne	ma	dimanîya	bimabe	maye	mabû
xwarin "eat"	dixwe	bixwe	xwar	dixwar	bixwaribe	xwarîye	xwaribû

TABLE 7.6

Irregular Forms of Some Verbs and Their Inflection in Kurmanji -2

Ew (3SG)	Present/prog.	Present subjunctive	Simple past	Past prog.	Past subjunctive	Present perfect	Past perfect
şûştin "wash"	dişû	bişû	şûşt	dişûşt	bişûştibe	şûştîye	şûştibû
avetin "throw"	davêje	bavêje	avêt	davêt	bavêtibe	avêtîye	avêtibû
dîtin "see"	dibîne	bibîne	dît	didît	bidîtibe	dîtîye	dîtibû

2.3 NEGATION IN KURMANJI

Negation in Kurmanji appears on verb stems as a prefix that has three phonological shapes as *n(a)-*, *n(e)-*, and *ni-*, as previously mentioned (Bedirxan & Lescot 2000, Thackston 2006). In the present/progressive tense it is realized as *n(a)-* while in all other tenses it is *n(e)-*. *Ni-* is only used with the auxiliary *karin* 'to be able to' in the present and past tenses. The interesting point about negation is that the appearance of the negation prefix precludes the appearance of other tense prefixes *di-* and *bi-*. To illustrate, when *diçim* 'I'm going' is negated the tense prefix *di-* is replaced by the negation marker, as in *naçim* 'I'm not going,' and similarly *biçim* 'that I go' turns into *neçim* 'that I do not go.' The same blocking applies to the irregular cases in which tense prefixes appear as infixes on some verbs such as *vedikim* ' I am opening' and its negated form is *venakim* 'I am not opening.'[7] Again, the *na-* precludes the *di-* prefix. However, negation does not preclude the prefix *di-* used in the past progressive tense but it only precedes it, so when *diçûm* 'I was going' is negated, the tense prefix *di-* follows the negation marker as in nediçûm 'I was not going.' Moreover, the TAM markers in the form of suffixes are not affected by this blocking phenomenon; the suffixes remain in their positions and the negation marker is prefixed to the verb stem. For instance, a verb gets inflected by a suffix in the past perfect tense so when *çûbûn* 'we/you/they had gone' is negated, we get *neçûbûn* 'we/you/they had not gone.' Lastly, this blocking phenomenon does not influence verb stems in the simple past because there is not an overt marker on the verb stem in this tense. The conjugation of the verb *çûn* 'to go' in all tenses for all persons and their negated forms are presented in Table 7.7 and Table 7.8.

TABLE 7.7

The Conjugation of the Verb *çûn* 'to go' – 1

çûn "to go"	Present/prog.	Present subjunctive	Future tense	Simple past	past Prog.
Ez (1SG)	diçim/naçim	biçim/neçim	biçim/neçim	çûm/neçûm	diçûm/nediçûm
Tu (2SG)	diçî/naçî	biçî/neçî	biçî/neçî	çûy/neçûy	diçûy/nediçûy
Ew (3SG)	diçe/naçe	biçe/neçe	biçe/neçe	çûØ/neçûØ	diçûØ/nediçûØ
Em /Hûn /Ew (1/2/3 PL)	diçin/naçin	biçin/neçin	biçin/neçin	çûn/neçûn	diçûn/nediçûn

[7] These are called compound verbs (Thackston 2006) or complex predicates (Haig 2002) in Kurmanji. The properties of these verbs will be explained in section 4.

TABLE 7.8
The Conjugation of the Verb *çûn* 'to go' – 2

çûn "to go"	Past subjunctive	Present perfect	Past perfect	Past conditional	Past Perfect conditional
Ez (1SG)	biçûbim/	çûme/	çûbûm/	biçûma/	biçûbûma/
	neçûbim	neçûme	neçûbûm	neçûma	neçûbûma
Tu (2SG)	biçûbî/	çûye/çûyî	çûbûyî/	biçûya/	biçûbûya/
	neçûbî	neçûye/î	neçûbûyî	neçûya	neçûbûya
Ew (3SG)	biçûbe/	çûye/	çûbûØ/	biçûya/	biçûbûya/
	neçûbe	neçûye	neçûbûØ	neçûya	neçûbûya
Em /Hûn /Ew	biçûbin/	çûne/	çûbûn/	biçûna/	biçûbûna/
(1/2/3 PL)	neçûbin	neçûne	neçûbûn	neçûna	neçûbûna

3. Premises of Template Morphology

How affixes are ordered or organized in a certain language is one of the central questions in morphological theory. Some linguists try to find the answer to this question in layered morphology, which assumes that affix ordering in many languages is stipulated by the constituent structure, while others (Simpson & Withgott 1986, Inkelas 1993, Stump 1997, 2006, and Nordlinger 2010) consider affix ordering in certain languages to have a templatic organization, since it appears in the form of arbitrary position classes. Inkelas (1993) discusses templatic or position-class systems as "morphological systems in which morphemes or morpheme classes are organized into a total linear ordering that has no apparent connection to syntactic, semantic or even phonological representation" (p. 560). Rice (2011) also defines templates or position classes as the system in which affix ordering is considered to be arbitrary from the perspective of general grammatical principles (semantic, syntactic, or phonological).

Simpson and Withgott (1986) propose several criteria that define template morphology and set it apart from layered morphology:

i. *Zero morphemes* are a characteristic of template morphology, but not of layered morphology.
ii. Layered morphology is subject to the *adjacency constraint*, whereas template morphology is not.
iii. Template morphology is subject to the *lookahead constraint*, whereas layered morphology is not.
iv. Template morphology lacks *headed structure*, while layered morphology has headed structure.
v. Template morphology allows a predicate to *encode more than one of its arguments*, while layered morphology encodes at most one argument.

In the next section, I will explain how Kurmanji Kurdish exhibits all of these properties of template morphology excluding the last one, and I will analyze how affixes are ordered and how negation operates in this language from a templatic structure perspective.

4. Accounting for Affix Ordering and Negation in Kurmanji Kurdish

I will explain the criteria identified by Simpson and Withgott (1986), and try to show in what ways affix ordering and negation in this language conform to these properties. Recall that the verbal categories and the verb structure were introduced in section 2. I assert that Kurmanji Kurdish exhibits the first four properties of template morphology claimed by Simpson and Withgott (1986) and also by others (Stump 1998, 2006, Inkelas 1993) in the literature. The last criterion, *encoding more than one of its arguments*, is not observed in Kurmanji since the subject and the object cannot agree simultaneously with the verb in this language.

(i) zero morphemes

Zero morphemes are common in template morphology (Simpson & Withgott 1986). What is meant by zero morpheme is "the meaningful use of absence of information" (p. 156). Stump (1997) refers to it as "significative absence," because it is the absence of an affix that denotes a formal difference between two forms. According to this criterion, the absence of an affix may signal a property or a set of properties in a templatic system. In Kurmanji, the absence of third-person agreement suffix in the past, past perfect, and past progressive tenses (see Tables 7.1, 7.3, 7.7, and 7.8) distinguishes the third-person singular form from the other persons inflected in these tenses. For instance, it is the zero morpheme that differentiates *çû-Ø* 's/he went' from *çû-y* 'you (sg) went' or *çû-n* 'we/you/they went.' The absence of overt agreement marking in the suffix position signals the property set "agreement."

(ii) no adjacency constraint

Template morphology is not subject to the adjacency constraint, which requires that the addition of an affix to a stem be sensitive to the stem's outermost layer of structure, because in template morphology the addition of an affix to a stem is mostly sensitive to the internal morphology of the stem. Similarly, in Kurmanji, the affixes are sensitive to the tense information encoded in the verb stem (see Tables 7.3, 7.5, 7.6, 7.7, and 7.8). To illustrate, in the past conditional the conditional suffix *-a* can only join with a past verb stem, but cannot join with a present verb stem. Likewise, the present perfect suffix *-e* can only be attached to the past verb stem, but it can never show up on the present verb stem in this language. The non-existing forms in (8b) prove that the appearance of affixes is strictly sensitive to the verb stem in Kurmanji.

(8) a. *bi-* <u>past verb stem</u>- agreement-*a*
bi-çû-m-a
SBJV-GO.PST -1SG- COND
'if I went'

past verb stem- agreement-*e*
çû- m-e
go.PST -1SG-PRF
'I have gone.'
b. * present verb stem- agreement-*e*
**ç- im-e*
go.PRS -1SG-PRF
* *bi*-present verb stem- *bû*- agreement-*a*
**bi*-ç-bû-m-a
SBJV-go.PRS –PST PRF-1SG- COND
* *bi*-past verb stem- agreement
**bi*-çû-m
SBJV-go.PST –1SG

The tense information encoded in the verb stem is not the outermost layer of the structure but it is relevant to the internal structure, which implies that the prefixes and the suffixes in this language are sensitive to the internal morphology of the verb stem.

(iii) lookahead constraint

The lookahead constraint, also known as discontinuous dependencies, is considered as a typical property of template morphology. It is closely related to the already discussed *no adjacency* constraint in *(ii)* and says that there may be dependencies between pairs of positions that are far away from each other in the word form; for instance, a particular affix in one position may necessitate a certain type of affix in another position to which it is not neighboring (Spencer 1991). In parallel with this constraint, we observe that in Kurmanji three constructions such as past subjunctive, past conditional, and past perfect conditional are formed through circumfixation in which the prefix and suffix positions are dependent, though they are not adjacent (see Table 7.3, also examples in (5) and (6)). To illustrate, past conditional form of the verb is created by prefixing the subjunctive prefix *bi*- and suffixing the conditional suffix *-(y)a* to the past verb stem. This clearly shows that the conditional suffix *-(y)a* requires the appearance of the subjunctive prefix *bi*- to express past conditional meaning in this language, and the conditional suffix does not have an independent meaning without the subjunctive prefix. Likewise, the past subjunctive is formed through the subjunctive prefix *bi*-and the past subjunctive suffix *-b* (see past subjunctive forms in Table 7.3, 7.5, 7.6, and 7.8). Again, the appearance of the past subjunctive suffix *-b* in this language is conditioned by the subjunctive prefix *bi*-. In both instances, the verb is clearly constituted by two discontinuous affixes, and these affixes together can express the intended meaning (9).

(9) a. ***bi**-çû-m-**a***
SBJV- go.PST - 1SG - suffix (conditional)
'If I went.'

b. ***bi**-çû-**b**-im*
SBJV-GO.PST –PST SBJV-1SG
'. . . that I went'

(iv) no headed structure

Simpson and Withgott (1986) consider headed words as typical of hierarchical layered morphology. In a headed structure, inflectional and derivational morphology are interspersed and inflectional morphology occurs outside of derivational morphology (Nordlinger 2010). Stump (2001) also refers to the same property of headed structures by pointing out that a word's inflectional marking is expected to be linear outside of any derivational or compounded formatives it contains. Unheaded words that do not conform to this property are typical of templatic systems. In Kurmanji, compound verbs behave like unheaded words when they are inflected by the TAM and negation prefixes—which provides further support for the existence of templatic system in this language. Compound verbs in this language consist of a nonverbal element (preverbal or complement) such as *ve*, *hil*, or *ra* plus a verb like *kirin* 'to do,' *dan/dayin* 'to give,' or *ketin* 'to fall,' and they are written as one word as in *vekirin* 'to open,' *hildan* 'to lift,' *raketin* 'to sleep,' and so on. There are also compound verbs consisting of two separate words: the first element is a nominal word such as *alîkarî* 'help (noun)' and the second element is a verb like *kirin* 'to do,' which are written separately as in *alikari kirin* 'to help' (Thackston 2006, Bedirxan & Lescot 2000). When these verbs are inflected, the TAM and negation prefixes are realized on the verbal element inside the compound as in *ve**di**kin-ve**na**kin, alîkarî **di**kin - alîkarî **na**kin.*

Moreover, template morphology predicts that functionally distinct elements occupying the same slot in the template must be mutually exclusive in that position. Mutual exclusivity is another property of templatic ordering that we detect in Kurmanji. As previously discussed in (1), although some TAM and person-number affixes occupy the same slot, they cannot occur together: either the verb stem (depending on the tense information it carries- past vs. present) or the selection of one affix precludes other potential candidates from occupying the same slot. The same blocking effect is observed in the negation phenomena discussed in section 2.4. I propose that the template given in (1), which is also repeated in (10), is the verbal template of Kurmanji.

(10) *[prefix 2] + [prefix 1] + [verb stem] + [suffix 1] +[suffix 2] + [suffix 3]*
negation + TAM + lexical stem + TAM + agreement + TAM TAM

Which slots will be filled and which one will remain empty changes depending on the verb stem. For instance, if we have a present verb stem, then only Prefix 2 and Suffix 2 positions are filled up, while Prefix 1, Suffix 1, and Suffix 3 positions are always empty; thus when one omits the empty slots, the result is the template in (11); which is Prefix2 -present stem-Suffix2.

(11) *[prefix 2]* + *[prefix 1]* + *[present verb stem]* + *[suffix 1]*+*[suffix 2]* + *[suffix 3]*
negation (*na-*, *ni-*) + empty + lexical stem + empty + agreement + empty TAM (*di-*, *bi-*)

Both the negation and TAM markers are potential candidates for the Prefix 1 position of the present verb stem; however, only one of the markers can occupy this slot. Therefore, if the progressive prefix *di-* emerges in this position, it precludes the present subjunctive prefix *bi-* and negation prefixes *na-*, *ni-*. Likewise, the appearance of the present subjunctive prefix *bi-* blocks the occurrence of *di-* and *na-*, *ni-*. In negative sentences, the negation prefixes *na-*, *ni-* block the emergence of the TAM prefixes in this slot. That's why we have verbs like *diçim*, *biçim*, and *naçim* but not forms such as **dibiçim*, **bidiçim*, **nadiçim*, **nebiçim*, **nebidiçim*, or **nadibiçim*. As for the suffix position, it is filled by the person-number agreement markers, the selection of which is conditioned by the agreeing argument of the sentence, either the subject or the object.

On the other hand, if we have a past stem, all prefix and suffix positions may have potential candidates (see Tables 7.5–7.8).

(12) [*prefix* 2] +[*prefix* 1] + [*past verb stem*] + [*suffix* 1] + [*suffix* 2] + [*suffix* 3]
negation (*ne-*) + TAM (*di-*) + lexical stem + TAM + agreement + TAM TAM (*bi-*)

As shown in (12), the first prefix position of the past verb stem has one candidate: the progressive prefix *di-*, and we do not observe any blocking effect in this slot even in negative sentences, so we have *diçûm* 'I was going' in affirmative sentences and *nediçûm* 'I was not going' in negative sentences. As for the second prefix position, it is reserved for the negation prefixes *ne-*, *ni-* and the subjunctive prefix *bi-*. Again, what happens is that in negative sentences, the negation prefix precludes the emergence of *bi-* in this slot. Thus, we have either *biçûbim* or *neçûbim* but not **bineçûbim* or **nebiçûbim*. The crucial point here is that although there are two prefix positions where TAM markers can appear, the past verb stem cannot have these two slots filled with two TAM markers simultaneously, hence we do not have forms such as **dibiçûm* or **bidiçûm*. However, the negation marker occupying the second prefix position and the TAM marker occupying the first prefix position can co-occur together. This is the reason that the emergence of the negation prefix does not block the appearance of the progressive prefix *di-* on the past verb stem, so we have *nediçûm* when we negate ***di**çûm*. The motivation behind why two TAM markers cannot co-occur may be semantic, since the appearance of the progressive aspect and the subjunctive mood markers together on the same verb does not seem to be semantically possible. In contrast, the co-occurrence of the negation and the TAM markers is semantically possible, and also this co-occurrence may be required by the wellformedness conditions of Kurmanji morphology. Note that when *çûm* 'I went' is negated we get *neçûm*, and when *diçûm* 'I was going' is negated we have *nediçûm*. If the co-occurrence of negation and *di-* were not possible on the past verb stem *-çû-* 'go,' then we would have the same negated form for the past progressive and the simple past tense verb, which is *neçûm*, and this might cause some misunderstanding for speakers during communication. This, in turn, implies that Kurmanji morphology allows this co-occurrence to distinguish between these two forms. As for suffix positions on the past verb stem, the first suffix position is occupied either by the past subjunctive suffix *-b* or by the past perfect suffix *–bû,* but not by both at the same time. The second suffix position is filled by person–number agreement markers whose selection is conditioned by the agreeing argument of the sentence, either the subject or the object. Lastly, either the present perfect suffix *-e* or the past/perfect conditional suffix *-a* fills the third suffix position. In fact, the suffixes occupying the first and the third positions (slots for the TAM markers) are also sensitive to what appears in the prefix and suffix positions on the past verb stem. For instance, if it is the progressive prefix *di-* that occupies the first prefix position, then the first and the third suffix positions must remain empty. On the other hand, if it is the subjunctive prefix *bi-* that fills the second prefix position, then the first suffix position must be filled either by the past subjunctive suffix *-b* or by the past perfect suffix *-bû*. Should *-b* occupy the first suffix position, then the third suffix position remains empty, but if *-bû* fills the first suffix position, then the past/perfect conditional suffix *-a* must appear in the third suffix position. Likewise, if in the third suffix position the present perfect suffix *-e* emerges, then the prefix positions must remain empty. To sum up, it can be asserted that in Kurmanji, not only the verb stem but also affixes occupying the same and the different prefix/suffix positions seem to be sensitive to each other's occurrence.

Although the occurrence of the TAM markers emerging as prefixes or suffixes on the verb stem are sensitive to each other, in negative sentences the negation prefix always has priority over other TAM markers occupying the same slot; therefore, when the negation prefix emerges, other TAM markers are omitted. The selection of the appropriate prefix takes place semantically; that is, if one needs to express negation, a negation marker is selected, if not a TAM marker is attached.

In fact, this blocking phenomenon may be considered to be similar to Inkelas's (1993) position-class morphology analysis of affix ordering in the Nimboran language. Inkelas points out that verbal morphemes in the Papuan language Nimboran occupy slots organized in an order and that certain morphemes are in complementary distribution, hence the morpheme that fills a certain slot precludes all the other possible morphemes relevant to that slot. The relative ordering of these morphemes is schematized in Inkelas (1993) as follows:

(13)	0	1	2	3	4	5	6	7	8
	root	PlSubj	DuSubj	MObj	IncDuSubj	Loc	Iter	Tense	SubjPers
			PlObj	Dur					
				Part					

(Inkelas 1993: 561)

The morphemes occupying the same slot—for example the morphemes occupying the second slot—can never co-occur in this language, thus the appearance of the one morpheme -DuSubj- automatically blocks the occurrence of the other morpheme -PlObj-. Inkelas (1993) names this morphological blocking phenomenon *symmetrical blocking*. What we observe in the Prefix 2 position of the Kurmanji verbal template that hosts the negation and TAM prefixes is similar to the *symmetrical blocking* phenomenon, because when the negation prefix appears on the verb, the TAM prefixes occupying this slot are blocked. However, apart from *symmetrical blocking*, one observes much more complex blocking cases in the Nimboran verbal system, where a particle in one slot blocks the appearance of an affix in another slot. Inkelas (1993) accounts for such blocking phenomenon through the position of class analysis, which assumes a hierarchical ordering of slots. She proposes that verbal morphemes in this language occupy slots organized in a strict hierarchical order and that these morphemes belong to position classes, each of which permits only one morpheme to surface. Each finite verb in Nimboran may have a maximum of seven morphemes including the verb root, and in certain cases a co-occurring verbal particle. She argues that there is no single position for particles, but rather they occupy some vertical segments in this hierarchical ordering. Therefore, a verbal particle such as *–ta* may block both position 2 and position 5 simultaneously, as it is hierarchically prior over these positions. An analysis in terms of position classes may also be possible for the Kurmanji data. Assume that we have three prefix positions in the verbal template of Kurmanji ordered as Prefix 3 - Prefix 2 - Prefix 1, and Prefix 2 position is only reserved for the negation marker, while Prefix 3 and 1 host the TAM markers. The position–class analysis will produce the right forms if affixes in these prefix positions attach hierarchically in the following order; Prefix 1 attaches first, then Prefix 2 follows and both Prefix 1 and 2 block Prefix 3 position. With the present form of the verb stem Prefix 1 position empty, we have *na-Ø-çi-m* "I am not going," as in (14a); on the other hand, if we have a past verb stem, Prefix 1 position is occupied by the progressive prefix *di-* and we have *ne-di-çû-m* "I was not going," illustrated in (14b).

(14) a. *[prefix 3]* + *[prefix 2]* + *[prefix 1]* + *[present verb stem]*
TAM (*bi-*, *di-*) + negation (*na-*, *ni-*) + empty + lexical stem

b. *[prefix 3]* + *[prefix 2]* + *[prefix 1]* + *[past verb stem]*
TAM (*bi-*) + negation (*ne-*) + TAM (*di-*) + lexical stem

The Prefix 2 and Prefix 1 positions are hierarchically prior over the TAM prefixes occupying Prefix 3 position, thus in negative sentences the negation prefix shown as [A] blocks the TAM prefixes represented as [B]. This shows that the former hierarchically precedes the latter, so when [A] enters into the competition, it wins by blocking [B], (15).

(15)

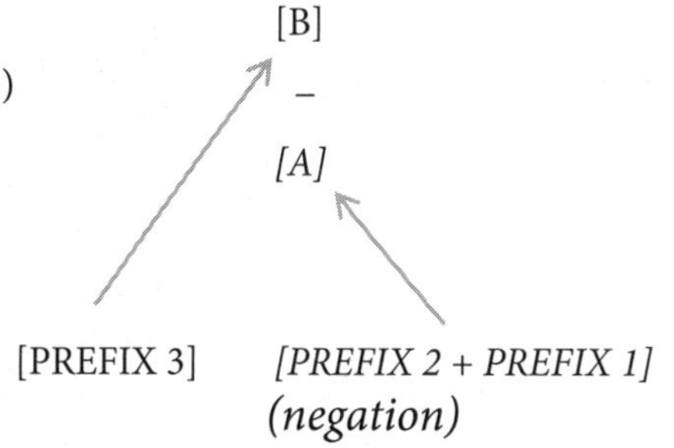

A template morphology analysis may be superior in our case because negation in Kurmanji takes scope semantically over the whole verb form, thus the negation markers are expected to be outside of the TAM markers—that is, to attach the last. Therefore, we assign negation and TAM markers to the same slot in the templates in (10), (11), and (12). The selection of the appropriate prefix takes place semantically, which means that if one needs to express negation a negation marker is selected; if not, a TAM marker is attached. However, as this study does not investigate the exact order in which the affixes attach, Kurmanji data cannot be seen as evidence against position-class morphology.

5. Conclusion

The fact that Kurmanji verbs have both prefixes and suffixes on them gives them a complex structure that makes them hard to analyze with only one morphological model. Although I propose that affix ordering in this language displays templatic features, there are also some examples exhibiting the properties of layered morphology. For instance, morphosyntactic properties of a Kurmanji verb can be deduced from the sum of its parts, verb stem and affixes; this compositionality is a property of layered morphology. For example, the past perfect conditional category is a combination of the subjunctive prefix *bi-* plus the past verb stem *-çû-* plus the past perfect suffix *-bû* plus the person/number agreement marker plus the conditional suffix *–a*, as in *bi-çû-bû-m-a* 'If I had gone.' Muysken (1986) proposes that "affix ordering, just like word order, results from a number of different components" (p. 637), thus a single morphological approach may not be enough to account for affix ordering in a language. According to Manova and Aranoff (2010), a particular language can benefit from a combination of two morphological organizations, basically templatic and layered morphology. Adopting their views, it may be proposed that these two models may account for the complex structure of verbs in Kurmanji. However, the whole internal structure of the verb and the nature of affix ordering are issues beyond the goals of this study. The current study is an introductory step in analyzing the ordering of affixes and negation phenomenon in this language. I have proposed that the morphology of Kurmanji is organized templatically, as evidenced by the fact that the negation prefix has priority over TAM prefixes occupying the same position on the verb stem; therefore when a verb is negated, the negation marker precludes TAM markers from occupying the same slot. In fact, semantics and scope may have a role in the organization of affixes and the blocking effects of negation in this language (see Rice 2000, Elly van Gelderen 2008, Croft 1991). For instance, in the literature, Athabaskan verbs are assumed to exhibit templatic properties but Rice (2000) argues against the traditional analyses showing that affix ordering in this language is regulated by scope-based principles/semantic composition. However, Kurmanji verbs display some morphological properties that are regarded as typical of templatic systems, such as nonheadedness, zero morpheme, no adjacency, lookahead, and mutual exclusivity of morphemes in the same slot. Note that Nordlinger (2010) considers mutual exclusivity problematic for a scope based account such as Rice's (2000), because she observes no motivation for such blocking on this account. In this study, we assumed that negation takes scope semantically over all other markers; the exact investigation of scopal principles that could regulate the order of affixes in Kurmanji is, however, left for future research.

Abbreviations

1	first person
2	second person
3	third person
COND	conditional
OBJ	object
PL	plural
PRF	perfect
PRS	present
PROG	progressive
PST	past
SBJ	subject
SBJV	subjunctive
SG	singular
TAM	tense-aspect-mood

References

Bedirxan, Celadet A. and Roger Lescot. 2000. *KürtçeDilbilgisi (Kurmanci)*. Istanbul: Dos Basım Yayın.

Croft, William. 1991. "The evolution of negation". *Journal of Linguistics* 27: 1–27.

Gelderen, Elly V. 2008. "Cycles of negation in Athapaskan". *Working Papers in Athabaskan Languages* 7: 49–64.

Gündoğdu, Songül. 2010. "Morphosyntactic Representations and Negation in Kurmanji Kurdish." Boğaziçi University. Unpublished Ms.

Gündoğdu, Songül. 2011a. "The Phrase Structure of Two Dialects of Kurmanji Kurdish: Standard Dialect and Muş Dialect." MA thesis, Boğaziçi University.

Gündoğdu, Songül. 2011b. "Case and Ergativity in Kurmanji." Paper Presented at the Fourth International Conference on Iranian Linguistics, Uppsala University, Sweden, June 17–19.

Haig, Geoffrey. 2002. "Noun-plus-verb Complex Predicates in Kurmanji Kurdish: Argumentsharing, Argument Incorporation, or What?" *Sprachtypologie und Universalienforschung* 55: 15–48.

Haig, Geoffrey. 2004. *Alignment in Kurdish: A Diachronic Perspective.* Unpublished Habilitationsschrift: Philosophische Fakultät der Christian-Albrechts-Universität zu Kiel. (Available as pdf at: http://www.linguistik.uni-kiel.de/KurdAlignment.pdf.)

Inkelas, Sharon. 1993. "Nimboran Position Class Morphology." *Natural Language and Linguistic Theory* 11: 559–624.

Manova, Stela and Mark Aranoff. 2010. "Modelling Affix Order." *Morphology* 20: 109–131.

Muysken, Pieter. 1986. "Approaches to affix order." *Linguistics* 24: 629–643.

Nordlinger, Rachel. 2010. "Verbal Morphology in Murinh Pantha: Evidence for Templates." *Morphology* 20: 321–341.

Rice, Keren. 2000. *Morpheme Order and Semantic Scope. Word Formation in the Athapaskan Verb.* Cambridge: Cambridge University Press.

Rice, Keren. 2011. "Principles of Affix Ordering: An Overview." *Word Structure* 4: 169–200.

Simpson, Jane. and Meg Withgott. 1986. "Pronominal Clitic Clusters and Templates." In *The Syntax of Pronominal Clitics: Syntax and Semantics*, edited by H. Bohrer, 149–174. New York: Academic Press.

Spencer, Andrew. 1991. *Morphological Theory: An Introduction to Word Structure in Generative Grammar.* Oxford: Blackwell.

Stump, Gregory T. 1997. "Template Morphology and Inflectional Morphology." In *Yearbook of Morphology*, edited by G. Booij and Jaap van Marle, 217–241. Dordrecht: Kluwer.

Stump, Gregory T. 1998. "Inflection." In *The Handbook of Morphology*, edited by Andrew Spencer and Arnold M. Zwicky, 13–43. Oxford: Blacxkwell.

Stump, Gregory T. 2001. *Inflectional Morphology*. Cambridge: Cambridge University Press.

Stump, Gregory T. 2006. "Template Morphology." In *Encyclopedia of language and Linguistics*, edited by K. Brown, 559–563. Oxford: Elsevier.

Thackston, Wheeler M. 2006. *Kurmanji Kurdish: A Reference Grammar with Selected Readings*. Harvard University. www.fas.harvard.edu/~iranian/Kurmanji/kurmanji_1_grammar.pdf.

PART III:

Psycholinguistic and Cognitive Ordering

8

Suffix Combinations in Italian: Selectional Restrictions and Processing Constraints

Luigi Talamo

1. Introduction

How affixes combine in a language largely coincides with the description of constraints on word-formation rules.[1] Focusing on affixation, these constraints may apply to just one affix (individual constraint) or be valid for a set of affixes, to the extreme point of a constraint that could account for all the affix combinations in the language. This paper deals with Italian affixation and aims to verify both types of constraints.

For Italian affixation, the pertinent literature describes individual constraints that are based on grammatical principles encompassing all levels of linguistic analysis. For instance, in Italian the suffix combination *-ale+-(z)ione* is ruled out by the individual constraint that rejects nonverbal bases to the suffix *-(z)ione*. As they select a given feature, in this case a word class, this type of constraints is called a 'selectional restriction.' Much of the morphological investigation into Italian has been devoted to describe selectional restrictions. As indeed partially shown in this paper, this type of constraint can explain a significant portion of Italian derivational affix combinations.

The only attempt to systematically establish a constraint that could be valid for a set of affixes was the application of a theory that is based on extragrammatical, psycholinguistic principles known as Complexity-Based Ordering (Hay 2002). This theory was first tested for English, and it is based on the hypothesis that affix hierarchy relates to parsing complexity and strongly correlates with the potential productivity of affixes. The present paper challenges the Complexity-Based Ordering in Italian, taking more data into account than previous related work (Gaeta 2005).

---

[1]The author wishes to thank Pier Marco Bertinetto, Stela Manova, Davide Ricca and an anonymous reviewer for their insightful comments on an earlier version of this paper. Some parts of this work were presented at the third Vienna Workshop on Affix Order and at the fifth Mediterranean Morphology Meeting.

This paper is organized as follows. In section 2, I present the terminology, notifications, and assumptions of this theory. In section 3, I discuss the other approaches to affix combinations in Italian. In section 4, I describe the data and method employed in my research. In section 5, I discuss exceptions to the hierarchies and how hierarchies are correlated to productivity values. In the same section, I analyze the role of selectional restrictions. Finally, in section 6, I summarize my results and propose some interlinguistic remarks and psycholinguistic insights.

2. Complexity-Based Ordering: An Overview

The Complexity-Based Order hypothesis (henceforth, CBO: Hay 2002) was originally conceived and tested for English (Hay & Plag 2004, Plag & Baayen 2009, and Zirkel 2010). It was then tentatively extended to Bulgarian (Manova 2010), German (Zirkel Hilkenbach 2011), and Italian (Gaeta 2005). Based on psycholinguistic evidence, CBO tries to give a parsing account of affix ordering, stating that affixes closer to the base (i.e., in an inner position) in a derivative are harder to parse than more external affixes. CBO aims to be valid for all affixation instances, being that its main output is a hierarchy that accounts for and predicts all affix combinations. For instance, if we take ABCDE as an hypothetical parsing hierarchy where affix A is less parsable than affix B, combinations such as BASE-A-B or BASE-C-E are predicted, while combinations such as *BASE-B-A or *BASE-E-D are ruled out.

This hypothesis assumes a dual-route processing model (see, for instance, Baayen 1992), such that complex words can be either parsed out; for example, *leggibile* 'readable' = *leggere* 'to read' + *-bile*, or whole-word retrieved from the mental lexicon, for example, *mobile* 'movable' is not decomposed into its morphological constituents *mo-* 'to move (suppletive base)' and *-bile*. In her 2002 paper, Hay claims that phonotactics and lexical frequency influence the choice between these concurrent processing strategies and argues that low-probability phonotactic transitions may posit morphological boundaries in complex words, thus easing the parsing. For instance, the word *pipeful* is probably decomposable, as it contains the low-probability transition /pf/. Furthermore, Hay argues that high-frequency forms are most likely to be accessed as whole words rather than decomposed. In particular, a derivative that is more frequent than its base is expected to be retrieved from the mental lexicon, while a derivative that is less frequent than its base is probably decomposed. This is called 'relative frequency,' and accounts the reason for which the same suffix *-ment* is probably accessed in a word like *discernment*, which is less frequent than its base *discern*, whereas it is not parsed out in a word such as *government*, which is more frequent than its base *govern*.

To assess the affix parsability, two measures are established based on relative frequency: the type- and token-parsing ratios. These measures are simple ratios between nonparsable and parsable types and tokens. If the affix A appears in 2 parsable types and in 5 nonparsable types, its type-parsing ratio is 0.4 (2/5). Parsing ratio values below 1 indicate that the given affix is not parsed out; by contrast, affixes that are parsed out show parsing ratio values above 1 (Hay & Baayen 2002). Using these measures, it is possible to order affixes into a parsability hierarchy. However, in the relevant CBO literature affix hierarchies were established in an alternative way, using diagonalized affix combination matrices. More precisely, in a matrix of zeros and units, the zeros represents the absence of affix combinations, and their presence is represented by units. As shown in Figure 8.1, the diagonalization is obtained by rearranging, manually or computationally, the matrix such that the fewest number of units stand below the diagonal.

In the computational procedure, the computer rearranges the matrix several times by moving rows from left to right (forward search) and from right to left (backward search) to reach the minimum number of combinations below the diagonal (the exceptions). As different matrices show the minimum number of exceptions, the ordering obtained is the average of these matrices. The final hierarchy is thus the mean of all hierarchies with the minimum number of exceptions. After this reorganization, the units still sitting below the diagonal are exceptions to the CBO hypothesis. One type of affix combination could not be ruled out by diagonalization is the one that occurs both as BASE-A-B and BASE-B-A. This is known as the "inside-outside problem" (Plag & Baayen 2009:114), and these combinations are dubbed "mirror-image combinations" (Manova 2010).

The actual affix combinations are obtained from several data sources, including dictionaries, corpora, and the web. The hierarchies are compared to those whose ordering depends on parsing ratio values and potential productivity measures, namely the category-conditioned degree of productivity (Baayen 2009). Not every language tested for the CBO possesses the wealth of information on parsability and potential productivity that is available for English. Works on Bulgarian and Italian are thus only able to investigate the ordering obtained by rearranging the matrix.

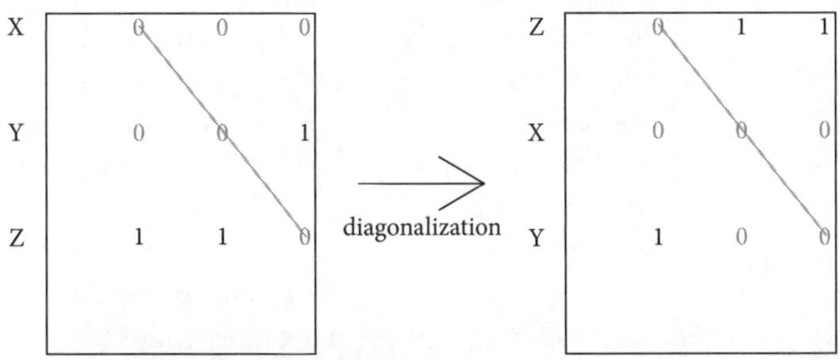

FIGURE 8.1 Example of a matrix diagonalization

3. Affix Combinations in Italian: History of Research

Studies on affix combinations in Italian closely follow those on the English language. Earlier papers, which I discuss in section 3.1 and 3.2, either draw from the lexicalist proposals or present a formal ordering based upon the affix position in derivatives, which resembles the one proposed by Fabb against lexical strata in English (Fabb 1988). These studies make the attempt, although unsuccessful, to establish constraints that could be applied to more than one affix. In his paper on CBO hypothesis, Gaeta claims that such a constraint could be established for Italian affixation. Section 3.3 deals with this. Finally, selectional restrictions represent the much argued, well-investigated part of the literature on Italian affix constraints. Section 3.4 is devoted to the discussion of these constraints and their prominent role in Italian affixation.

3.1 LEXICAL STRATA IN ITALIAN?

The stratalist hypothesis considers the idea that many loaned words from a foreign language may split the derivational system of the borrowing language into two parts, or strata: a native and a foreign stratum. According to Saciuk (1969), this was exactly the case for English, as it was strongly influenced by Norman French. Following this hypothesis, affixes that belong to native stratum do not cause phonological or morphophonological phenomena, such as resyllabification or deletion. Moreover, in the derivative, native affixes are predicted to stay further from the base than foreign affixes, which are supposed to be closer to the base. According to Spencer (1991), the native stratum is commonly called the second stratum, while the foreign stratum is the first stratum.

As previously stressed (Gaeta 2005:231), a stratalist approach to Italian morphology cannot be justified using such etymological terms. Indeed, Nespor (1993) relies only on phonological phenomena in applying what she calls "a rule of the Italian lexical phonology." The phenomenon involved in this rule is the affrication of the phoneme [t(:)] before certain suffixes, most notably *-ione*: *sottra[t:]o* 'subtracted' >*sottra[t:s]ione* 'subtraction.' Given the same segmental context, this phenomenon does not apply to nonderivatives: for example, *tra[t:]iene* 'hold (3sg)' or, most importantly, before other suffixes, for example, *-iere*: *gelato* 'ice cream' >*gela[t]iere* 'one who sells ice cream,' not **gela[ts]iere*. Suffixes that cause affrication belong to a sort of first stratum, while suffixes that do not to a second stratum. This rule predicts the ordering BASE-*ione-iere*, as in *spedizioniere* 'customs broker,' *munizioniere* 'munitioner,' and denies the specular suffix combination, BASE-*iere-ione* (Nespor 1993: 146–147).

However, as the suffix *-iere* does not output verbs and the suffix *-ione* only accepts verbs as bases, the combinations between these suffixes may be easily ruled out by positing a selectional restriction. Moreover, as the literature discusses no other Italian lexical phonology rules, this proposal could not be extended to other affix combinations.

3.2 AFFIX POSITION AND FORMAL ORDERING

The proposal tries to describe Italian affix ordering as templatic (Manova & Aronoff 2010: 112), that is, an ordering that is ordered but inexplicable, being unmotivated by grammatical and extragrammatical principles. On the position of the suffix in the derivative (Table 8.1), four combination schemas are identified by Scalise (1994) in order to account for the combination of 30 Italian suffixes.

The first schema describes initial/closing suffixes, namely suffixes that are always contiguous to the base and never followed by other suffixes: for example, *-eto* as in *ros-eto* 'rosegarden' >**roset-izzare* '? rosegarden-ize.' The second schema describes suffixes that are always contiguous to the base and can be followed by other suffixes (initial suffixes), for example *-evole* in *piacevole* 'pleasure' >*piacevol-mente* 'in a pleasure manner.' The third schema involves intermediate suffixes, where suffixes can go before and after other suffixes: for example *-ale*, *nazion-ale* 'national' >*nazional-izzare* 'nationalize' and *conversa-zione* 'conversation' >*conversazion-ale* 'conversational.' Finally, suffixes that cannot be followed by other suffixes belong to the last schema, for example *capital-ismo* 'capitalism' >**capitalism-izzare* '? capitalism-ize.'

The first and last combination schemas resemble the recent "mono-suffix constraint" proposed for English (Aronoff & Fuhrhop 2002). Although some linguistic data support this claim for Italian (e.g., the suffix *-ismo* is never followed by other suffixes in my data; see the Appendix), the other two schemas are problematic for two reasons. First, the second schema captures some suffixes that may also appear after other suffixes; that is, the suffix *-tore*. This suffix appears several times in my data after another suffix: *bors-eggiare* 'to pick-pocket' >*borseggia-tore* 'a pick-pocket' and *cement-ificare* 'to cover something with concrete' >*cementifica-tore* 'person, especially a politician who ruins natural landscapes with concrete covering' (cfr. Gaeta 2005: note 5). Second, as already discussed by Scalise (1994), the third schema is not further sortable; that is, further hierarchies between intermediate suffixes are hard to assume. Take the three suffixes *-ale*, *-izzare*, and *-(z)ione*. Starting from complex forms such as *commerci-al-izzare* 'to market' and *industrial-izza-zione* 'industrialization,' one may assume the following rule: 'The suffix *-ale* precedes the suffix *-izzare*, which in turn is followed by the suffix *-(z)ione*.' However, this rule is easily falsified by complex forms

TABLE 8.1

Four Combination Schemas for the Italian Suffixation[2]

1.	Base	Suffix	#	
2.	Base	Suffix	Other suffixes	#
3.	Base	Other suffixes	Suffix	Other suffixes
4.	Base	Other suffixes	Suffix	#

[2] # indicates end of word.

such as *conversa-zion-ale* 'conversational' or, worse, *costitu-zion-al-izzare* 'to make something constitutional (law).' In addition, as already stressed by Gaeta (2005), the third schema describes over half of the investigated Italian suffixes, making this constraint proposal useless.

3.3 COMPLEXITY-BASED ORDERING IN ITALIAN

Gaeta (2005) investigates whether the CBO hypothesis could account for the combinations of a 16 Italian suffix set in a newspaper corpus of 75 million tokens (informally, corpus La Stampa, three years from the daily *La Stampa*, Gaeta & Ricca 2002); his suffix set yields 46 total combinations from the 240 virtually possible; that is, not considering any constraints, such as selectional restrictions.

Gaeta reorganizes the combination matrix manually, finding 17 combinations standing below the diagonal. He then recognizes opacity and high relative frequency values as the two factors that appear to favor whole-word access, justifying these exceptions to the CBO ordering. For instance, combinations like *-bile+-izzare* and *-ico+-ista* are justified exceptions, because their first derivative forms show high relative frequency values. At the same time, combinations such as *-(t)ore+-ista*, *-bile+-ista* and *mento+-ista* are again justified exceptions, because of the morphotactic and morphosemantic opacity shown by their first derivative form; that is, *ascensorista* 'liftman,' containing the morphotactic opaque derivative form *ascensore* 'lift,' which has the verb *ascendere* 'to ascend' as its base and *risorgimentista* 'related to an Italian historical period,' containing the morphosemantic opaque derivative form *Risorgimento*, which is loosely connected to the verb *risorgere* 'to rise.' (Gaeta 2005: 241)

This leads to the conclusion that a hierarchic ordering is possible for the investigated suffixes. However, two issues may lessen the validity of this claim: the first is peculiar to the linguistic structure of Italian, and the other is methodological. Starting from the linguistic issue, as previously indicated (Gaeta 2005: 236), Italian suffixation lacks phonotactic encounters that, in Hay's original claim, ease parsing in derivatives. Several suffixes that begin with a vowel and stress-changing suffixes do not exist. When these phonotactic encounters exist, as in prefixes, they may posit morphological boundaries or not: consider /sg/ as in s-gridare 'to scold' versus sgabello 'stool,' /sn/ as in s-naturare 'to denature' versus s- versus snello 'slender' and /skw/ as in s-qualificare 'disqualify' versus squalo 'shark' (Dressler & Dziubalska-Kołaczyk 2006). For the methodological issue, Gaeta (2005) makes no attempt to correlate the matrix obtained hierarchy with a productivity based hierarchy; that is, to validate the hierarchy with productivity data.

By using linear regression analysis, it is possible to verify whether a correlation exists between the hierarchy established by Gaeta and productivity values employed in previous studies (Gaeta & Ricca 2003 and Gaeta & Ricca 2006), which investigate the productivity of a large set of Italian affixes in the corpus La Stampa;

incidentally, this corpus is the same used by Gaeta (2005). Productivity values are obtained using the "variable-corpus approach" (Gaeta & Ricca 2006), an improvement of Baayen's category-conditioned degree of productivity (Baayen 2009). Following this approach, productivity values are calculated at equal token numbers (N) for different affixes; this assumes that hapax are counted for each affix in subcorpora—that is, chunks of the corpus. In the correlation I employ productivity values with N fixed to 50,000 tokens, which covers the largest number of affixes. Since productivity information for five suffixes is missing (Gaeta & Ricca 2003), I must refer to only two-thirds of the suffixes.

Before discussing the results of linear regression, I will briefly explain the meaning of two values used in this kind of statistical analysis:[3] the *p*-value and the R-Squared value. Quoting Baayen (2008: 97), "the *p*-value tells us whether the coefficient is significantly different from zero. If the coefficient for a predictor is zero, there is no relation at all between the predictor and the dependent variable, in which case it is worthless as a predictor." In the correlation discussed here, the dependent variable is represented by the suffix rank which, according to the CBO hypothesis, is dependent by the category-conditioned productivity, representing the predictor. A *p*-value of 0.05 is traditionally considered the threshold under it is safe to assume a correlation. The R-Squared value "quantifies the proportion of the variance in the data that is captured and explained by the regression model" (Baayen 2008: 96). The variance to be captured here is the variability of the suffix rank, which is dependent by the category-conditioned productivity. The more the R-Squared value is closer to 1—that is, 100% of the variance—the better the correlation between the two variables.

The correlation between Gaeta's original ordering and productivity (Table 8.2) is not significant; I obtain a *p*-value of 0.274 (27%) and a very low proportion of explained variance, just the 4% (R-Squared = 0.040).

TABLE 8.2

Original Ordering for Gaeta's 2005 Suffix Combinations, with Productivity Values from Gaeta and Ricca 2003, and Gaeta and Ricca 2006

Suffix	Rank	$P(N)*10^3$	Suffix	Rank	$P(N)*10^3$
-(z)a	1	0.3	-(t)orio	9	?
-izzare	2	3.8	-aggio	10	?
-(z)ione	3	5.1	-bile	11	6.3
-ista	4	6.2	-mento	12	4.9
-ico	5	?	-oso	13	1.6
-ezza	6	1.3	-trice	14	?
-eggiare	7	?	-ità	15	6.3
-(t)ore	8	5	-iere	16	?

[3]For an introduction to statistical analysis in linguistics, see, e.g., Baayen 2008.

TABLE 8.3

Alternative Ordering for Gaeta's 2005 Suffix Combinations, with Productivity Values from Gaeta and Ricca 2003, and Gaeta and Ricca 2006

Suffix	Rank	P(N)*10³	Suffix	Rank	P(N)*10³
-ico	1	?	-ità	7.29	6.3
-eggiare	2.06	?	-(t)ore	8.05	5
-bile	4.87	6.3	-oso	10.22	1.6
-aggio	6.51	?	-ista	12.29	6.2
-(z)a	6.6	0.3	-iere	12.75	?
-(z)ione	6.65	5.1	-izzare	13.96	3.8
-ezza	6.71	1.3	-(t)orio	15.13	?
-mento	6.78	4.9	-trice	15.14	?

Table 8.3 gives an alternative ordering using the computational procedure discussed in section 2 and implemented using a script written by Baayen for the R statistical program. This script performs a rearrangement similar to that in Plag and Baayen (2009) and Zirkel (2010);[4] the corank values in Table 8.3 are decimal because this hierarchy is the mean of all hierarchies with the minimum number (10) of exceptions. The alternative ordering was achieved after considering 43 alternative orderings in a forward search and 189 in a backward search.

By further correlating the alternative ordering with productivity, the correlation is even less significant: $p = 0.730$, R-Squared = 0.015.

It seems that suffix productivity does not support the existence of a hierarchical ordering of the suffixes. However, Gaeta (2005) provides a hierarchy in which exceptions are justified by internal CBO assumptions; that is, morphotactic/morphosemantic transparency and relative frequency. This paper considers whether this hypothesis could be applied to a larger set of affixes and data.

3.4 SELECTIONAL RESTRICTIONS

As it is encoded in morphemes, this type of constraint is morphological (Manova & Aronoff 2010: 117–118). However, whether the direction of derivation proceeds from the right to the left (affix-driven approach) or from the left to the right (base-driven approach; Plag 1999) is discussed. Or, put into simpler terms, in the derivative BASE-A-B is it the suffix B that selects the suffix A, or the suffix B that selects the suffix A?

The pertinent literature on selectional restrictions in Italian frequently discuss the interaction between lexical base and one affix; that is, BASE-A, assuming that it is the affix that selects the base. This could be easily extended to the affix

[4]The R script available at Baayen's website does not perform exactly the same rearrangement used in those works. More precisely, this script also performs backward searches instead of forward searches. I slightly modified this script to adapt it to the Italian data.

combination BASE-A-B, claiming that it is the suffix B that selects the affix A. Thus, for instance, in the suffix combination *-osità*, (e.g., *golosità* 'gluttony') it is the suffix *-ità* which, according to its selectional restrictions that denies nonadjectival input, selects the suffix *-oso*, whose output are adjectives.

The base-driven approach to Italian suffix combinations is briefly discussed by Gaeta in his paper on CBO hypothesis (Gaeta 2005: 234); however, since other studies do not provide evidence for a base-driven approach to Italian affix combination, I employ the affix-driven approach to describe the selectional restrictions of the suffixes involved in this study (see Table 8.5).

4. Data and Methods

I investigated the combinations yielded by a 34-suffix set in two different linguistic data collections: the *Grande Dizionario Italiano dell'Uso* and the La Repubblica corpus. Suffix combinations that show at least one type in either one source were considered. This means that archaic or literary formations, which are mostly found in the dictionary, were taken into account as well as several nonce-formations, which are scattered through the corpus. Take the form *limon-at-iere* 'one who sells lemonade,' which is no longer in use and probably replaced by *limon-aio* or the odd derivative *brasil-ian-ata*, 'act or thing typical of Brazilian people'; they both count as types of the combination *-ata+-iere* and *-ano+-ata*, respectively. These types are sometimes signs of a low availability of these combinations, showing fewer than two types (see the Appendix).

In section 4.1, I discuss the nature of the two linguistic data sources. In section 4.2, I describe the series of criteria that a complex form must fulfill to be counted as an actual example of a suffix combination. In section 4.3, I give a list of suffixes investigated in this study.

4.1 LINGUISTIC DATA COLLECTIONS USED IN THIS WORK

The *Grande Dizionario Italiano dell'Uso* (henceforth GRADIT 2003) is a vast Italian dictionary, containing over 250,000 lemmas. As the name of the dictionary suggests, each lemma is annotated as to its 'use,' helping identify suffix combinations that are scarcely used (the label 'BU,' Italian *Basso Uso*: Low Use) or obsolete (the label 'OB,' Italian *Obsoleto*: 'Archaic'). Published in seven volumes, this dictionary is also available in electronic form, with a simple but effective search engine.

The La Repubblica corpus (Baroni et al. 2004) is a corpus of approximately 330 million tokens, one of the largest available for the Italian language. It contains all articles published by the national daily *La Repubblica* between 1985 and 2000, and it is annotated regarding parts of speech, lemmatized and encoded in the Corpus Query Processor System (Christ 1994), which allows for the execution of sophisticated queries. Unlike English corpora such as CELEX, this corpus lacks a morphological annotation. Suffix combination queries are thus hand-crafted, and results are checked for erroneous forms, including homographic sequences, mistyped forms, and the criteria discussed in section 4.2. Given the nature of the collected texts (i.e., journalistic texts), the language of this corpus is monodialectal with the following sociolinguistic features: well educated, middle-class and residents of Northern and Central Italy.

4.2 CRITERIA USED IN THE ASSESSMENT OF A SUFFIX COMBINATION

To be counted as an actual example, a suffix combination (i) must be linear and continuous, and (ii) each suffix involved must obey certain morphosemantic and morphotactic transparency criteria. Several productivity-oriented studies present similar criteria for investigating Italian word formation (Gaeta & Ricca 2003, Gaeta & Ricca 2006, Štichauer 2009).

A combination is linear and continuous when a morphological process does not interrupt it. In a suffix combination, this can occur whenever conversion is involved, such as with noun-to-verb conversion, adjective-to-noun conversion, and so on. Accordingly, the following complex forms were excluded: *mobilitazione* 'campaigning,' with (base) *mobilità* 'mobility' >(noun-to-verb conversion) *mobilitare* 'mobilize' > (suffixation in *-(z)ione*) *mobilitazione*; *occhialeria*, with (base) 'eyewear shop,' *occhiale* 'eye-related' >(adjective-to-noun conversion) *occhiale* 'glasses' > (suffixation in -^{2}eria) *occhialeria*. In the latter case, the conversion process involves a semantic drift; *occhiale* is a relational adjective no longer available, being synchronically replaced by *oculare*. Furthermore, I only considered the outmost derivational cycles (Plag 1999: 28–29). For example, I did not consider *auto-real-izza-zione* 'self-fulfillment' and *real-izza-zione* 'fulfillment' as two distinct types of the combination *-izzare* + *-zione*.

For morphosemantic and morphotactic transparency, I employed two different scales: the scale of morphosemantic transparency (Talamo & Celata 2011, Talamo, Celata, & Bertinetto 2014) and the "universal scale of morphotactic transparency" (Dressler 1985, 2005).

Similarly to the morphosemantic scale proposed by Dressler (2005: 271–272) for English compounds, the scale of morphosemantic transparency (Table 8.4) represents the derivative meaning as the *composition* between the base meaning,

TABLE 8.4
The Scale of Morphosemantic Transparency

Level	Example	Base	Suffix
1	*sparizione* 'disappearing'	+	+
2a	*passione* 'passion'	+/–	+
2b	*stazione* 'station'	+	+/–
3	*potabile* 'drinkable'	–	+

which is basically a lexical meaning, and the affix meaning, which is a word-formation meaning (German *Wortbildungsbedeutung*).

With respect to Dressler's compounding scale, suffixes are parallel to compound heads. Complex forms characterized by the highest transparency level, showing full transparency in both base and suffix, are classified at the first level. For instance, in the derivative *sparizione* 'disappearing,' an Italian speaker can easily recognize the base verb *sparire* 'to disappear,' and semantics of suffix *-zione*: *disappearing* is an action noun. The second level of morphosemantic transparency can be further divided into two sublevels: sublevel 2a and sublevel 2b. Complex forms in sublevel 2a retain the *typical* word-formation meaning of their suffixes, but present a lexicalized base; derivatives in sublevel 2b retain the lexical meaning of their base, but may show a non*typical* suffix meaning. As an example of 2a, consider the Italian word *passione* 'passion,' which has one of the *typical* meaning of the suffix *-(z)ione*—that is, that of a result noun; however, the original, metaphorical relationship to the verb *patire* 'to suffer' is lost. As for 2b, consider *stazione* 'station'; if the suffix *-zione* had retained its *typical* word-formation meaning, *stazione* would mean 'act of staying' (Gaeta & Ricca 2003: 70). However, in contemporary Italian, *stazione* is a locative rather than an action noun. The third level of morphosemantic transparency is glossed as "base-less" in Gaeta and Ricca (2003: 71); the *Wortbildungsbedeutung* of their suffixes is still present, but their base is not an autonomous form in Italian. As an example, for an Italian speaker *potabile* 'drinkable' is an adjective that means 'that can be **pota*-ed,' but **pota-* is not an autonomous form in Italian.

In the present study, I only considered complex forms showing the first and second morphosemantic transparency levels in the first suffixation cycle. Indeed, it is rare to find a second or lower transparency level in the second suffixation cycle. According to Gaeta and Ricca (2003: 71), there might be good theoretical and psycholinguistic reasons to count base-less forms as genuine complex forms, considering that their further suffixation gives rise to a genuine suffix combination. However, following the methodology of previous CBO works (Zirkel 2010: 248–249), these combinations were not included. Thus, forms such as *pota-bil-izzare* 'to make something drinkable' were not counted as a type for the combination *-bile* + *-izzare*.

Turning to the morphotactic aspect, I adopted the universal scale of morphotactic transparency (Dressler 1985: 330–331, 2005: 272–273). All sorts of complex forms displaying any transparency degrees pertaining to Italian morphotactics were considered: from the highest transparency degree, where no morphophonological phenomena are involved in the word-formation process (e.g., *Italia* 'Italy' > *itali-ano* 'Italian'), to the lowest degree, where a lexical phenomenon such as strong suppletion obfuscates the derivation (e.g., *Basilicata* 'name of a southern Italian region' > *luc-ano* 'inhabitant of Basilicata'). This is not the place to discuss every detail of the morphotactic scale; the reader can consult Gaeta (2002: 64–65), who presents a detailed account of this scale regarding Italian word-formation, although restricted to action nouns. Finally, as in semantic transparency, this paper considered the morphotactic transparency of the first suffixation cycle. Indeed, outer suffixes (i.e., second or third cycles) usually show a higher degree of morphotactic transparency (i.e., they are more transparent). This is in agreement with Hay's claim.

4.3 THE SUFFIX SET

The set previously established by Gaeta (2005) contains the following suffixes:

- deverbal suffixes *-mento*, *-(z)ione*, *-aggio*, and *-(z)a*, derivating action nouns;
- agentive and relational suffixes *-(t)ore*, *-trice*, *-iere*, and *-ista*;
- deadjectival suffixes *-ità* and *-ezza*, derivating quality nouns;
- adjectivizing suffixes *-bile*, *-(t)orio*, *-oso*, and *-ico*;
- verbalizing suffixes *-eggiare* and *-izzare*.

This set was expanded by adding the following suffixes, to reach a total of 34 suffixes:

- deverbal and denominal suffixes *-(t)ura* and *-ata*, derivating action nouns and *-evole*, derivating adjectives;
- nominalizing suffixes *-iera*, *-iero*, and *-ario*, derivating relational nouns and adjectives as well as instrumental nouns;
- nominalizing suffix *-ismo*, derivating abstract nouns and nominalizing suffix -^{2}*eria*, derivating collective nouns;
- deadjectival/denominal suffixes *-aggine* and -^{1}*eria*, both derivating quality nouns;
- adjectival suffix *-ale*, derivating relational adjective;
- denominal suffixes *-(i)ano* and *-ese*, both derivating relational adjective;
- verbalizing suffix *-ificare*;
- evaluative suffixes *-one*, *-ume*, *-etto*, *-ino*.

The suffix set could have been extended by counting the adjectival and nominal output of *-iera*, *-iero*, *-ista*, and *-ario* as different affixes, as is in Plag and Baayen (2009) for the nominal *-ary*, adjectival *-ary*, nominal (but not adjectival) *–ist*, and so on. This has not been done for two reasons: one methodological (i) and the other theoretical (ii). For reason (i), the previous CBO hypothesis test in Italian (section 3.3 and Gaeta 2005) did not distinguish between the adjectival and nominal *-ista*. This study is conceived as an extension of Gaeta's previous findings, and thus uses the same methodology. For reason (ii), I claim that when identifying affixes, morphosemantic criteria, similar to those outlined in the previous section, should be employed rather than output category (for a similar position, see Plag 1998 on the English suffix *-ize*). Accordingly, the four Italian suffixes listed above are not to be split into nominal and adjectival, as they express a relational meaning

TABLE 8.5

Selectional Restrictions of the 34 Suffixes Investigated in the Present Study: "?" Indicates an Untrustworthy Restriction

Suffix	Example	Input word class	Output word class	Phonetic restrictions (input)	Semantic restrictions (input)	Typical meaning
-(V)ale	*nuziale* 'wedding'	N/?Adj	Adj		?[+technical]	relational adjectives
-(i)ano	*gramsciano* 'follower of Antonio Gramsci'	N	Adj/N		anthroponyms, toponyms	relational adjectives
-(t)ore	*evidenziatore* 'marking pen'	V/?N	N			agent and instrument nouns
-(t)orio	*amatorio* 'amatory'	V	N/Adj			relational adjectives
-(z)a	*arroganza* 'arrogance'	V/Adj	N			action nouns
-(z)ione	*sparizione* 'disappearing'	V	N	base ending is not /tsV/	?[+learnèd]	action nouns
-aggine	*stupidaggine* 'foolishness'	Adj/N			qualitative adjectives, ethonyms	quality nouns
-aggio	*boicottaggio* 'boycott'	V/N	N	base does not contain /ʤa/	?[+foreign] and/or [+technical]	action nouns
-ario	*autoritario* 'authoritarian'	N/V	N/Adj		?[+technical] verbs	relational noun and adjectives
-ata	*passeggiata* 'a walk'	V/N	N		continuative verbs, concrete nouns	action and collective nouns
-bile	considerabile '*considerable*'	V/N/?Adj	Adj	? not monosyllabic base	?[+learnèd]	qualitative adjectives
-eggiare	*asineggiare* 'to behave in a stupid manner'	N/Adj	V			similative and performative verbs
-^{1}eria	*meschineria* 'pettiness'	Adj/N	N		qualitative adjectives, animal nouns, ethnonyms	quality nouns
-^{2}eria	*panetteria* 'bakery'	N	N			collective and locative nouns
-ese	*francese* 'french'	N/?Adj	Adj/N		toponyms, anthroponyms, simplex nouns	relational adjectives and nouns

-etto	*bacetto* 'little kiss'	Adj/N/V	Adj/N/V	base does not contain /tV/		evaluative (diminutive, ameliorative)
-evole	*maneggevole* 'handy'	V/N/?Adj	Adj		?[-learnèd]	qualitative adjectives
-ezza	*bellezza* 'beauty'	Adj	N		qualitative adjectives, ?[-learnèd]	quality nouns
-ico	*filmico* 'related to a movie'	N/?Adj	Adj		?[+technical]	relational adjectives
-iera	*bistecchiera* 'steak pan'	N	N			instrumental and locative nouns
-iere	*giardiniere* 'gardener'	N/?V	N			relational nouns
-iero	*guerriero* 'warrior'	N	N/Adj			relational adjectives and nouns
-ificare	*burrificare* 'to butter'	N/Adj	V			causative and inchoative verbs
-ino	*uccellino* 'little bird'	Adj/N/V	Adj/N/V		? base does not contain /inV/	evaluative (diminutive, ameliorative, pejorative)
-ismo	*anarchismo* 'anarchism'	N/Adj/V	N			abstract nouns
-ista	*socialista* 'socialist'	N/Adj/V	N/Adj	last syllable coda is not /j/ or /s/		adjective and relational nouns
-ità	*abilità* 'skill'	Adj	N		?[+learnèd]	quality nouns
-izzare	*realizzare* 'to fulfil'	N/Adj	V			causative verbs
-mento	*movimento* 'movement'	V	N	not monosyllabicbase	?[-learnèd]	action nouns
-one	*librone* 'big book'	Adj/N/V	Adj/N/V			evaluative (augmentative)
-oso	*goloso* 'greedy'	N	Adj			qualitative adjectives
-trice	*fotocopiatrice* 'photocopier'	V/?N	N			agent and instrumental nouns
-tura	*apertura* 'opening'	V/?N	N		?[-learnèd] and [+technical]	action nouns
-ume	borghesume 'petty bourgeois'	N/Adj	N			evaluative and collective nouns (pejorative)

between the base and its derivative, which is retained in both the nominal and adjectival forms. The only "true" homographic suffix in my set is the suffix *-eria*, which is distinguished on a semantic basis between *-¹eria*, whose derivational meaning is forming quality nouns, as in *ghiottoneria* 'gluttony' and *-²eria*, which instead forms collective nouns, as in *argenteria* 'silvery.'

Considering allomorphies, *-(V)ale* represents the suffix *-ale* and its allomorphies *-iale*, *-eale* and *-uale*, *-(t)ore* comprises the suffix *-tore* and its allomorph *-ore*, *-(t)orio* encompasses the suffix *-torio* and its allomorph *–orio*, and *-(z)ione* includes the suffix *-zione* and its allomorph *-ione*, while *-ità* comprises the suffix *-ità* and its allomorph *-età*.

Selectional restrictions are obtained from the pertinent literature, primarily, the collective book on Italian word-formation (Grossmann & Rainer 2004). I also used works on specific suffixes, including Gaeta's monograph on Italian action nouns (Gaeta 2002) and Bisetto's paper on Italian suffix *-(t)ore* (Bisetto 1995), as well as dictionaries, such as GRADIT (2003), and my own linguistic competence. These information are provided in Table 8.5, along with the *typical* suffix meaning.

In terms of productivity, according to Gaeta and Ricca (2003: 89), some suffixes in this set are scarcely productive, including *-evole*, *-ificare*, and *-(z)a*, while others belong to a productive group, though they show different productivity values: *-ezza*, *-oso*, *-(V)ale*, *-(t)ura*, *-(z)ione*, *-(t)ore*, *-trice*, *-ista*, *-ità*, *-bile*, and *-ismo* (suffixes are sorted from the less productive). Gaeta and Ricca recognize quite an inflectional behavior in the suffix *-(i)ano*, the most productive suffix in their set. For technical reasons, I subsumed two different suffixes under this affix: the ethnic *-(i)ano* as in *brasiliano* 'Brazilian' and the relational, deonomastic *-(i)ano*, as in *gramsciano* 'follower of Antonio Gramsci.' Gaeta and Ricca's productivity values only refer to the ethnic *-(i)ano*.

5. Constraining Italian Suffixation: Selectional Restrictions Versus the CBO Hypothesis

The 34 investigated suffixes yielded 184 combinations, 15.1% of the 1,122 potential combinations (34^2–34)). Examples of attested suffix combinations are given in the Appendix.

Figure 8.2 shows that there are two cases of recursiveness (combinations on the diagonal): *-ino+-ino* and *-one+-one*. Following Zirkel (2010: 251), this type of combination is not considered in the total number of 184 combinations. The potential number of combinations is also computed without the 34 recursive combinations. Let us briefly consider here the matter of recursiveness in Italian. As my collected linguistic data suggest, combinations of two identical suffixes are rare, usually involve allomorphy, and, most important, are restricted only to the evaluation domain, similarly to what is claimed by Manova (2010: 293) for Bulgarian diminutive domain. The combination *-ino+-ino*, which infringes the phonological

	(V)ale	(i)ano	(t)ore	(t)orio	(z)a	(z)ione	aggine	aggio	ario	ata	bile	eggiare	(1)eria	(2)eria	ese	etto	evole	ezza	ico	iera	iere	iero	ificare	ino	ismo	ista	ità	izzare	mento	one	oso	trice	tura	ume
(V)ale	0	0	0	0	0	0	0	0	0	0	0	1	1	0	1	0	0	1	0	0	0	0	0	0	1	1	1	1	0	0	0	0	0	0
(i)ano	0	0	0	0	0	0	1	0	0	1	0	1	1	0	0	1	0	0	1	0	0	0	0	1	1	1	1	1	0	1	0	0	0	0
(t)ore	1	0	0	0	0	0	0	0	0	0	0	0	0	0	1	0	0	0	0	0	0	0	0	1	1	1	0	1	0	1	0	0	0	0
(t)orio	0	0	0	0	0	0	0	0	0	0	0	0	0	0	0	0	0	0	0	0	0	0	0	0	0	0	1	0	0	0	0	0	0	0
(z)a	1	0	0	0	0	0	0	0	0	0	0	0	0	0	0	1	0	0	0	0	1	1	0	1	1	0	0	0	0	0	1	0	0	0
(z)ione	1	0	0	0	0	0	0	0	1	0	0	0	0	0	1	0	1	0	0	0	1	0	0	0	1	1	0	0	0	0	0	0	0	0
aggine	0	0	0	0	0	0	0	0	0	0	0	0	0	0	0	0	0	0	0	0	0	0	0	0	0	0	0	0	0	0	1	0	0	0
aggio	0	0	0	0	0	0	0	0	0	0	0	0	0	1	0	1	0	0	0	0	0	0	0	0	1	1	0	0	0	0	0	0	0	0
ario	1	0	0	0	0	0	0	0	0	0	0	0	0	0	1	1	0	0	0	0	0	0	0	0	1	1	1	1	0	1	1	0	0	0
ata	0	0	0	0	0	0	0	0	0	0	0	0	0	0	0	0	0	0	0	0	1	0	0	1	0	0	0	0	0	1	0	0	0	0
bile	0	0	0	0	0	0	0	0	0	0	0	0	0	0	0	0	0	1	0	0	0	0	0	0	1	1	1	1	0	0	0	0	0	0
eggiare	0	0	1	1	0	0	0	0	0	0	1	0	0	0	0	0	1	0	0	0	0	0	0	0	0	0	0	0	1	0	0	1	1	0
(1)eria	0	0	0	0	0	0	0	0	0	0	0	0	0	0	0	0	0	0	0	0	0	0	0	0	0	0	0	0	0	0	0	0	0	0
(2)eria	1	0	0	0	0	0	0	0	0	0	0	0	0	0	0	1	0	0	0	0	0	0	0	0	0	0	0	0	0	0	0	0	0	0
ese	0	0	0	0	0	0	0	0	0	0	0	1	1	0	0	0	0	0	0	0	0	0	0	1	1	1	1	1	0	1	0	0	0	1
etto	0	0	0	0	0	0	0	1	1	1	0	0	0	1	0	0	0	0	0	1	1	0	0	1	1	1	0	1	0	1	1	0	0	1
evole	0	0	0	0	0	0	1	0	0	0	0	1	0	0	0	0	0	1	0	0	0	0	0	0	1	1	0	1	0	0	0	0	0	0
ezza	0	0	0	0	0	0	0	0	0	0	0	0	0	0	0	0	0	0	0	0	0	0	0	1	0	0	0	0	0	0	1	0	0	0
ico	1	0	0	0	0	0	0	0	0	0	0	0	1	0	0	0	0	0	0	0	0	0	0	0	1	1	1	1	0	0	0	0	0	1
iera	0	0	0	0	0	0	0	0	0	0	0	0	0	0	0	0	0	0	0	0	0	0	0	1	0	0	0	0	0	0	0	0	0	0
iere	0	0	0	0	0	0	0	1	0	0	0	0	0	0	1	1	0	0	0	0	0	0	0	1	1	0	1	0	0	1	0	0	0	0
iero	0	0	0	0	0	0	0	0	0	0	0	0	0	0	0	0	0	0	0	0	0	0	0	0	1	0	0	0	0	0	0	0	0	0
ificare	0	0	1	1	0	1	0	0	0	0	0	0	0	0	0	0	0	0	0	0	0	0	0	0	0	0	0	0	0	0	0	1	1	0
ino	0	0	0	0	0	0	0	0	0	0	0	0	0	1	0	0	0	0	0	0	0	0	0	1	0	1	0	0	0	1	0	0	0	0
ismo	0	0	0	0	0	0	0	0	0	0	0	0	0	0	0	0	0	0	0	0	0	0	0	0	0	0	0	0	0	0	0	0	0	0
ista	1	0	0	0	0	0	0	0	1	0	0	1	0	1	1	0	0	0	1	0	0	0	0	0	0	0	0	1	0	1	0	0	0	1
ità	0	0	0	0	0	0	0	0	1	0	0	0	0	0	0	0	0	0	0	1	0	1	0	0	0	0	0	1	0	0	1	0	0	0
izzare	0	0	1	1	0	1	0	0	0	1	0	0	0	0	0	0	0	0	0	0	0	0	0	0	0	0	0	0	1	0	0	1	1	0
mento	1	0	0	0	0	0	0	0	1	0	0	0	0	0	0	0	0	0	0	0	0	0	0	1	1	1	0	0	0	0	1	0	0	0
one	0	0	0	0	0	0	0	0	0	1	0	1	0	0	0	1	0	0	0	1	1	0	0	1	0	1	0	0	0	1	0	0	0	0
oso	0	0	0	0	0	0	1	0	0	0	0	0	1	0	0	1	0	0	0	0	0	0	0	1	1	1	1	0	0	1	0	0	0	1
trice	0	0	0	0	0	0	0	0	0	0	0	0	0	0	0	0	0	0	0	0	0	0	0	0	0	0	1	0	0	0	0	0	0	0
tura	1	0	0	0	0	0	1	0	0	0	1	1	0	0	0	1	0	0	0	1	1	1	0	1	1	1	0	1	0	1	0	0	0	0
ume	0	0	0	0	0	0	0	0	0	0	0	0	0	0	0	0	0	0	0	0	0	0	0	0	0	0	0	0	0	0	0	0	0	0

FIGURE 8.2 Adjacency matrix for a set of 34 Italian suffixes. Suffixes are listed in alphabetical order. Combinations attested are marked by the black 1, unattested combinations by a light grey 0.

restrictions described in Table 8.5, is attested by just one type: *violinino* 'little violin,' whose first suffixation instance, *violino* 'violin,' is not morphosemantically transparent. In the other case of recursiveness, the second suffix *-one* occurs as its allomorph *-cione* (/ʧone/), seen in two types: *scarponcione* 'big boot' and *maglioncione* 'big sweater.' These are again two cases of morphosemantic opaque forms, *scarpone* 'boot' <*scarpa* 'shoe' and *maglione* 'sweater' <*maglia* 'shirt.'

5.1 THE ROLE OF SELECTIONAL RESTRICTIONS

Figure 8.3 shows that selectional restrictions rule out 469 combinations (41.8% of the potential combinations). Not all restrictions presented in Table 8.5 were actually used. More specifically, restrictions labeled with a question mark (?) are not applied. Considering the morphosyntactic input category, constraints are more severe, while conversely, semantic restrictions are more permissive. Let us discuss the nominalizing suffix *-tura*. According to Gaeta (2004: 336), this suffix may also accept nouns as its input: for example, *bottinatura* 'act of collecting pollen seeds' < *bottino* 'bounty.' However, this position is problematic, as (i) one must explain the vowel /a/ between lexical root and suffix and (ii) the complex form may be derived from a verbal base, say ? *bottinare* 'to make a bounty.' Nevertheless, as it entails the existence of 'virtual' verbs, even this last claim is questionable. The issue also extends to other suffixes, including *-trice*, *-(t)ore* and *-(t)orio*. Accordingly, Figure 8.3 shows that these suffixes all display a selectional restriction that denies combinations with suffixes that output nouns.

The literature proposes several criteria for semantic restrictions, including [±technical], [±good] or [±learnèd]. According to Gaeta (2004), the suffix *-(z)ione* seems to prefer [+learnèd] bases/suffixes, as in the combination *-izzare+-(z)ione*. Notwithstanding, the combination *-ificare+-(z)ione* is attested with many types in both dictionary and corpus. I thus suggest that criteria such as those mentioned above should be taken as preferences rather than true constraints. Other semantic restrictions have instead been used, including restrictions that select a specific semantic type of a word class: for instance, the suffix *-(i)ano* only selects two specific type of nouns, that is, anthroponyms and toponyms.

	(V)ale	(i)ano	(t)ore	(t)orio	(z)a	(z)ione	aggine	aggio	ario	ata	bile	eggiare	(1)eria	(2)eria	ese	etto	evole	ezza	ico	iera	iere	iero	ificare	ino	ismo	ista	ità	izzare	mento	one	oso	trice	tura	ume
(V)ale	0	sr	sr	sr	0	sr	sr	sr	sr	sr	sr	1	SR	0	SR	0	sr	SR	sr	sr	sr	sr	0	0	1	1	1	1	sr	0	sr	sr	sr	0
(i)ano	0	0	sr	sr	0	sr	1	0	0	1	0	1	1	0	0	1	0	sr	1	0	0	0	0	1	1	1	1	1	sr	1	0	sr	sr	0
(t)ore	1	sr	0	sr	sr	sr	sr	0	0	0	0	0	sr	0	SR	0	0	sr	0	0	0	0	0	1	1	1	sr	1	sr	1	0	sr	sr	0
(t)orio	0	sr	sr	0	sr	sr	sr	0	0	0	0	0	sr	0	sr	0	0	sr	0	0	0	0	0	0	0	0	1	0	sr	0	0	sr	sr	0
(z)a	1	sr	sr	sr	0	sr	sr	0	0	sr	0	0	sr	0	sr	1	0	sr	0	0	1	1	0	1	1	0	sr	0	sr	0	1	sr	sr	0
(z)ione	1	sr	sr	sr	sr	0	sr	0	1	sr	0	0	sr	0	SR	0	1	sr	0	0	1	0	0	0	1	1	sr	0	sr	0	0	sr	sr	0
aggine	0	sr	sr	sr	sr	sr	0	sr	0	sr	0	0	sr	0	sr	0	0	sr	0	0	0	0	0	0	0	0	sr	0	sr	0	1	sr	sr	0
aggio	0	sr	sr	sr	sr	sr	sr	0	0	sr	0	0	sr	1	sr	1	0	sr	0	0	0	0	0	0	1	1	sr	0	sr	0	0	sr	sr	0
ario	1	sr	sr	sr	sr	sr	sr	0	0	0	0	0	sr	0	SR	1	0	sr	0	0	0	0	0	0	1	1	1	1	sr	1	1	sr	sr	0
ata	0	sr	sr	sr	sr	sr	sr	0	0	0	0	0	sr	0	sr	0	0	sr	0	0	1	0	0	1	0	0	sr	0	sr	1	0	sr	sr	0
bile	0	sr	sr	sr	0	sr	sr	sr	sr	sr	0	0	0	sr	sr	0	0	1	sr	sr	sr	sr	0	0	1	1	1	1	sr	0	sr	sr	sr	0
eggiare	sr	sr	1	1	0	sr	sr	0	0	0	1	0	sr	sr	sr	sr	1	sr	0	sr	sr	sr	sr	0	0	0	sr	sr	1	0	sr	1	1	sr
(1)eria	0	sr	sr	sr	sr	sr	sr	0	0	sr	0	0	0	0	sr	0	0	sr	0	0	0	0	0	0	0	0	sr	0	sr	0	0	sr	sr	0
(2)eria	1	sr	sr	sr	sr	sr	sr	0	0	0	0	0	sr	0	sr	1	0	sr	0	0	0	0	0	0	0	0	sr	0	sr	0	0	sr	sr	0
ese	0	sr	sr	sr	0	sr	sr	0	0	0	0	1	1	0	0	0	0	0	0	0	0	0	0	1	1	1	1	1	sr	1	0	sr	sr	1
etto	0	sr	sr	sr	sr	sr	sr	1	1	1	0	0	sr	1	sr	0	0	sr	0	1	1	0	0	1	1	1	sr	1	sr	1	1	sr	sr	1
evole	0	sr	sr	sr	0	sr	1	sr	sr	sr	sr	1	0	sr	sr	0	0	1	sr	sr	sr	sr	0	0	1	1	0	1	sr	0	sr	sr	sr	0
ezza	0	sr	sr	sr	sr	sr	sr	0	0	sr	0	0	sr	0	sr	0	0	0	0	0	0	0	0	1	0	0	sr	0	sr	0	1	sr	sr	0
ico	1	sr	sr	sr	0	sr	sr	sr	sr	sr	sr	0	SR	sr	sr	0	sr	sr	0	sr	sr	sr	0	0	1	1	1	1	sr	0	sr	sr	sr	1
iera	0	sr	sr	sr	sr	sr	sr	0	0	0	0	0	sr	0	sr	0	0	sr	0	0	0	0	0	1	0	0	sr	0	sr	0	0	sr	sr	0
iere	0	sr	sr	sr	sr	sr	sr	1	0	0	0	0	sr	0	SR	1	0	sr	0	0	0	0	0	1	1	0	SR	0	sr	1	0	sr	sr	0
iero	0	sr	sr	sr	sr	sr	sr	0	0	0	0	0	sr	0	sr	0	0	sr	0	0	0	0	0	0	1	0	0	0	sr	0	0	sr	sr	0
ificare	sr	sr	1	1	0	1	sr	0	0	0	0	sr	sr	0	sr	sr	0	sr	0	sr	sr	sr	0	0	0	0	sr	sr	0	0	sr	1	1	sr
ino	0	sr	sr	sr	sr	sr	sr	0	0	0	0	0	sr	1	sr	0	0	sr	0	0	0	0	0	1	0	1	sr	0	sr	1	0	sr	sr	0
ismo	0	sr	sr	sr	sr	sr	sr	0	0	sr	0	0	sr	0	sr	0	0	sr	0	0	0	0	0	0	0	0	sr	0	sr	0	0	sr	sr	0
ista	1	sr	sr	sr	sr	sr	sr	0	1	0	0	1	sr	1	SR	sr	0	sr	1	0	0	0	0	0	0	0	sr	1	sr	1	0	sr	sr	1
ità	0	sr	sr	sr	sr	sr	sr	0	1	sr	0	0	sr	0	sr	sr	0	sr	0	1	0	1	0	0	0	0	0	1	sr	0	1	sr	sr	0
izzare	sr	sr	1	1	0	1	sr	0	0	0	1	sr	sr	0	sr	sr	0	sr	0	sr	sr	sr	sr	0	0	0	sr	0	1	0	sr	1	1	sr
mento	1	sr	sr	sr	sr	sr	sr	0	1	sr	0	0	sr	0	sr	0	0	sr	0	0	0	0	0	1	1	1	sr	0	0	0	1	sr	sr	0
one	0	sr	sr	sr	sr	sr	sr	0	0	1	0	1	sr	0	sr	1	0	sr	0	1	1	0	0	1	0	1	sr	0	sr	SR	0	sr	sr	0
oso	0	sr	sr	sr	0	sr	1	sr	sr	sr	sr	0	1	sr	sr	1	sr	0	sr	sr	sr	sr	0	1	1	1	1	0	sr	1	0	sr	sr	1
trice	0	sr	sr	sr	sr	sr	sr	0	0	0	0	0	sr	0	sr	0	0	sr	0	0	0	0	0	0	0	0	SR	0	sr	0	0	0	sr	0
tura	1	sr	sr	sr	sr	sr	SR	0	0	sr	1	1	sr	0	sr	1	0	sr	0	1	1	1	0	1	1	1	sr	1	sr	1	0	sr	0	0
ume	0	sr	sr	sr	sr	sr	sr	0	0	0	0	0	sr	0	sr	0	0	sr	0	0	0	0	0	0	0	0	sr	0	sr	0	0	sr	sr	0

FIGURE 8.3 The role of selectional restrictions for a 34 Italian suffix set. Suffixes are listed in alphabetical order. Attested combinations are marked by '1,' combinations denied by selectional restrictions by 'sr' and attested combinations that are denied by selectional restrictions by 'SR.' Unattested combinations are marked by light grey zeros.

Twelve combinations, marked by 'SR' in Figure 8.3, violate selectional restrictions. Most violators infringe on semantic restrictions and the suffix *-ese* commits half of the violations. Suffix *-ese* has a semantic restriction that limits its input to anthroponyms, toponyms and simplex nouns, theoretically preventing it from combining with other suffixes. However, violators are justified to a certain extent either by being nonce-formations: *tradu-zion-ese* 'language of awkward translations,' *giudizi-ari-ese* 'specialized legal language,' or by showing morphosemantic opacity in the first derivative: *carabin-ier-ese* 'specialized language of Italian military policeman,' where *carabiniere* 'Italian military policeman' is no more related to *carabina* 'carbine' and could be considered a simplex, meeting the restriction of the suffix *-ese*. Finally, two suffix combinations violate constraints on word classes: *-iere+-ità* and *-trice+-ità*. The suffix *-ità* should select only adjectives as input, while both *-iere* and *-trice* yield nouns. Both suffix combinations show only one type: *ragion*-ier-*ità* 'accountant-ness' and *mo-tric-ità* 'motility.' The first complex form is a hapax in the La Repubblica corpus and should be again regarded as a nonce-formation, referring to some Italian movies starring an accountant (*ragioniere*). Considering *motricità*, this derivative is a loanword from the French *motricité*.

5.2 CHALLENGING THE CBO HYPOTHESIS

As introduced by Plag and Baayen (2009), a suffix combination matrix showing no exceptions is represented in graph theory as an acyclic graph; that is, a graph whose nodes (the suffixes) are connected by downward-pointing edges (the suffix combinations). This is connected to the idea that the CBO hypothesis implicates, to a certain extent, acyclicity in suffix combinations. In Figure 8.5, which represents the matrix in Figure 8.4, there are several edges pointing both upward and downward, corresponding to the suffix combinations that can both occur as BASE-A-B and BASE-B-A (mirror-image combinations; section 2). In addition to these edges, which are called 'mutual edges' in graph theory, there are a certain number of upward-pointing edges. Both mutual and upward-pointing edges introduce cyclicity, thus violating the CBO hypothesis.

In trying to reduce the cyclicity and minimize the exceptions by using the computational procedure described in sections 2 and 3.3, I found a minimum number of 30 violators, which comprises the 21 mirror-image combinations that could not be removed with further matrix rearrangements. As I mentioned before, recursiveness cases (2) are not included among the violators. The minimum number of violators was found by the computer after performing 54 alternative orderings in a forward search and 656 alternative orderings in a backward search, with a maximum number of 72 exceptions.

As outlined above, the minimum number of exceptions may be shown by different matrices. The matrix in Figure 8.4 thus displays one possible best arrangement, and the established hierarchy in Table 8.6 is the mean of all the best hierarchies; that is, the hierarchies with a minimum number of exceptions of 30.

	ificare	(i)ano	tura	(z)a	(z)ione	mento	etto	ario	(t)ore	ico	ata	(V)ale	bile	evole	iere	ezza	aggio	oso	ese	one	ista	izzare	eggiare	trice	(t)orio	ità	iera	iero	ino	ume	(1)eria	(2)eria	aggine	ismo
ificare	0	0	1	0	1	0	0	0	1	0	0	0	0	0	0	0	0	0	0	0	0	0	0	1	1	0	0	0	0	0	0	0	0	0
(i)ano	0	0	0	0	0	0	1	0	0	1	1	0	0	0	0	0	0	0	0	1	1	1	1	0	0	1	0	0	1	0	1	0	1	1
tura	0	0	0	0	0	0	1	0	0	0	0	1	1	0	1	0	0	0	0	1	1	1	1	0	0	0	1	1	1	0	0	0	1	1
(z)a	0	0	0	0	0	0	1	0	0	0	0	1	0	0	1	0	0	1	0	0	0	0	0	0	0	0	0	1	1	0	0	0	0	1
(z)ione	0	0	0	0	0	0	0	1	0	0	0	1	0	1	1	0	0	0	1	0	1	0	0	0	0	0	0	0	0	0	0	0	0	1
mento	0	0	0	0	0	0	0	1	0	0	0	1	0	0	0	0	0	1	0	0	1	0	0	0	0	0	0	0	1	0	0	0	0	1
etto	0	0	0	0	0	0	0	1	0	0	1	0	0	0	1	0	1	1	0	1	1	1	0	0	0	0	1	0	1	1	0	1	0	1
ario	0	0	0	0	0	0	1	0	0	0	0	1	0	0	0	0	0	1	1	1	1	1	0	0	0	1	0	0	0	0	0	0	0	1
(t)ore	0	0	0	0	0	0	0	0	0	0	0	1	0	0	0	0	0	0	1	1	1	1	0	0	0	0	0	0	1	0	0	0	0	1
ico	0	0	0	0	0	0	0	0	0	0	0	1	0	0	0	0	0	0	0	0	1	1	0	0	0	1	0	0	0	1	1	0	0	1
ata	0	0	0	0	0	0	0	0	0	0	0	0	0	0	1	0	0	0	0	1	0	0	0	0	0	0	0	0	1	0	0	0	0	0
(V)ale	0	0	0	0	0	0	0	0	0	0	0	0	0	0	0	1	0	0	1	0	1	1	1	0	0	1	0	0	0	0	1	0	0	1
bile	0	0	0	0	0	0	0	0	0	0	0	0	0	0	0	1	0	0	0	0	1	1	0	0	0	1	0	0	0	0	0	0	0	1
evole	0	0	0	0	0	0	0	0	0	0	0	0	0	0	0	1	0	0	0	0	1	1	1	0	0	0	0	0	0	0	0	0	1	1
iere	0	0	0	0	0	0	1	0	0	0	0	0	0	0	0	0	1	0	1	1	0	0	0	0	0	1	0	0	1	0	0	0	0	1
ezza	0	0	0	0	0	0	0	0	0	0	0	0	0	0	0	0	0	1	0	0	0	0	0	0	0	0	0	0	1	0	0	0	0	0
aggio	0	0	0	0	0	0	1	0	0	0	0	0	0	0	0	0	0	0	0	0	1	0	0	0	0	0	0	0	0	0	0	1	0	1
oso	0	0	0	0	0	0	1	0	0	0	0	0	0	0	0	0	0	0	0	1	1	0	0	0	0	1	0	0	1	1	1	0	1	1
ese	0	0	0	0	0	0	0	0	0	0	0	0	0	0	0	0	0	0	0	1	1	1	1	0	0	1	0	0	1	1	1	0	0	1
one	0	0	0	0	0	0	1	0	0	0	1	0	0	0	1	0	0	0	0	1	1	0	1	0	0	0	1	0	1	0	0	0	0	0
ista	0	0	0	0	0	0	0	1	0	1	0	1	0	0	0	0	0	0	1	1	0	1	1	0	0	0	0	0	0	1	0	1	0	0
izzare	0	0	1	0	1	1	0	0	1	0	0	0	1	0	0	0	0	0	0	0	0	0	0	1	1	0	0	0	0	0	0	0	0	0
eggiare	0	0	1	0	0	1	0	0	1	0	0	0	1	1	0	0	0	0	0	0	0	0	0	1	1	0	0	0	0	0	0	0	0	0
trice	0	0	0	0	0	0	0	0	0	0	0	0	0	0	0	0	0	0	0	0	0	0	0	0	0	1	0	0	0	0	0	0	0	0
(t)orio	0	0	0	0	0	0	0	0	0	0	0	0	0	0	0	0	0	0	0	0	0	0	0	0	0	1	0	0	0	0	0	0	0	0
ità	0	0	0	0	0	0	0	1	0	0	0	0	0	0	0	0	0	1	0	0	0	1	0	0	0	0	1	1	0	0	0	0	0	0
iera	0	0	0	0	0	0	0	0	0	0	0	0	0	0	0	0	0	0	0	0	0	0	0	0	0	0	0	0	1	0	0	0	0	0
iero	0	0	0	0	0	0	0	0	0	0	0	0	0	0	0	0	0	0	0	0	0	0	0	0	0	0	0	0	0	0	0	0	0	1
ino	0	0	0	0	0	0	0	0	0	0	0	0	0	0	0	0	0	0	0	1	1	0	0	0	0	0	0	0	1	0	0	1	0	0
ume	0	0	0	0	0	0	0	0	0	0	0	0	0	0	0	0	0	0	0	0	0	0	0	0	0	0	0	0	0	0	0	0	0	0
(1)eria	0	0	0	0	0	0	0	0	0	0	0	0	0	0	0	0	0	0	0	0	0	0	0	0	0	0	0	0	0	0	0	0	0	0
(2)eria	0	0	0	0	0	0	1	0	0	0	0	1	0	0	0	0	0	0	0	0	0	0	0	0	0	0	0	0	0	0	0	0	0	0
aggine	0	0	0	0	0	0	0	0	0	0	0	0	0	0	0	0	0	1	0	0	0	0	0	0	0	0	0	0	0	0	0	0	0	0
ismo	0	0	0	0	0	0	0	0	0	0	0	0	0	0	0	0	0	0	0	0	0	0	0	0	0	0	0	0	0	0	0	0	0	0

FIGURE 8.4 Adjacency matrix for a complexity-based ordering for 34 Italian derivational suffixes with a minimum of 30 exceptions.

The question that now arises is whether the minimum number of exceptions found is likely to arise by chance. Given a graph with 34 nodes (suffixes), what is the minimum number of violators (= combinations below the matrix) if these nodes were allowed to combine randomly in 186 edges (184 combinations + 2 recursiveness cases)? As it is impossible to consider all possible graphs, I employed the following method; using the Erdos-Renyi model implemented in the igraph package for the R statistical program (Csárdi & Nepusz 2006), I generated 10,000 graphs with 34 nodes and 186 randomly assigned edges. The graphs included cycles (violators) and loops (recursiveness cases). For each generated graph I computed the number of violators and mirror-image combinations, the latter including recursiveness cases (loops). Next, I converted each graph into its adjacency matrix and I minimized the violator numbers using the same procedure employed to reorganize the original matrix. I refer to this latter procedure as 'the optimization procedure.'

The mean number of violators found in these 10,000 graphs was 90.26. After the optimization procedure, the mean number of violators found in generated graphs is still higher (41.48) than the violators found in the actual graph (30). However, a lower number of violators (27) was observed once. This means that

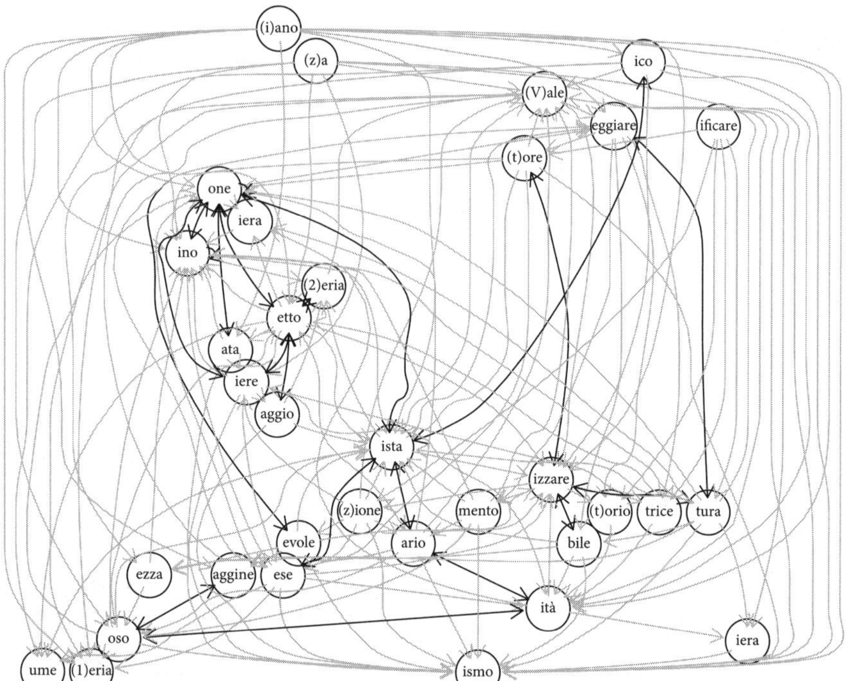

FIGURE 8.5 Directed graph for 34 Italian derivational suffixes (nodes) and 186 suffix combinations (edges). Mirror-image combinations are highlighted by black arrows, pointing in both directions.

there is a small probability (0.0001) that the actual number of violators could be due to chance. Furthermore, the number of the actual mirror-image combination (21) is higher than the mean number found in 10,000 graphs (17.15) and, although it was observed only once, it was four times the smallest number (5) (Table 8.7).

Since mirror-image combinations show few types and, in many cases, less than three types, this type of exceptions do not posit a serious issue against the CBO hypothesis. For instance, the high-frequency suffix combination *-eggiare+-evole*, as in *maneggevole* 'handy' is mirrored by *-evole+-eggiare*, whose only type is the literary form *piacevoleggiare* 'to behave in a pleasing manner,' which is a hapax in the La Repubblica corpus. In some cases, both suffix orderings show few types; that is, *-ario+-etto*, as in *glossarietto* 'little vocabulary,' and *-etto+-ario*, as in *fascettario* 'small band container.' However, the established hierarchy (Table 8.6), while accounting for low frequency combinations, such as *-aleria*, *-alese*, *-anico*, *-evolismo*, *-bilezza*, and many others, fails to predict several high frequency combinations, such as those with verbalizing suffixes: *-izzamento*, *-izzabile*, *-izzazione*, *-eggiamento*, and *-eggiabile*.

As earlier discussed (section 4.1), low frequency combinations often show obsolete or nonce-formations types. These combinations may not play the same role in processing as those counting a greater number of types. To obtain a more balanced hierarchy, an alternative matrix is established using the suffix combinations

TABLE 8.6

Suffixes with Their Mean Ranks and Productivity Values Computed Using the Corpus-Variable Approach at Two Different Values of N (19,000 and 50,000). All 184 Combinations Are Considered, with a Minimum of 30 Exceptions. All Productivity Values Are Intended *10,000

Affix	Rank	N=19	N=50	Affix	Rank	N=19	N=50
ificare	1.311	0.6		ese	18.111		
(i)ano	2.922			oso	18.360	3.7	1.6
(z)a	3.407	0.7	0.3	ista	20.456	11.3	6.2
tura	3.498	6.6	3.5	one	20.518		
mento	4.992	10.4	4.9	izzare	22.450	7.6	3.8
(z)ione	5.805	13.4	5.1	eggiare	22.549	4.1	
etto	6.755			trice	24.453	10.8	
ario	8.230			(t)orio	24.547		
(t)ore	9.166	9.4	5.0	ità	26.0	13.4	6.3
ata	10.120			iera	27.024		
ico	10.148			ino	29.169		
(V)ale	12.412			iero	29.692		
bile	13.334	11.3	6.3	aggine	31.502		
iere	13.680			ume	31.544		
evole	14.288	0.3		[1]eria	31.611		
ezza	16.412	2.7	1.3	[2]eria	31.634		
aggio	17.076	1.5		ismo	31.823	15.2	8.2

displaying more than three types, combining the GRADIT dictionary and the La Repubblica corpus. The alternative matrix contains 100 combinations and a minimum number of 11 exceptions, which was obtained after performing 54 alternative orderings in forward search and 3675 alternative orderings in backward search. Using the same procedure described above, I challenge the minimum number of violators under probabilistic terms. Table 8.7 displays results that I obtained after generating 10,000 cyclic graphs with 34 nodes and 100 randomly assigned edges.

TABLE 8.7

Number of Violators in 10,000 Generated Graphs with 34 nodes and 186 Edges (All) and with 34 Nodes and 100 edges (Partial)

	Before optimization				
All	90.26	115	1	69	3
Partial	49.97	70	1	31	1
	After optimization				
All	41.84	54	1	27	1
Partial	15.18	23	4	5	1
	Mirror-image combinations				
All	17.15	30.5	1	5	1
Partial	4.4	14	2	0	69
	Mean	Biggest	Observed (times)	Smallest	Observed (times)

However, since the alternative matrix has no cases of recursiveness, loops were not included in the generated graphs. The mean number of violators found in this experiment was 49.97, higher than the violator number found in the actual graph (11). Although after optimization the mean violator number found is higher (15.18) than the actual violator number (10), a smaller number (5) was observed once. Thus, a small probability (0.0001) exists that the actual minimum number of violators could arise by chance. More surprisingly, a graph without mirror-image combinations was observed 69 times (0.0069), and the mean mirror-image combinations found in 10,000 graphs is lower (4.4) than the actual mirror-image combinations (5).

The mirror-image combinations are *-one+-etto* and *-etto+-one*, *-ico+-ista* and *-ista+-ico*, *-one+-ata* and *-ata+-one*, *-izzare+-bile* and *-bile+-izzare*, *-ità+-ario*, and *-ario+-ità*. Some of these combinations may not be exceptions to the CBO hypothesis because their bases show high relative frequency values, as in *-itario* (mean relative frequency: 0.909, type parsing ratio: 1) or morphosemantic opacity. Consider *-onata*, as in *pallonata* 'blow with a ball,' where the first suffixed word *pallone* 'ball' is not fully transparent. However, several other mirror-image combinations are problematic. Combinations such as *-icista*, *-onetto*, and, above all, *-izzabile* have many types showing high transparency, both morphotactic and morphosemantic.

The hierarchy (Table 8.8) that is established on the alternative matrix appears different from the hierarchy comprising all the suffix combinations. Intriguingly,

TABLE 8.8

Suffixes with Their Mean Ranks and Productivity values Computed with the Corpus-Variable Approach at Two Different Values of N (19,000 and 50,000). Only Combinations with at Least Three Types Are Considered. All Productivity Values Are Intended *10,000

Affix	Rank	N=19	N=50	Affix	Rank	N=19	N=50
mento	1.442	10.4	4.9	aggio	17.143	1.5	
(i)ano	3.059			ino	17.530		
ico	3.063			ista	20.0	11.3	6.2
ificare	3.085	0.6		izzare	21.7	7.6	3.8
(z)ione	5.493	13.4	5.1	evole	24.571	0.3	
eggiare	5.787	4.1		trice	24.694	10.8	
ario	7.979			[2]eria	28.347		
oso	8.094	3.7	1.6	(t)orio	28.386		
iere	9.880			aggine	28.438		
ese	10.327			ume	28.489		
etto	10.973			(z)a	28.540	0.7	0.3
(t)ore	12.787	9.4	5.0	iera	28.601		
tura	13.297	6.6	3.5	iero	28.671		
ata	13.524			ità	28.745	13.4	6.3
one	13.725			[1]eria	28.754		
bile	16.184	11.3	6.3	ismo	28.834	15.2	8.2
(V)ale	16.628			ezza	28.922	2.7	1.3

some suffix ranks are preserved, namely those in the starting positions of the hierarchy. In both hierarchies, the suffixes *-(i)ano*, *-mento*, and *-(z)ione* occupy quite the same positions. Unsurprisingly, the biggest variation in rankings is observed in suffixes that show a low number of suffix combinations.

Generally speaking, both hierarchies make wrong predictions on both low frequency suffix combinations and combinations showing hundreds of types, namely those that feature verbalizing suffixes *-izzare* and *-eggiare*. These suffixes are ranked after suffixes *-bile*, *-tore*, and *-mento*, but in two cumulated sources (dictionary and corpus), suffix combination *-izzabile* displays 171 types, *-izzatore* shows 187 types, the types of *-eggiamento* total 132, and *-izzazione* is likely the highest frequency combination in my set, counting more than one thousand types. Moreover, CBO hierarchies should also be able to predict noncontiguous combinations, such as forms like BASE-A-X-B. Again, some of the highest frequency combinations violate both hierarchies; for example, the suffix *-(z)ione* is meant to occupy a more inner position than affix *-(V)ale*, but 195 types attest to the suffix combination *-alizzazione*.

As discussed in section 2, one claim of the CBO hypothesis is the correlation between suffix ranks and other values, namely productivity and the parsing ratio. Plag and Baayen (2009) have introduced another type of value, lexical processing data, to correlate suffix rank. As there are no morphological annotated corpora such as CELEX for Italian,[5] parsing ratio values can only be obtained through a careful and time-consuming analysis of complex forms. Moreover, as far as I know, there is no lexical processing database for Italian, such as the English Lexicon Project for the English language. As noted in section 3.3, Gaeta and Ricca (2003, 2006) provide a source of data on affix productivity in Italian, which partially permits correlating the suffix rank with the category-conditioned degree of productivity. Unfortunately, productivity data for several suffixes considered in my set are missing. Due to technical reasons explained in Gaeta and Ricca (2003: note 16), there are no productivity values for *-iere*, *-iero*, or *-iera*. Furthermore, due to the variable-corpus approach employed in Gaeta and Ricca's works, it is not possible to compare all affixes and some data may not be reliable (Gaeta & Ricca 2006), because (i) some suffixes do not display fixed token values (N) and (ii) some suffixes reach the given N quickly. For reason (i), the N value embraced by most suffixes is 50,000. However, some low frequency suffixes, such as *-evole* or *-aggio*, do not reach the threshold. For reason (ii), consider how the suffix *-(z)ione* scores 19,000 tokens in the La Stampa corpus after just 1,300,000 tokens. According to Gaeta and Ricca (2006), one should not take words attested once in such a small subcorpus as hapax legomena.

---

[5] An annotated lexicon of Italian derivatives has been developed at Scuola Normale Superiore, Pisa. It is available at the website http://derivatario.sns.it. See Talamo, Celata, and Bertinetto 2014 for further information.

Given the above-mentioned problems, the suffix ranks are correlated with two orders of productivity: one at N = 19,000 and the other at N = 50,000. The lower value aims to cover low frequency suffixes (e.g., *-trice*, *-evole*, and *-aggio*), while the higher is meant to provide more reliable data for middle and high frequency suffixes (e.g., *-(i)ano*, *-(z)ione*, *-bile*, . . .). The straight line present in each of the four diagrams in Figure 8.6 shows graphically the correlation obtained through a linear regression analysis, in which the dependent variable is represented by the suffix rank and the independent variable by the category-conditioned productivity. If a correlation is found between the two variables, many of the data points in the diagram will be concentrated to the straight line, which is called a 'regression line.' As it is clear from Figure 8.6, there is a lack of correlation in all four of the diagrams. The lack of correlation is also confirmed by the two correlation coefficients *p* and R-Squared, which I have presented in section 3.3.

I begin with the discussion of the hierarchy that assume all 184 combinations (Table 8.6 and the upper panel in Figure 8.6). When the N value is fixed to 19,000, covering the largest number of suffixes (17 out of 34), I observe a *p*-value of 0.136 with only 14% of the variance explained by the regression model (R-Squared = 0.141). When the N value is fixed to 50,000, there is a slightly worse

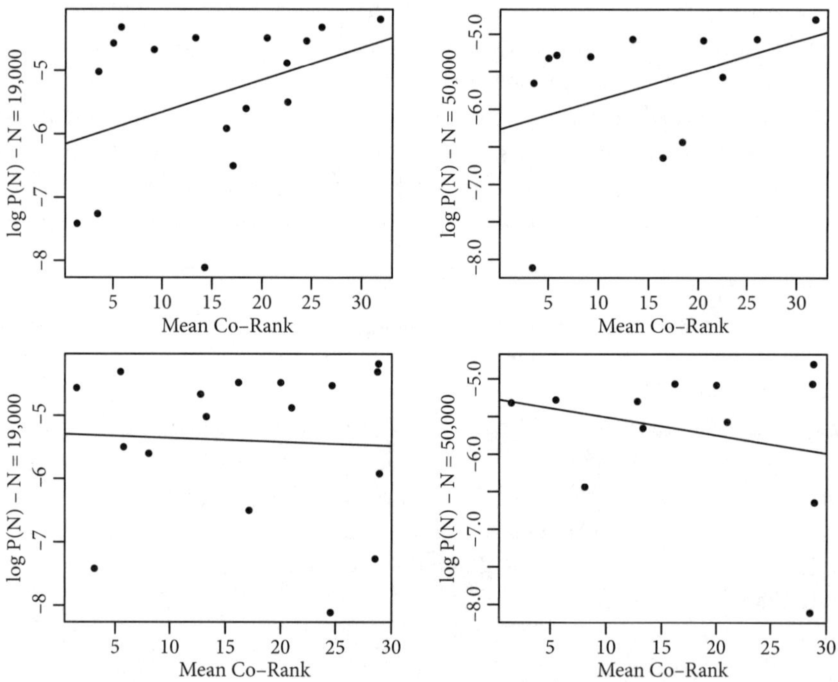

FIGURE 8.6 Log-transformed productivity at N=19,000 (left) and N=50,000 (right) as a function of mean corank obtained with all combinations (upper panels) and combinations with more than two types (lower panels).

degree of correlation, with a p-value of 0.202 and a similar percentage of variance (16%: R-Squared = 0.157).

For the alternative hierarchy (Table 8.8 and Figure 8.6), the correlation is even less significant, with $p = 0.850$ for the first order of productivity (N = 19,000) and $p = 0.433$ for the second (N = 50,000). In both the cases, the proportion of the suffix rank variability that is predicted by the productivity is very low: <1% (R-Squared = 0.002) and 6% (R-Squared = 0.062), respectively.

Finally, there is a third element in the linear regression analysis that deserves a brief discussion: the slope of the regression line. According to Zirkel (2010: 258), if the regression line has a positive slope, this indicates that "the general pattern in the data is that less productive prefixes tend to have lower mean ranks than more productive prefixes." In the correlation between the productivity and the hierarchy assuming all combinations, the slopes are positive: 0.0509 (N = 19,000) and 0.0393 (N = 50,000), whereas in the correlation involving the alternative hierarchy are negative: -0.0062 (N = 19,000) and -0.0239 (N = 50,000). However, since all slopes are very close to zero, there is not much to speculate: these values simply confirm the absence of correlation.

Let us now turn to a comparison with other languages. A measure introduced by Baayen (2010) and known as exception rate is suitable for comparing the CBO hypothesis across different languages. The exception rate is the ratio between violator number and total combination number. Table 8.9 displays the exception rate in different languages and data sets.

The CBO hypothesis shows the lowest exception rates when it is tested on the English language. More precisely, the suffix hierarchies for English are particularly robust. The highest exceptions rate is shown by the earlier study that tested the hypothesis in Italian, especially when using manual reordering, which is employed by Gaeta (2005). The hierarchies discussed in this study show an average exception rate. The table below also shows the probabilistic exception rate; that is, the ratio between the mean violator number observed in the random runs and total

TABLE 8.9

The Exception Rate in Different Languages and Data Sets

	Exception rate	
Language (no. of affixes)	Actual	Probabilistic
English (15 suffixes)	0	?
English (31 suffixes)	0.062	0.439
English (15 prefixes)	0.101	0.449
Bulgarian (22 suffixes)	0.139	?
Italian (16 suffixes, manual reordering)	0.369	?
Italian (16 suffixes, automatic reordering)	0.217	?
Italian (34 suffixes, total)	0.161	0.224
Italian (34 suffixes, partial)	0.110	0.151

combination number. In this paper and in papers challenging the CBO hypothesis in English, the graphs were generated using different models and the violator number is optimized using slightly different algorithms. However, what is important is that the ratio between the actual and probabilistic exception rate is much higher in English than in Italian, when the actual exception rate is very close to the exception rate that arises by chance.

6 Conclusion

The present work expands on earlier studies on CBO in Italian (Gaeta 2005) by doubling the suffix set and employing larger linguistic data sources for the suffix combination search.

It has been shown that selectional restrictions rule out more than one-third of the combinations. Through computational means similar to those used by Plag and Baayen (2009), a suffix hierarchy can be established. Although lower, the number of violators of this hierarchy is close to that which would occur if the suffix were allowed to combine randomly. Moreover, some combinations could not be removed by matrix rearrangement. These are the mirror-image combinations, involving half of the exceptions. Another unavoidable exception is represented by two cases of recursiveness, although these were limited to evaluative suffixes.

As low frequency combinations represent a conspicuous part of the total, an alternative hierarchy was established that discards combinations with two or fewer types. The minimum number of violators found was again close to that which was observed probabilistically, while the mirror-image combinations, amounting to one half of the total, increased. In probabilistic terms, this indicates that a hierarchy can only be partially established for Italian suffixation.

Furthermore, hierarchies obtained in both ways are challenged by exceptions showing hundreds of types, such as *-izzazione* and *-izzabile*. In both hierarchies, the suffix *–izzare* (ranks 22 and 21) is ranked well after the suffixes *-(z)ione* (ranks 6 and 5) and *–bile* (ranks 13 and 16).

Due to the scarcity of data, it turned out to be impossible to establish a thorough correlation to productivity. Regrettably, the productivity values of several Italian suffixes are not available. However, even in their partiality, my results suggest a lack of correlation between the suffix hierarchies and productivity; that is, highly ranked suffixes do not turn out to be more productive than lower ranked suffixes.

In their 2009 paper on CBO, Plag and Baayen surmised that the balance of storage and computation would shift dramatically in favor of parsing in languages with more productive morphology. Their claim was that in languages such as Italian or Russian, "parsing constraints would play a minor role at best" (Plag & Baayen 2009: 146). As there are neither parsing nor lexical decision data available for Italian, this study could not directly verify this claim. However, if an affix hierarchy depends on parsing principles, the restricted role of a parsing constraint in Italian appears to be confirmed by my findings. Finally, a question recently raised (Baayen 2010) is whether affix order is really dependent on parsing or is motivated by other principles. In the latter case, further research has to discover which principles motivate the establishment of partial hierarchies in Italian. Until then, it is safe to assume that Italian suffix combinations are mostly ruled out by grammatical, affix-pertinent constraints; that is, selectional restrictions rather than constraints accounting for a large set of suffixes.

Appendix

Examples of attested suffix combinations; if not indicated otherwise, the suffix combination is attested at least in GRADIT.

Combinations with at least three types: *sentimentaleggiare, naturalezza, fondamentalismo, occidentalista, esistenzialità, commercializzare, americaneggiare, napoletaneria, italianetto* (La Repubblica), *persianino, africanismo, americanista, cristianità, cubanizzare, conservatoriale, conservatorismo, bruciatorista, composizionale, concessionario, abolizionismo, creazionista, arbitraggista, impresariale, glossarietto, autoritarismo, documentarista, abitudinarietà, bancarizzare, camminatina, frittatone, sensibilista, possibilismo, addestrabilità, contabilizzare, borseggiatore, palleggiatorio, tratteggiabile, maneggevole, amoreggiamento, arpeggiatrice, tinteggiatura, milaneseria, giapponesino* (La Repubblica), *piemontesismo, portoghesista, inglesità, cinesizzare, svedesona* (La Repubblica), *spaghettata, cavetteria, barchettina, bozzettismo, librettista, quadrettone, arrendevolezza, anarchicheggiante, burocraticismo, bellicista, cosmicità, economicizzare, giustizierismo, portierone* (La Repubblica), *cementificatore, pacificatorio, diversificazione, falsificabile, saponificatrice, carrettinista, cubisteggiare, ebanisteria, stilnovistico, marxistizzare, pubblicitario, privatizzatore, lottizzatorio, realizzazione, strumentalizzabile, ipnotizzamento, democratizzatrice, comportamentale, medicamentario, sentimentino, tradimentismo, comportamentista, medicamentoso, pallonata, pallonetto, palloncino, aquilonista, spiritosaggine, fumoseria, capricciosetto, virtuosismo, delittuosità, noiosone, scritturale, fritturina, letturista*.

Combinations with less than three types: *radicaleria, aziendalese* (La Repubblica), *venezianaggine, brasilianata* (La Repubblica), *romanico, indianone* (La Repubblica), *traduttorese, motorino, riflettorizzare, giocatorone, contradditorietà, influenzale, influenzetta, scadenziere, vacanziero, vacanzina, vacanzismo, speranzoso, traduzionese, passionevole, spedizioniere, seccagginoso, ingranaggeria* (La Repubblica), *sondaggetto* (La Repubblica), *sondaggismo, giudiziarese* (La Repubblica), *documentarione* (La Repubblica), *volontarioso, limonatiere, mobilezza, osteriale* (La Repubblica), *osterietta* (La Repubblica), *franceseggiare, inglesume, picchettaggio, fascettario, spaghettiera, trombettiere, fumettizzare* (La Repubblica), *polpettoso* (La Repubblica), *macchiettume* (La Repubblica), *sazievolaggine, piacevoleggiare, colpevolismo, colpevolista, consapevolizzare, sciocchezzina* (La Repubblica), *altezzoso, angelicale, arcadicheria, burocraticume* (La Repubblica), *camerierina* (La Repubblica), *pompieraggio* (La Repubblica), *carabinierese* (La Repubblica), *ragionieretto* (La Repubblica), *carabinierino* (La Repubblica), *ragionierità* (La Repubblica), *salottierismo, rettificatura, cestineria, violinino* (La Repubblica), *paninone* (La Repubblica), *apprendistale, vocabolistario, giornalistese* (La Repubblica), *capitalistone, ereditiera, veritiero, felicitizzante, dignitoso, marmorizzatura, falconeggiare, falconiera, banconiere, maglioncione* (La Repubblica), *spiritosino* (La Repubblica), *virtuosista* (La Repubblica), *tifosume* (La Repubblica), *misturaggine, scritturabile, caricatureggiare, sculturetta* (La Repubblica), *fatturiera, calzaturiere, calzaturiero, caricaturismo, miniaturizzare, cinturone.*

References

Aronoff, Mark, and Nanna Fuhrop. 2002. "Restricting Suffix Combinations in German and English: Closing Suffixes and the Monosuffix Constraint." *Natural Language and Linguistics Theory* 20:451–490.

Baayen, Harald R. 1992. "Quantitative Aspects of Morphological Productivity." In *Yearbook of Morphology 1991*, edited by Geert Booij and Jaap Van Marle, 109–149. Dordrecht: Kluwer.

Baayen, Harald R. 2008. *Analyzing Linguistic Data. A Practical Introduction to Statistics.* Cambridge: Cambridge University Press.

Baayen, Harald R. 2009. "Corpus Linguistics in Morphology: Morphological productivity." In *Corpus Linguistics. An International Handbook*, edited by Anke Lüdeling and Meria Kytö, 900–919. Berlin: De Gruyter Mouton.

Baayen, Haarald R. 2010. "The Directed Compound Graph of English. An Exploration of Lexical Connectivity and Its Processing Consequences." In *New Impulses in Word-formation (Linguistische Berichte Sonderheft 17)*, edited by Susan Olsen, 383–402. Hamburg: Buske.

Baroni, Marco, Silvia Bernardini, Federica Comastri, Lorenzo Piccioni, Guy Aston, and Marco Mazzoleni. 2004. "Introducing the La Repubblica Corpus: A Large, Annotated, TEI(XML)-Compliant Corpus of Newspaper Italian." In *Proceedings of LREC 2004*, 1771–1774. Lisbon: European Language Resource Association.

Bisetto, Antonietta. 1995. "Il suffisso –tore." *Quaderni Patavini di Linguistica* 14:39–71.

Christ, Oliver. 1994. "A Modular and Flexible Architecture for an Integrated Corpus Query System." In *Proceedings of COMPLEX'94. 3rd Conference on Computational Lexicography and Text Research (Budapest, 7–10/7/1994)*, 23–32.

Csárdi, Gábor, and Tamás Nepusz. 2006. "The Igraph Software Package for Complex Network Research." *InterJournal*, Complex Systems, 1695.

Dressler, Wolfgang. U. 1985. "On the Predictiveness of Natural Morphology." *Journal of Linguistics* 21(2):321–337.

Dressler, Wolfgang. U. 2005. "Word-formation in Natural Morphology." In *Handbook of Word-Formation (Studies in Natural Language and Linguistics Theory, volume 64)*, edited by Pavol Štichauer and Rochelle Lieber, 267–284. Dordrecht: Springer.

Dressler, Wolfgang U., and Katarzyna Dziubalska-Kołaczyk. 2006. "Proposing Morphonotactics." *Italian Journal of Linguistics/Rivista di linguistica* 18(2):249–266.

Fabb, Nigel. 1988. "English Suffixation is Constrained Only by Selectional Restrictions." *Natural Language and Linguistics Theory* 6(4):527–539.

Gaeta, Livio. 2002. *Quando i verbi compaiono come nomi: un saggio di Morfologia Naturale*. Milano: Franco Angeli.

Gaeta, Livio. 2004. "Nomi d'azione." In *La formazione delle parole in italiano*, edited by Maria Grossmann and Franz Rainer, 314–349. Tübingen: Niemeyer.

Gaeta, Livio. 2005. "Combinazioni di suffissi in italiano. *La formazione delle parole." In Atti del XXXVII Congresso Internazionale di Studi della Società di Linguistica Italiana (L'Aquila, 25–27/9/2003)*, edited by Maria Grossmann and Anna Maria Thornton, 229–247. Roma: Bulzoni.

Gaeta, Livio, and Davide Ricca. 2002. "Corpora testuali e produttività morfologica:i nomi d'azione italiani in due annate de La Stampa." In *Parallela IX. Testo variazione informatica/Text Variation Informatik. Atti del IX Incontro italo-austriaco dei linguisti*, edited by Roland Bauer and Hans Goebl, 223–249. Wilhelmsfeld: Egert.

Gaeta, Livio, and Davide Ricca. 2003. "Frequency and Productivity in Italian Derivation: A Comparison between Corpus-based and Lexicographical Data." *Italian Journal of Linguistics/Rivista di linguistica* 15(1):63–98.

Gaeta, Livio, and Davide Ricca. 2006. "Productivity in Italian Word Formation: A Variable-corpus Approach." *Linguistics* 44(1):57–91.

GRADIT (2003). *Grande dizionario italiano dell'uso, edizione su CD-ROM. Ideato e diretto da Tullio De Mauro*, Torino: Unione Tipografico-Editrice Torinese.

Grossmann, Maria, and Franz Rainer, eds. 2004. *La formazione delle parole in italiano*. Tübingen: Niemeyer.

Hay, Jennifer. 2002. "From Speech Perception to Morphology: Affix Ordering Revisited." *Language* 78(3):527–555.

Hay, Jennifer, and Haarald R. Baayen. 2002. "Parsing and Productivity." *Yearbook of Morphology 2001*, edited by Geert Booij and Jaap Van Marle, 203–255. Dordrecht: Kluwer.

Hay, Jennifer, and Ingo Plag. 2004. "What Constrains Possible Suffix Combinations? On the Interaction of Grammatical and Processing Restrictions in Derivational Morphology." *Natural Language and Linguistics Theory* 22(3):565–596.

Manova, Stela. 2010. "Suffix Combinations in Bulgarian: Parsability and Hierarchy-based Ordering." *Morphology* 20(1):267–296.

Manova, Stela, and Mark Aronoff. 2010. "Modeling Affix Order." *Morphology* 20(1):109–131.

Merlini Barbaresi, Lavinia and Wolfgang U. Dressler. 1994. *Morphopragmatics: Diminutives and Intensifiers in Italian, German, and Other Languages*. Berlin: De Gruyter Mouton.

Nespor, Marina. 1993. *Fonologia*. Bologna: il Mulino.

Plag, Ingo. 1998. "The Polisemy of *-ize* Derivatives: On the Role of Semantics in Word Formation." *Yearbook of Morphology 1997*, edited by Geert Booij and Jaap Van Marle, 219–242. Dordrecht: Kluwer.

Plag, Ingo. 1999. *Morphological Productivity: Structural Constraints in English Derivation*. Berlin: De Gruyter Mouton.

Plag, Ingo, and Haarald R. Baayen. 2009. "Suffix Ordering and Morphological Processing." *Language* 85(1):109–152.

Saciuk, Bohdan. 1969. "The Stratal Division of the Lexicon." *Papers in Linguistics* 1(3):464–532.

Scalise, Sergio. 1994. *Morfologia*. Bologna: il Mulino.

Spencer, Andrew. 1991. *Morphological Theory. An Introduction to Word Structure in Generative Grammar*. Oxford: Blackwell.

Štichauer, Pavol. 2009. "Morphological Productivity in Diachrony: The Case of the Deverbal Nouns in -mento, -zione and -gione in Old Italian from the 13th to the 16th Century." In *Selected proceedings of the 6th Décembrettes: Morphology in Bordeaux*, edited by Fabio Montermini, Gilles Boyé, and Jesse Tseng, 138–147. Somerville, MA: Cascadilla Proceedings Project.

Talamo, Luigi, and Chiara Celata. 2011. "Toward a Morphological Analysis of the Italian Lexicon: Developing Tools for a Corpus-based Approach." *Quaderni del Laboratorio di Linguistica* 11(1).

Talamo, Luigi, and Chiara Celata, and Pier Marco Bertinetto. 2014. "derIvaTario: An Annotated Lexicon of Italian Derivatives." Submitted.

Zirkel, Linda. 2010. "Prefix Combinations in English." *Morphology* 20(1):239–266.

Zirkel Hilkenbach, Linda. 2011. "Can Complexity-based Ordering be Extended from English to German?" Paper presented at the 3rd workshop on Affix Order. Vienna, January 14–16.

9

Affix Order and the Structure of the Slavic Word

Stela Manova

1. Introduction

This article investigates the structural properties of the Slavic word in terms of affix ordering in three Slavic languages, the South Slavic Bulgarian, the East Slavic Russian, and the West Slavic Polish and thus covers all three subgroups of the Slavic family.[1] The discussion is with a focus on suffixation, in particular on suffixation in derivation.[2]

Recently much research has been carried out on affix ordering in lesser-studied languages; see the overviews in Manova and Aronoff (2010) and in Rice (2011). There has been much research on the ordering of the English derivational affixes as well, especially on the order of the suffixes, and a number of specific proposals have been formulated (in chronological order): level ordering or stratal approach (Siegel 1974; Allen 1978; Selkirk 1982; Kiparsky 1982, Mohanan 1986; Giegerich 1999); selectional restrictions (Fabb 1988; Plag 1996, 1999); the monosuffix constraint (Aronoff and Fuhrhop 2002), and the parsability hypothesis (Hay 2001, 2002, 2003) or complexity-based ordering (Plag 2002; Hay and Plag 2004; Plag and Baayen 2009). In this list of approaches, every following approach was formulated in response to its predecessor; that is, every following approach demonstrates that the predecessor

---

[1]The author was supported by the Austrian Science Fund (FWF), grant V64-G03, and the European Science Foundation (ESF), NetWordS-09-RNP-089 / Individual Grant 5566. Portions of this study were presented at the Fifth Annual Meeting of the Slavic Linguistics Society, Chicago, October 2010; the Linguistic Seminars of the Universitat Autònoma de Barcelona, December 2011; the University of Sofia, March 2013; the Scuola Normale Superiore di Pisa, May 2013; as well as within the CogSci Talk Series of the Research Platform Cognitive Science, University of Vienna, June 2013. I am grateful to the audiences of all these events for stimulating discussion. I am particularly grateful to A. Bagasheva, P. M. Bertinetto, W. U. Dressler, V. Pirrelli, and L. Talamo.

[2]Abbreviations used in the text: ADJ–adjective, ASP–aspect, C–consonant, DEF–definite, DIM–diminutive, FEM–feminine, GEN–genitive, GEND–gender, MASC–masculine, N–noun, NEUT–neuter, NUM–number, PER–person, PL–plural, PREF–prefix, SG–singular, SUFF–suffix, TM–thematic marker, TNS–tense, V–verb.

approach makes incorrect predictions. Therefore in the present article we will pay attention only to the most recent proposal, that of complexity-based ordering. Despite the many recent studies on affix ordering in various languages, the ordering of the Slavic affixes, especially that of the derivational suffixes, has been neglected so far. The goal of the present paper is therefore to fill this gap. Unlike the derivational suffixes, the Slavic verbal prefixes have been subject of intensive research recently (Svenonius 2004; Tatevosov 2008, among others), which is maybe due to the following facts: prefixes are much less numerous than the suffixes; attach primarily to verbs to mark the perfective aspect, while the suffixes derive nouns, adjectives, and verbs; and prefixes are easier to identify in dictionaries and corpora because of their initial position in the word form, while the derivational suffixes are often followed by inflection. So far, the ordering of the Slavic prefixes has been explained as primarily due to syntactic and semantic factors (see section 6.4). Nevertheless, the most recent approach to affix ordering in English, the complexity-based ordering, has been tested against data from suffixation in Slavic. In a paper on the order of the Bulgarian derivational and inflectional suffixes Manova (2010) shows that the complexity-based ordering cannot account for the order of the Bulgarian derivational suffixes, and in Manova (2011b) she makes a novel proposal, the so-called cognitive approach. Manova's (2011b) approach to affix order differs from the other proposals in the literature as it is based on the following two assumptions: (1) affixes are not organized in the same way in the whole word, but the latter has different domains with respect to affixation; and (2) suffix order is best analyzed in terms of binary combinations of suffixes of the type SUFF1-SUFF2 where SUFF1 and SUFF2 are any two neighboring suffixes in a word (sub)domain. This article provides novel data, from Russian, in support of Manova's (2011b) proposal. The paper is with a focus on derivational suffixation and for the ordering of the nonevaluative derivational suffixes, following Manova (2011b) it is claimed that the lexical-category specification of the suffix (whether a noun (N), an adjective (ADJ), or a verb (V)), the suffix-particular semantics (whether the suffix derives persons, objects, etc.), and the notion of default are the factors responsible for suffix combinability. The lexical- and semantic-category specifications of a suffix are seen as being cognitive in nature. As the majority of the suffix combinations under scrutiny in this paper are either fixed or predictable, it is claimed that they are (most probably) rote learned and represent instances of entrenched morphological structure in the sense of cognitive grammar. To gain a typological perspective, the paper also compares the structure of the Slavic word with that of the English word.

The article is organized as follows. Section 2 introduces the basic theoretical assumptions of the analysis. Section 3 lists and explains the sources of data used. Section 4 discusses the structure of the Slavic word. Section 5 briefly presents the most recent approach to affix ordering in English, the complexity-based ordering, and shows with data from Bulgarian, Russian, and Polish why this approach fails to account for affix ordering in Slavic. Section 6 defines the affix-order domains in the structure of the Slavic word and illustrates them with numerous examples from the three Slavic languages under scrutiny. Section 7 accommodates the discussion, and in section 8 conclusions are drawn.

2. Theoretical Assumptions

The theoretical framework of this study mixes principles and assumptions from cognitive grammar (Langacker 1987, 1991; Taylor 2002; and Geeraerts 2006), including resent research in cognitive neuroscience (Mestres-Missé et al. 2010, and the references therein), and Natural morphology (Dressler et al. 1987; Dressler 2005).

With cognitive grammar, it is assumed that there are only three objects of study: language in its perceptible form, symbolized content, and symbolic associations between phonological and semantic structures. Thus grammar is an inventory of units (phonological, semantic, and symbolic structure) that has been established or entrenched in the speaker's mind through frequency of previous use (Taylor 2002). We speak of entrenchment if an affix combination does not need to be assembled (compositionally) from its parts on each occasion of its use. Following cognitive grammar, this study is also strictly usage-based; that is, bottom-up, and the analysis takes place from data to theory.

The lexical category specification of a suffix plays a pertinent role in the analysis, and we define the major lexical categories in the spirit of cognitive grammar; that is, with Langacker (1987) and Croft (2001). Langacker (1987), based on *relationality* (i.e., +/– relational) and *way of scanning* (whether summarily scanned, i.e., conceived statistically and holistically, or sequentially scanned, i.e., mentally scanned through time), recognizes *things* (N), *processes* (V), and *modifiers* (ADJ). Croft (2001) defines objects, properties, and actions in terms of four semantic properties: *relationality, stativity, transitoriness*, and *gradability*. Thus prototypically, nouns name things or objects, verbs denote processes or actions, and adjectives are modifiers and express properties.

Additionally, we see research in cognitive neuroscience showing that nouns and verbs have different representations in the brain (Mestres-Missé et al. 2010, among many others) as evidence for the correctness of an analysis that pays attention to the lexical category specification of an affix.

As for the cognitive nature of the semantic categories used in the analysis, following cognitive semantics (e.g., Fillmore's 1982 frame semantics) and conceptual semantics (Jackendoff 1990), we assume that there is no principle difference between meaning and conceptualization.

We also refer to Natural morphology, a semiotic and cognitively oriented theory of morphology compatible with cognitive grammar (Dressler 1990). According to the naturalness parameter of iconicity (constructional diagrammaticity), there are different types of affixation (see also Manova 2011a): affixation by addition in which addition of meaning is reflected by addition of form, and the less iconic affixation by substitution (truncation in Aronoff 1976). The English derivation *beauti-ful* → *beautiful-ness* and its Russian equivalent *kras-ivyj* → *krasiv-ost'* (*-yj* is inflection) are examples of affixation by addition, while the derivation *Marx-ism* → *Marx-ist* is an instance of affixation by substitution. Although *beautifulness/krasivost'* and *Marxist* are analyzable as compositional units, only affixation by addition involves affix ordering. Thus in this study we will always control how two suffixes interact, and make a clear distinction between addition and substitution (see Table 9.4 in 6.2).

3. Data

As already mentioned in the previous section, this study is usage-based and thus data are of particular importance for the analysis. The data cited herein were collected for the project *(De)composing the Slavic Word* (grant V64-G03 from the Austrian Science Fund-FWF) and come from various sources such as dictionaries, corpora, and the Internet. The collected material was also checked by native speakers. The role of the native speakers was to ensure the usage-based character of the study; that is, to consider suffix combinations that are really in use. The Internet was particularly helpful for the verification of the existing combinations of the diminutive suffixes.

The following dictionaries and corpora are the major sources of data for this study. For Bulgarian: Stankov (2002), *New Spelling Dictionary of Modern Bulgarian*; Andrejčin (1975), *Reverse Dictionary of Modern Bulgarian*, and *the Bulgarian National Corpus*. For Russian: Kuznecova and Efremova (1986), *A Morpheme Dictionary of Russian*; electronic version of Zaliznjak (1977), *A Grammatical Dictionary of Russian*, the *Yandex Dictionaries*, as well as *The Russian National Corpus*. For Polish: Doroszewski (1997), electronic version of *Słownik Języka Polskiego* (SJP); Dubisz (2008), electronic version of *Uniwersalny Słownik Języka Polskiego* (USJP); Saloni (2007), electronic version of *Słownik Gramatyczny Języka Polskiego* (SGJP); Jadacka (2001), *Słownik gniazd słowotwórczych współczesnego języka polskiego* (SGSWJP); and *the Polish National Corpus*.

4. The Structure of the Slavic Word

According to the traditional morphological typology (Skalička 1979), the Slavic languages represent the inflecting-fusional type and make a clear distinction between derivational and inflectional suffix slots. This is best visible in the inflectional paradigms of derived words and is illustrated in Table 9.1 with the paradigm of the Russian noun *učitel'* 'teacher': the latter contains the derivational suffix *-tel'* written in bold italic. The suffixes that follow -***tel'*** are inflectional.

Based on the principle of constructional iconicity (diagrammaticity) mentioned in section 2 and the notion of prototype;—that is, that the easiest way of accessing a phenomenon is via that manifestation of it that is most salient (cf. Langacker 1987)—I postulate the generalized form of the structure of the Slavic word in (1) (cf. Manova 2005, 2010):

TABLE 9.1

Declension of R. *učitel'* 'teacher'

NUMBER CASE	SG	PL
NOM	*uči-**tel**'*	*uči-**tel**-i*
ACC	*uči-**tel**-ja*	*uči-**tel**-ej*
GEN	*uči-**tel**-ja*	*uči-**tel**-ej*
DAT	*uči-**tel**-ju*	*uči-**tel**-jam*
INSTR	*uči-**tel**-em*	*uči-**tel**-jami*
LOC	*uči-**tel**-e*	*uči-**tel**-jax*

(1) (PREFIX)-BASE-(DERIVATIONAL SUFF)-(THEMATIC MARKER)-(INFLECTIONAL SUFF)

As indicated by the brackets, the slot BASE is always occupied (in Table 9.1, the base is *uči-*), whereas the other slots may be empty; for example, the prefix slot of the noun *učitel'* in Table 9.1 is empty. Additionally, the BASE can be a root, a stem, or a word: *uči-* in is a stem (consists of the root *uč-* and the suffix *-i-* that is a TM); see the explanations in Manova (2011a). Prototypically, derivation takes place in the derivational slot of the word, whereas inflection operates in the inflectional slot. Thematic markers are assumed only in verbal morphology, where they have inflectional status (cf. Manova 2005, 2011a). Crucially, the different word slots can be occupied by more than one affix. Table 9.2 lists the inflectional forms of the adjective *uči-**tel'**-**sk**-ij* that has two derivational suffixes, namely ***-tel'-sk-***.

Since in Bulgarian, in contrast to Russian (and Polish), there is no category of case but definiteness is morphologically marked by suffixes, the definite form of the Bulgarian equivalent of the Russian *uči-tel'-sk-ij* in (2) contains not only more than one derivational suffix but also two inflectional suffixes:

(2) Bg. *učitelski* 'teacher's'
MASC SG *uči-**tel-sk**-i* (like in Russian, two derivational suffixes, *-tel* and *-sk-*)
MASC SG DEF *uči-tel-sk-**i-jat*** (two inflectional suffixes)
PL *uči-tel-sk-i* (is syncretic with the MASC SG, but there is no gender in the PL)
PL DEF *uči-tel-sk-**i-te*** (two inflectional suffixes)

TABLE 9.2

Declension of R. *učitel'skij* 'teacher's' (Derivational Suffixes *-tel'* and *-sk-*)

GENNUM CASE	MASC SG	PL
NOM	*uči-**tel'-sk**-ij*	*uči-**tel'-sk**-ie*
ACC	*uči-**tel'-sk**-ij /-ogo*	*uči-**tel'-sk**-ie /-ix*
GEN	*uči-**tel'-sk**-ogo*	*uči-**tel'-sk**-ix*
DAT	*uči-**tel'-sk**-omu*	*uči-**tel'-sk**-im*
INSTR	*uči-**tel'-sk**-im*	*uči-**tel'-sk**-imi*
LOC	*uči-**tel'-sk**-om*	*uči-**tel'-sk**-ix*

The examples in (3) illustrate the use of more than one prefix:

(3) Russian
   a. ***pre-pod**-nesti* 'to present' (from *nesti* 'to carry' → *pod-nesti* 'to serve')
   b. ***pere-pere**-delat'* 'to remake again' (from *delat'* 'to make' → *pere-delat'* 'to remake')
   c. ***pere-pere**-delka* 'sth that has been remade twice'

The noun *pereperedelka* in (3c) is derived from the verb *pereperedelat'* in (3b) and was found in the Russian National Corpus and on the Internet.

With respect to multiple prefixation, Bulgarian and Polish behave like Russian, and therefore I do not give Bulgarian and Polish examples here.

5. Complexity-Based Ordering (CBO) and Why It Fails to Account for Slavic Data

In this section, I briefly review the most recent proposal in the literature on affix ordering, the so-called complexity-based ordering (CBO), and explain why it cannot account for the structure of the Slavic word.

CBO is based on the so-called parsability hypothesis (Hay 2001, 2002, 2003). According to Hay's proposal, affixes order in such a way that more parsable affixes always follow less parsable ones in the word form because this order is easier to process. Parsability correlates with a number of factors such as productivity, semantic transparency, phonotactics across the morpheme boundary, and relative frequency. It occurs by gradations; parsability can be measured, and affixes can be ordered hierarchically according to their ability to parse. The parsabilty hypothesis assumes a double route access of morphologically complex words—that is, a derived word can be accessed either as a whole unit (whole-word route) or as a decomposable unit (decomposition route). Which route is preferred depends on relative frequency. If the derivative is more frequent than its base, the relative frequency is high and the whole-word route is more probable, as in *whiteness* versus *white*. If the base is more frequent than the derivative, the relative frequency is low and the decomposition route is more probable, as in *green* versus *greenness*. As parsable affixes add structure, affix ordering based on parsability has been termed complexity-based ordering (CBO). The label indicates that complexity increases from the innermost to the outermost affix. There has been much research on affix ordering in CBO: see Plag (2002), Hay and Plag (2004), Plag and Baayen (2009), and Zirkel (2010), to mention just a few articles. CBO also relies on a hierarchy, which can be illustrated with the following hypothetical example. If the suffixes A, B, C, D, and E form a hierarchy, they occur in a complex word in such a way that if we attach, let us say, C to a base, then we can attach only suffixes that follow C on the hierarchy; that is, the combination CDE is possible while the combination *CAD is not. This is how CBO predicts and restricts the order of the English suffixes.

Manova (2010) applies CBO to data from Bulgarian and shows that it cannot account for the order of the derivational suffixes in this language. In Manova's (2010) paper, the ordering of 12 out of the 22 derivative suffixes discussed is incompatible with CBO. The major problems for CBO pose permutations and repetitions of suffixes, as in (4) through (17) below.

(I) AB - BA order of affixes

Bulgarian

(4) ***-(l)iv & -ost***
   a. *mil-ost* 'mercy' → *mil-ost-iv* 'merciful'
   b. *săn-liv* 'sleepy' → *săn-liv-ost* 'sleepiness'

(5) ***-en & -ota***
   a. *măč-en* 'difficult' → *măč-n-ota* 'difficultness'
   b. *sam-ota* 'loneliness' → *sam-ot-en* 'lonely

Russian

(6) ***-ost' & -(l)ivyj***
   a. *mil-ost'* 'mercy' → *mil-ost-ivyj* 'merciful'
   b. *son-livyj* 'sleepy' → *son-liv-ost'* 'sleepiness'

(7) ***-ota & -nyj***
   a. *dobr-ota* 'goodness' → *dobr-ot-nyj* 'good'
   b. *tem-n-y* 'dark' → *tem-n-ota* 'darkness'

Polish

(8) ***-ość & -iwy/-ywy***
   a. *lit-ość* 'mercy' → *lit-ośc-iwy* 'mercyful'
   b. *robacz-ywy* 'wormy' → *robacz-yw-ość* 'wormyness'

(9) ***-ny & -ota***
   a. *dusz-ny* 'stuffy' → *dusz-n-ota* 'stuffiness'
   b. *rob-ota* 'work' → *rob-ot-ny* 'hard-working'

(II) ABA order of affixes

(10) Bulgarian
   a. *revn-iv* 'jealous' → *revn-ost'* 'jealousy'→ *revn-ost-en* 'devoted' → *revn-**ost-n-ost**'* 'devotedness'
   b. *verojaten* 'probable' → *verojatn-ost* 'probability' → *verojatn-ost-en* 'related to probability' → *verojatn-**ost-n-ost*** '(greater) probability'

(11) Russian
   a. *revn-iv-yj* 'jealous' → *revn-ost'* 'jealousy'→ *revn-ost-n-yj* 'devoted' → *revn-**ost-n-ost**'* 'devotedness'

b. *verojatn-yj* 'probable' → *verojatn-ost'* 'probability' → *verojatn-ost-n-yj* 'related to probability' → *verojatn-**ost-n-ost**'* '(greater) probability'

(III) ABAB order of affixes

Bulgarian

(12) *lice* 'face' → *lič-en* 'personal' → *lič-n-ost* 'person, personality' → *lič-n-ost-en* 'related to personality' → *lič-**n-ost-n-ost*** '(greater) personality'

Russian

(13) *lico* 'face' → *lič-n-yj* 'personal' → *lič-n-ost'* 'person, personality' → *lič-n-ost-n-yj* 'related to personality' → *lič-**n-ost-n-ost'*** '(greater) personality'

Examples such as these in (II) and (III) are less typical of Polish.

(IV) AA order of affixes in double diminutives

Bulgarian

(14) *kuče* 'dog' → DIM1 *kuč-ence* → DIM2 *kuč-**enc-ence***

Russian

(15) *kartina* 'picture' → DIM1 *kartin-ka* 'small picture'
→ DIM2 *kartin -**oč-ka*** 'very small picture'

Polish

(16) *dom* 'house' → DIM1 *dom-ek* 'small house'
→ DIM2 *dom-**ecz-ek*** 'very small house'

(V) AAA order of affixes in triple diminutives

(17) Bulgarian

dete 'child' → DIM1 *det-ence*
→ DIM2 *det-enc-ence*
→ DIM3 *det-**enc-enc-ence***

Triple diminutives are less typical of Russian and Polish.

As already mentioned, examples such as these in (4) through (17) pose a real challenge to CBO as the latter is a hierarchy-based theory; that is, in CBO a suffix is in a fixed relationship to all other suffixes in the CBO hierarchy and can either precede or follow another suffix, but not both. Additionally, CBO allows a suffix to be accessed only once in the derivation of a word i.e., a suffix cannot be repeated. Problems with CBO due to affix permutations have been reported for English prefixes (Zirkel 2010) and for Italian (Talamo, this volume) and German derivational suffixes (Zirkel-Hilkenbach 2011). Therefore, in this paper I do not follow CBO but Manova's (2011b) cognitive approach.

6. Affix Order Domains in the Slavic Word

In section 4 we postulated a generalized form of the Slavic word (1) and illustrated that in Slavic the word can exhibit more than one derivational and more than one inflectional suffix. In (IV) and (V) in the previous section, we demonstrated that diminutive suffixes could be repeated. Thus, the schema in (18) below is a version of (1), and, through the separation of the derivational slot into a nonevaluative and evaluative subslots, it accommodates the fact that the Slavic word can have more than one diminutive suffix. In (18), every slot and subslot that can host more than one affix is associated with more than one arrow; that is, a single arrow means that within a word, only one single affix can occur in that slot; two arrows stand for two (types of) affixes; and three arrows mean that more than two affixes can co-occur in a particular slot. I refer to the slots that can host more than one suffix as word domains. I distinguish prefixational domain and suffixational domain, the latter with derivational and inflectional domains, and nonevaluative and evaluative subdomains within the derivational domain. In 6.1 through 6.3, I illustrate the order of the suffixes in the different domains, showing that (i) each domain has its own specific ordering, and (ii) fixed binary combinations of the type SIFF1-SUFF2, where SUFF1 and SUFF2 are any two neighboring suffixes in a domain, play an important role in affix ordering.

(18) Affix order domains in the Slavic word

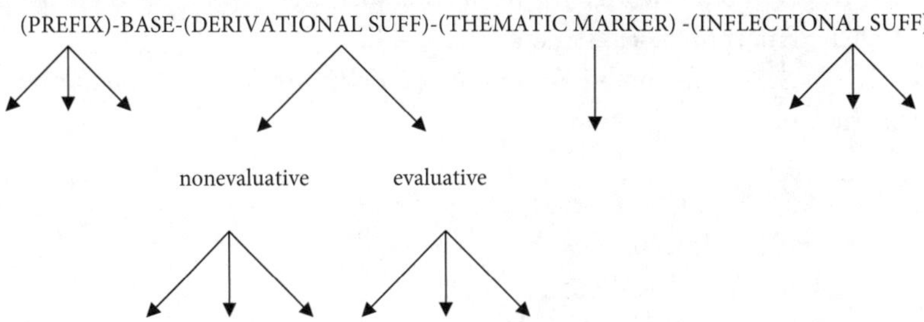

6.1. INFLECTIONAL DOMAIN

In this subsection, the order of the inflectional suffixes is illustrated with data from Bulgarian, as the other two languages under scrutiny in this paper, Russian and Polish, use a single inflectional suffix in noun and adjective morphology.

As regards the ordering of the inflectional suffixes in Bulgarian nouns, adjectives, and verbs, Manova (2010) shows that it is governed by phonological factors and is consonant with the principle of relevance (Bybee 1985) that may be interpreted as being cognitive in nature; see Muysken (1986). However, for the present study, it is more important that the order of the suffixes is templatic and can be described in terms of fixed (i.e., entrenched) binary combinations.

Actually, Bulgarian nouns and adjectives cannot have more than two inflectional suffixes, as the templates of the noun (19) and adjective inflection (20) indicates:
Bulgarian noun inflection:

(19) BASE-NUM-DEF
a. *măž-ø-ø* 'man'
măž -ø-a (for objects) & *măž-ø-ăt* (for subjects and predicatives) 'man-DEF'
măž-e-ø 'man-PL'
măž-e-te 'man-PL-DEF'
b. *žen-a-ø* 'woman-SG'
žen-a-ta 'woman-SG-DEF'
žen-i-ø 'woman-PL'
žen-i-te 'woman-PL-DEF'
c. *sel-o-ø* 'village-SG'
sel-o-to 'village-SG-DEF'
sel-a-ø 'village-PL'
sel-a-ta 'village-PL-DEF'

As only one single combination exists, namely -NUM-DEF, it is a fixed combination; native speakers use it every time when they express -NUM-DEF, which implies that they should know it by heart and most probably access it as an inseparable whole. In other words, the fixed combination -NUM-DEF is an example of entrenched morphological structure.

In contrast to nouns, adjectives mark GEND and NUM cumulatively:
Bulgarian adjective inflection:

(20) BASE–GEND/NUM–DEF
a. *mil-ø-ø* 'dear' (masculine)
mil-ø-ijat[3] 'dear-DEF'
b. *mil-a-ø* 'dear-FEM/SG'
mil-a-ta 'dear-FEM/SG-DEF'
c. *mil-o-ø* 'dear-NEUT/SG'
mil-o-to 'dear-NEUT/SG-DEF'
d. *mil-i-ø* 'dear-PL'
mil-i-te 'dear-PL-DEF'

---

[3]The suffix *-ijat* attaches as a single morpheme, and its initial *-i* is not the same as the *-i* that marks plural in (20d). The *-i* in *-ijat* is inherited from the long form of the Old Bulgarian adjectives. Modern Bulgarian does not distinguish between long and short adjectives, and although diachronically *-ijat* consist of two suffixes (*-i-* that made the adjective long, and *-jat* for definiteness), synchronically *-ijat* is a single suffix.

In adjective inflection, –GEND/NUM–DEF is a fixed (i.e., entrenched) combination that should be rote-learned by speakers.

In verb inflection, more than two suffixes are possible. The template for verbs is given in (21). Since Bulgarian verbs are seldom derived, the BASE coincides with the verb root by default. TNS/PER/NUM are cumulatively expressed by a single suffix.

(21) PREF–BASE–ASP–TM–TNS/PER/NUM

Crucially, the three suffixes that may follow the base participate in two fixed binary combinations: (i) aspectual suffixes, if present, are always followed by the thematic marker *-a-*; that is, the combination –ASP–TM– is always of the type -ASP-*a*-, as in *na-pis-**v-a**-m* '(I) write,' and (ii) as it is typical of inflectional morphology, thematic markers (the suffixes in the TM slot) govern the selection of the TNS/PER/NUM suffixes. In other words, *-a-*, whether preceded by an ASP-suffix or not, is always followed by the same TNS/PER/NUM suffix in 1 SG PRES; that is, by *-m*, as in *na-pis-v-**a-m*** and *slag-**a-m*** 'to put.' Such verbs are even called *am*-verbs in some descriptions of Bulgarian, as all verbs terminating in *-am* in 1 SG PRES have a conjugation pattern of their own. As the Bulgarian verb inflection is the most complex among Slavic and for the purposes of this study it is not necessary to give the full inflectional paradigm, for additional examples that illustrate the template in (21) I refer the curious reader to Manova (2008, 2010, 2011a), where the paradigm of the Bulgarian verbs is explained from different perspectives. Thus in verbal morphology we have two fixed combinations, -ASP-TM- and -TM-TNS/PER/NUM that should be learned by heart; that is, they are instances of entrenched morphological structure.

In sum, the order of the inflectional suffixes is fixed and can be described in terms of fixed binary combinations of suffixes that appear to be instances of entrenched morphological structure. As can be seen from the templates in (19) through (21), inflectional suffixes cannot be repeated. Thus, (i) templatic order in terms of entrenched binary combinations of suffixes and (ii) no repetition of suffixes define the inflectional domain. As we will see in the next subsection, the affixes in the evaluative domain are characterized by other properties.

6.2. EVALUATIVE DOMAIN

This subsection discusses the order of the diminutive suffixes in Bulgarian, Russian, and Polish double diminutives, as well as in triple diminutives in Bulgarian. Triple diminutives are less typical of Russian and Polish. Augmentative suffixes that are also evaluative suffixes are not considered, as they do not combine with each other.[4] Intriguingly, although all diminutive suffixes have the same semantics,

[4]A combination of an augmentative and a diminutive suffix is possible but extremely rare, and not all augmentative suffixes can be followed by a diminutive suffix. I therefore see nouns that exhibit a combination of an augmentative and a diminutive suffix as rote-learned, and do not discuss them herein.

TABLE 9.3

Combinability of the DIM Suffixes in Bulgarian (cf. Manova and Winternitz 2011; Manova, in press)

Nouns in	DIM1 suffixes	DIM2 suffixes	DIM3 suffixes
in -*C*	-*ec* (unproductive)		
	-*le* (unproductive) -*če*	-*ence*	-*ence*
	-*čica* (unproductive)		
in -*a*	-*ica*	-*ka*	
	-*ka*	-*ica*	-*ica*
	-*ička* (unproductive)		
in -*o*	-*ce*	-*ence*	-*ence*
in -*e*	-*ence*		
	-*ice* (unproductive)		

they do not combine freely, but only particular fixed combinations are allowed and those combinations often involve repetition of the same suffix form, although it would have been possible to combine the suffixes so that no repetition of form occurs. In Bulgarian, the suffixes -*ica* and -*ence* can be repeated (Table 9.3). In Russian ((22) through (24)) and Polish (Table 9.4), the repetitions are of the type -*ek*+-*ek*, -*ka*+ -*ka*, and -*ko*+-*ko*. Moreover, the combinations of diminutive suffixes either involve only two suffixes (the case of Bulgarian, Russian, and Polish double diminutives) or can be modeled in terms of two-suffix pairings (the case of Bulgarian triple diminutives).

With respect to the combinability of the diminutive suffixes, Russian resembles Polish (Table 9.4) to some extent. Examples (22) through (24) illustrate the combinations of diminutivizers found in Russian double diminutives (repeated suffixes are in bold italic):

(22) basic nouns in -*C* / *C'* → DIM1 → DIM2

a. -*ok* + -*ek*: *golos* 'voice'→ *golos-ok* → *golos-oč-ek*
b. ***-ek* + *-ek*:** *den'* 'day' → *den-ek* → *den-eč-ek*
c. -*ik* + -*ek*: *nož* 'knife' → *nož-ik* → *nož-ič-ek*
d. -*ica* + -*ka* (feminine basic nouns, unproductive): *čast'* 'part' → *čast-ica* → *čast-ič-ka*

(23) basic nouns in -*a* → DIM1 → DIM2

a. ***-ka* + *-ka*:** *igla* 'needle' → *igol-ka* → *igol-oč-ka*
b. -*ica* + -*ka*: *voda* 'water' → *vod-ica* → *vod-ič-ka;*
c. -*eška* + -*ka: ryba* 'fish' → *ryb-eška* → *ryb-ešeč-ka*

(24) basic nouns in -*o* and -*e* → DIM1 → DIM2

a. ***-ko* + *-ko*:** *sito*, sieve' → DIM1 *sit-ko* → DIM2 *sit-eč-ko*
b. -*yško* + -*ko: sol'nce* 'sun' → DIM1 *soln-yško* → *DIM2 soln-yšeč-ko* (found on the Internet)

TABLE 9.4

Combinability of DIM Suffixes in Polish (cf. Szymanek and Derkach 2005; Manova and Winternitz 2011)

Nouns in	DIM1 suffixes	DIM2 suffixes	
		Productive (attach by addition)	Unproductive (attach by substitution of a DIM1 suffix, i.e., do not combine with DIM1 suffixes)
-*C*	-*ek* -*ik* / -*yk* -*uszek* (unproductive)	-*ek*	-*uszek*, -*aszek*
	-*iszek* /-*yszek* (unproductive) -*aszek* (unproductive) -*ulek* (unproductive) -*ka* (unproductive, selects feminine nouns)		
-*a*	-*ka*	-*ka*	
	-*uszka* (unproductive) -*iczka* /-*yczka* (unproductive)		
-*o* / -*e*	-*ko*	-*ko*	
	-*uszko* (unproductive)		

In sum, the suffixes in the evaluative domain, like the inflectional suffixes, appear in a fixed order, but in contrast to the inflectional suffixes, evaluative suffixes can apply recursively and even repetition of the same suffix (phonological form) is possible.

6.3 NONEVALUATIVE DERIVATIONAL DOMAIN

Due to the fairly limited number of the inflectional and diminutive suffixes it is easy to list all their combinations, which is what we actually did in the two subsections above. However, listing all combinations of all nonevaluative derivational suffixes is virtually impossible. Therefore, in the literature the order of the derivational suffixes (in any language) is discussed with the help of a set of suffixes. This strategy is also employed in the present paper. We will analyze the combinations of 20 Russian derivational suffixes with all other suffixes in Russian. The analysis considers data from Russian that have not been examined as evidence for affix ordering in the literature so far, and follows Manova's (2011b) cognitive approach:

First, (i) as it is hard to formulate a single rule that accounts for all SUFF2 suffixes that may follow a particular SUFF1 (Table 9.5), the SUFF2 suffixes are distributed into three groups according to their lexical-category specifications; that is, into $SUFF2_N$, $SUFF2_V$, and $SUFF2_{ADJ}$, as illustrated in (25) and exemplified in Table 9.6, which is a version of Table 9.5.

(25)

SUFF1 → $SUFF2_N$
SUFF1 → $SUFF2_{ADJ}$
SUFF1 → $SUFF2_V$

TABLE 9.5

Combinability of the Russian Suffix *-tel'* Deriving Persons

No	SUFF1	SUFF1 lexical category & semantics	SUFF2	Examples
1.	*-tel'*	N person	*-stvo, -ščina, -skij*	*učitel'stvo* 'being a teacher; teachers (coll.)', *ljubitel'ščina* 'dilettantism', *učitel'skij* 'teacher's'

TABLE 9.6

Combinability of the Russian Suffix *-tel'* Deriving Persons, SUFF2 Suffixes Distributed According to Their Lexical Category

No	SUFF1	SUFF1 lexical category & semantics	SUFF2	Examples
1.	*-tel'*	N person	N: *-stvo, -ščina* (2)	*učitel'stvo* 'being a teacher; teachers (coll.)', *ljubitel'ščina* 'dilettantism'
			ADJ: *-skij*	*učitel'skij* 'teacher's'

As illustrated in (25), in Table 9.6 all SUFF2 suffixes in Russian that follow the SUFF1 *-tel'* for nouns denoting persons are distributed into two groups according to their lexical category specifications. Crucially, there is now one single SUFF2 that derives adjectives; that is, the combination *-tel'-skij* is **fixed**; and there are two SUFF2 suffixes, *-stvo* and *-ščina,* that both derive abstract nouns. However, the suffix *-ščina* forms only two types, which is indicated by the number in brackets after the suffix—that is, by (2). The information about the types formed by the suffix *-ščina* is based on counting of derivatives in the Morpheme dictionary of Russian (Kuznecova and Efremova 1986) that contains 52,000 words and the Russian National Corpus that has over 500 million tokens. (Intriguingly, the numbers of types derived by a particular SUFF1-SUFF2 combination do not differ significantly in both sources; see also the discussion in Manova and Talamo, accepted). There is no number after the suffix *–stvo*, since this suffix derives a great number of types. Actually, for all suffixes that derive more than 10 types no information about the derived number of types is provided. However, it is important for the analysis that of the combinations *-tel'-stvo* and *-tel'-ščina*, *-tel'-stvo* is the default option. Manova (2011b) refers to such defaults as **predictable** combinations.

Manova (2011b) assumes that in cases of suffixation by default, the exceptions (i.e., the competing suffixes) should derive up to five types, which are rote-learned and should be listed (i.e., ignored in the analysis), since if a particular suffix combination occurs only in five words, the speaker should know those words by heart irrespective of whether there is a rule that accounts for the derivation of those five words. The five types in Manova (2011b) were assumed based on the data analyzed and in order to have a threshold that is not very high and maximally challenges the theoretical framework. Recent research on suffixation in Italian (Manova and Talamo, accepted) provides evidence that the precise number of exceptions in cases of suffixation by default seems to be 10. Manova and Talamo count the exceptions to default suffixation patterns in Italian in two sources: the *derIvaTario* that is based on the 4 Mio CoLFIS corpus (Bertinetto et al. 2005) and annotated for research on derivational morphology, and the 380 Mio La Repubblica corpus. Crucially, the number of exceptions does not depend on the corpus size, and in both corpora SUFF2 suffixes that derive up to ten types (exceptions) accompany a default suffixation pattern. Therefore, in Table 9.8 no numbers of types are given for suffixes that derive more than 10 types. Nevertheless, as can be seen from Table 9.8, the combinations of the 20 Russian derivational suffixes under scrutiny in this article can be described with an even lower threshold.

Next, (ii) in cases where more than one SUFF2 of a particular lexical category is available, suffix-particular semantics differentiates among the suffixes. To illustrate, in Table 9.7 two nominalizing suffixes, *-ec* and *-ost'* follow the adjectivizer *-livyj,* but while *-ec* derives persons, the attachment of *-ost'* results in an abstract noun. Thus, both suffixes *-ec* and *-ost'* are clearly differentiable in terms of intentional semantics (or conceptualization); that is, in terms of what the speaker wants to say. In such instances, we also speak of **predictable** combinations.

In Table 9.8, the methodology introduced in (i) and (ii) above is applied to the description of the combinations of 20 derivational suffixes in Russian. Fixed combinations are marked by '1' in the last column of Table 9.8, whereas a ***bold italic*** *number* in the same column indicates a predictable combination. Thus, as can be seen from the last column in Table 9.8, all combinations of the 20 Russian derivational suffixes under scrutiny in this paper with all other suffixes in Russian are either fixed or predictable. Nevertheless, there are a few peculiar cases: see, for example, the suffix *-ar'* (no. 2 in Table 9.8) which combines with three nominalizing suffixes, *-stvo, -nik,* and *-nja.* In the last column of Table 9.8 it is written 'N: (3) ***2***.' This means that two of the three suffixes, namely *-nik* and *-nja,* derive the same semantics, which is then illustrated with *svin-ar-nik* and *svin-ar-nja,* both meaning 'piggery.' Similar to the suffix *-ar,'* to which two suffixes, *-nik* and *-nja,* with exactly the same semantics attach, is the suffix *-ost'* (see Table 9.9). Note, however, the number of types that *-ar-nik* (1) and *-ost-(l)ivyj* (4) derive, a single type and 4 types respectively (Table 9.9). Thus, of the combinations *-ar-nik* and *-ar-nja, -ar-nja* is the default; and then of *-ar-stvo* and *-ar-nja, -ar-stvo* derives abstract nouns, while *-ar-nja* forms nouns for places; that is, *-ar-stvo* and *-ar-nja* differ in terms of intentional semantics. Therefore, all combinations of *-ar'* with $SUFF2_N$ suffixes are classified as predictable. As for *-ost-nyj, -ost-noj,* and *ost-livyj* (Table 9.9), the first two derive the majority of types, *-ost-nyj* being the more usual combination, but there are still many adjectives with *-noj,* and one finds a number of doublets on the Internet—such as *skorostnoj poezd* and *skorostnyj poezd,* both

TABLE 9.7

The Role of the SUFF2 Particular Semantics

SUFF1	SUFF1 lexical category & semantics	SUFF2	Examples
-livyj	ADJ qualitative	N: *-ec, -ost'*	*son-liv-ec 'sleepy person'* *molča-liv-ost'* 'taciturnity'

TABLE 9.8

Combinability of 20 Russian Derivational Suffixes with All Other Suffixes in Russian

No	SUFF1	Lexical & semantic category	SUFF2 suffixes according to their lexical category	Examples	Translations	SUFF2 types in numbers
1.	*-tel'*	N person	N: *-stvo, -ščina* (2) ADJ: *-skij*	*pisa-tel'-stvo* *ljubi-tel'-ščina* *pisa-tel'-skij*	being a writer, writers (collect.) dilettantism writer's	N: **2** ADJ:1
2.	*-ar'*	N person	N: *-stvo*, *-nik* (1) / *-nja* ADJ: *-skij*	*aptek-ar-stvo* *svin-ar-nik* / *svin-ar-nja* *aptek-ar-skij*	being a chemist, chemists (collect.) piggery / piggery chemist's	N: (3) **2** ADJ: 1
3.	*-(j)ak*	N person	N: *-estvo* ADJ: *-eskij*	*zeml-jač-estvo* *zeml-jač-eskij*	countrymen (collect.) countryman's	N: 1 ADJ: 1
4.	*-ač*	N person	N: *-estvo* ADJ: *-eskij*	*cirk-ač-estvo* *cirk-ač-eskij*	being a circus actor, circus actors (collect.) circus actor's	N: 1 ADJ: 1
5.	*-un*	N person	N: *-stvo* ADJ: *-skij*	*opek-un-stvo* *opek-un-skij*	guardianship guardian's	N: 1 ADJ: 1
6.	*-(e)stvo*	N abstract	ADJ: *-ennyj*	*rod-stv-enn-yj*	kindred	ADJ: 1
7.	*-(iz)acija*	N abstract	ADJ: *-onnyj*	*inform-aci-onn-yj*	informational	ADJ: 1
8.	*-ost'*	N abstract	N: *-nik* ADJ: *-nyj* /*-noj*, *-(l)ivyj* (4)	*skor-ost-nik* *slad-ost-nyj* *skor-ost-noj* *mil-ost-ivyj*	fast worker delightful high-speed merciful	N: 1 ADJ: (3) **2**
9.	*-izm*	N abstract				
10.	*-tel'*	N object	ADJ: *-nyj*	*ukaza-tel'-nyj*	indicatory	ADJ: 1
11.	*-ar'*	N object	N: *-nik* ADJ: *-nyj*	*slov-ar-nik* *slov-ar-nyj*	lexicographer dictionary-	N: 1 ADJ: 1
12.	*-nik*	N place	ADJ: *-nyj* /*-ovyj*	*tajn-ič- nyj* = = *tajn-ik-ovyj*	secret	ADJ: (2) 1
13.	*-livyj*	ADJ qualitative	N: *-ec, -ost'*	*son-liv-ec* *molča-liv-ost'*	sleepy person taciturnity	N: **2**
14.	*-istyj*	ADJ qualitative	N: *-ost'*	*gor-ist-ost'*	mountainousness	N: 1
15.	*-skij*	ADJ possessive/ relational	N: *-ost'*	*det-sk-ost'*	childishness	N: 1
16.	*-abel'nyj*	ADJ qualitative	N: *-ost'*	*comfort-abel'n-ost'*	amenities	N: 1
17.	*-atel'nyj*	ADJ relational	N: *-ost'*	*dyh-atel'n-ost'*	respiration	N: 1
18.	*-astyj*	ADJ qualitative	N: *-ost'*	*zubast-ost'*	with sharp teeth	N: 1
19.	*-(n)ičat'*	V impf. inchoative	N: *-nie* / *-n'e*	*obezjann-iča-nie* = *obezjann-iča-n'e*	apery	N: (2) 1
20.	*-nut'*	V impf. inchoative	N: *-nie*	*sox-ne-nie*	drying	N: 1

TABLE 9.9

More Than One SUFF2 Deriving the Same Semantics (1)

No in table 9.8	SUFF1	Lexical & semantic category	SUFF2 suffixes according to their lexical category	Examples	Translations	SUFF2 types in numbers
2.	*-ar'*	N person	N: ***-stvo***, *-nik* (1) / *-nja* ADJ: ***-skij***	*aptek-ar-stvo* *svin-ar-nik* = *svin-ar-nja* *aptek-ar-skij*	being a chemist, chemists (collect.) piggery / piggery chemist's	N: (3) **2** ADJ: 1
8.	*-ost'*	N abstract	N: *-nik* ADJ: *-nyj* /*-noj*, *-(l)ivyj* (4)	*skor-ost-nik* *slad-ost-nyj* *skor-ost-noj* *mil-ost-ivyj*	fast worker delightful high-speed merciful	N: 1 ADJ: (3) **2**

meaning 'high-speed train.' *Skorostnyj* is registered only once in the Russian National Corpus, but used many times on the Internet.

The examples in Table 9.10 are to some extent similar to the *-nyj*/*-noj* case in Table 9.9.

The Russian examples in Table 9.9 and Table 9.10 present a real challenge to morphological theory, as the latter postulates blocking in derivation claiming that two derivatives never have exactly the same semantics (Aronoff 1976; Rainer 1988, and many others). Note that Manova (2011b) did not find such examples in English and Bulgarian, and, actually, used the principle of blocking to differentiate between two or more SUFF2 suffixes with the same lexical-category specification. We see this peculiarity of the Russian derivational morphology as evidence for semantically governed combinations of suffixes; that is, both SUFF2 suffixes with the same semantics in Russian are the same suffix at a semantic level, and therefore they do not violate Manova's (2011b) cognitive ordering. I will return to this issue in the discussion in section 7. All fixed and predictable suffix combinations are actually instances of entrenched morphological structure.

6.4. A NOTE ON PREFIXATION

The present article is focused on suffixation, but as I claim for the existence of word domains with respect to affix ordering, the goal of this note is to provide evidence for the existence of a prefixational domain.

TABLE 9.10

More Than One SUFF2 Deriving the Same Semantics (2)

No in table 9.8	SUFF1	Lexical & semantic category	SUFF2 suffixes according to their lexical category	Examples	Translations	SUFF2 types in numbers
12.	*-nik*	N place	ADJ: *-nyj* /*-ovyj*	*tajn-ič-nyj* = = *tajn-ik-ovyj*	secret	ADJ: (2) 1
19.	*-(n)ičat'*	V impf. inchoative	N: *-nie* / *-n'e*	*obezjann-iča-nie* = *obezjann-iča-n'e*	apery	N: (2) 1

If one looks at the studies on prefixation in the literature, it appears that cross-linguistically, the ordering of prefixes is much less restricted than that of suffixes. Thus in the literature on affix ordering so far, Chintang (Sino-Tibetan, Nepal) prefixes (Bickel et al. 2007) have been cited as a clear instance of 'free' affix order. Note, however, that Bickel's et al. (2007) use of 'free' is misleading, since the authors show that the order of the prefixes is governed by a phonological factor–prosodic subcategorization; that is, "prefixes occur in variable orders because each prefix and each stem element project a phonological word of their own, and each such word can host a prefix, at any position" (p. 43). Thus, 'free' only means that semantic or morphosyntactic factors cannot explain the ordering of the prefixes in Chintang.

Zirkel (2010), in a discussion of the CBO hierarchy of the English prefixes, comes to the conclusion that the ordering of the English prefixes is less restricted in comparison to that of the English suffixes, as more prefixes than suffixes participate in AB-BA mirror-image combinations and thus violate CBO.

Recently, much research has been carried out on the ordering of the Slavic verbal prefixes. The analyses are usually based on syntactic and semantic factors. Babko-Malaya (1999) and Svenonius (2004) explain the position of a prefix in the verb form with semantic compositionality: superlexical prefixes have fixed meaning and their attachment means semantic compositionality, lexical prefixes are highly idiosyncratic in meaning, and there are also purely perfectivizing prefixes (idiosyncratic prefixes are the closest to the root). Tatevosov (2008) adds to this classification a class of intermediate prefixes; that is, prefixes that are neither lexical nor superlexical. Markova (2011) proposes the following types of prefixes: lexical prefixes with a syntactic position below VP, inner prefixes that are above VP and below vP, and outer prefixes that are above vP. However, one and the same (phonological form of a) prefix often belongs to more than one group of prefixes; that is, in a sequence of prefixes, the same prefix can occupy different positions with respect to the root, which means permutation and repetition of prefixes. For examples of prefix combinations from Slavic, I refer the reader to the studies mentioned in this paragraph; recall also (3) above.

In sum, there is evidence in the literature on affix order that the ordering of the affixes in the prefixational word domain is less restricted than that of the affixes in the suffixational word domain. This fact is seen as supportive for the domain-specific analysis promoted in the present article.

7. Discussion

We started the description of the ordering of the suffixes in the Slavic word with the assumption of domains that accommodate more than one affix of the same type. We also demonstrated that the affixes in each domain order differently, in the sense that they exhibit different affix order peculiarities: (i) inflectional suffixes have a fixed order and cannot be repeated; (ii) evaluative suffixes also form fixed combinations but can be repeated; and (iii) the ordering of the nonevaluative derivational suffixes depends on the lexical-category specification of a suffix, the suffix-particular semantics, and the notion of default, but the 20 nonevaluative derivational suffixes from Russian under scrutiny in this paper form only fixed and predictable combinations—that is, in all free domains entrenchment of morphological structure (cf. on cognitive grammar in section 2) in terms of fixed (and predictable) suffix combinations seems to play an important role. We also discussed that the ordering of the prefixes is less restricted than that of the suffixes. Thus we are ready now to compare the structure of the Slavic word with that of the English word, and provide some additional explanation of why theories that are formulated to account for the ordering of the English derivational suffixes fail to account for the ordering of the Slavic suffixes. The structure of the Slavic word in terms of word domains, (18) above, is repeated in (26) for convenience.

(26) Affix order domains in the structure of the Slavic word

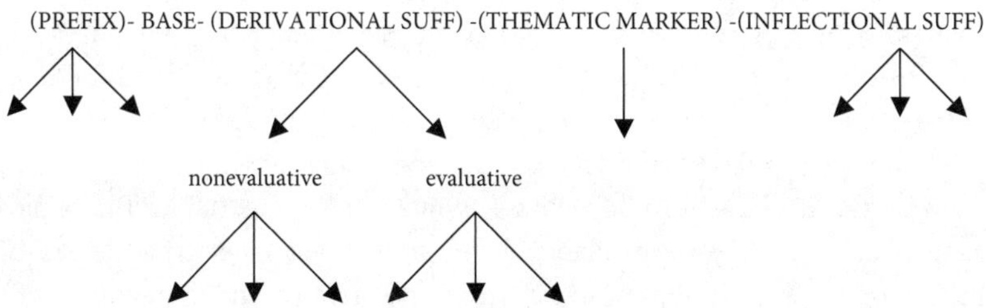

The schema in (27) gives the domains found in the structure of the English word.

(27) Affix order domains in the structure of the English word

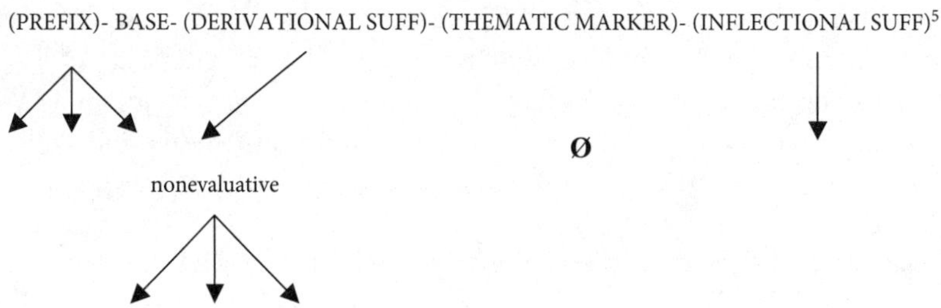

[5]Forms such as *children's,* with two inflectional suffixes, are rather exceptional and not considered here. Note that the second suffix does not attach in the way the other inflectional suffixes do but requires a special marking (').

Ø means that there are no thematic markers in English, or at least English does not possess affixes that could be seen as parallel to the thematic markers in Slavic languages. Of course, English also has evaluative suffixes; that is, suffixes that derive the meaning 'small X,' for example the suffix *-ette*, as in *cigar* → *cigar-ette*, *opera* → *operette*. However, it does not make sense to put *-ette* in a different suffix slot. There is nothing related to affix order that would motivate such a decision for English.

A comparison between (26) and (27) shows that the Slavic word has a greater number of suffix slots and domains than the English word. These facts thus explain why affix order approaches (CBO and others) designed to account for the ordering of the English suffixes has difficulty in accounting for the order of the Slavic suffixes. Manova's (2011b) approach assumes the existence of domains in suffixation, which thus makes it more flexible and suitable to explain the affix ordering in the Slavic word. Moreover, Manova's approach allows for typological comparisons. Thus, Manova (2011b) demonstrates that the type of suffix ordering we illustrated with the combinations of the 20 nonevaluative derivational suffixes in Russian is also operative in English and Bulgarian—the only difference being that in English and Bulgarian the principle of morphological blocking holds (better), whereas in Russian it is often the case that two nonevaluative derivational suffixes derive the same semantics (recall the examples in Table 9.9 and Table 9.10).

Suffix-particular semantics plays an important role in this study, and it was claimed that in cases in which there is more than one SUFF2 that derives the same lexical category, the available SUFF2 suffixes are expected to derive different meanings (or exactly the same meaning in Russian). The importance of suffix-particular semantics can be further illustrated with the behavior of homophonous (Table 9.11) and synonymous (Table 9.12) SUFF1 suffixes. Homophonous suffixes, such as the suffix *-tel'* that derives concepts such as persons and objects (examples in Table 9.11), have the same form but different semantics, and are therefore followed by different SUFF2 suffixes while synonymous SUFF1 suffixes have different form but the same semantics (all suffixes in Table 9.12 derive nouns denoting persons), and are therefore followed by the same SUFF2 suffixes.

Examples such as these in Table 9.11 (of homophonous suffixes) and Table 9.12 (of synonymous suffixes), together with the examples in Table 9.9 and Table 9.10 in

TABLE 9.11

Combinability of Homophonous Suffixes in Russian, Based on Table 9.8

No in table 9.8	SUFF1	Lexical & semantic category	SUFF2 suffixes according to their lexical category	Examples	Translations	SUFF2 types in numbers
1.	*-tel'*	N person	N: *-stvo, -ščina* (2) ADJ: *-skij*	*pisa-tel'-stvo* *ljubi-tel'-ščina* *pisa-tel'-skij*	being a writer, writers (collect.) dilettantism writer's	N: **2** ADJ:1
10.	*-tel'*	N object	ADJ: *-nyj*	*ukaza-tel'-nyj*	indicatory	ADJ: 1

TABLE 9.12

Combinability of Synonymous Suffixes in Russian, Based on Table 9.8

No in table 9.8	SUFF1	Lexical & semantic category	SUFF2 suffixes according to their lexical category	Examples	Translations	SUFF2 types in numbers
1.	*-tel'*	N person	N: *-stvo, -ščina* (2) ADJ: ***-skij***	*pisa-tel'-stvo* *ljubi-tel'-ščina* *pisa-tel'-skij*	being a writer, writers (collect.) dilettantism writer's	N: **2** ADJ:1
2.	*-ar'*	N person	N: *-stvo,* *-nik* (1) / *-nja* ADJ: *-skij*	*aptek-ar-stvo* *svin-ar-nik* / *svin-ar-nja* *aptek-ar-skij*	being a chemist, chemists (collect.) piggery / piggery chemist's	N: (3) **2** ADJ: 1
3.	*-(j)ak*	N person	N: *-estvo* ADJ: *-eskij*	*zeml-jač-estvo* *zeml-jač-eskij*	countrymen (collect.) countryman's	N: 1 ADJ: 1
4.	*-ač*	N person	N: *-estvo* ADJ: *-eskij*	*cirk-ač-estvo* *cirk-ač-eskij*	being a circus actor, circus actors (collect.) circus actor's	N: 1 ADJ: 1
5.	*-un*	N person	N: *-stvo* ADJ: *-skij*	*opek-un-stvo* *opek-un-skij*	guardianship guardian's	N: 1 ADJ: 1

which two derivational suffixes derive exactly the same semantics, seem to speak for a model of grammar in which (1) suffixes have semantics; that is, are word-like, and (2) that suffixes that express the same semantics should be treated as the same suffix, which is consonant with Distributed morphology (DM) (Halle and Marantz 1993). However, as the fixed and predictable suffix combinations we define do not depend on lexical bases (roots in DM), DM would not be able to derive them, as derivation in DM takes place step by step, starting from the root. Moreover, fixed and predictable combinations, since entrenched pieces of morphological structure are most probably not always attached as derived structure but as semantic wholes where SUFF2 gives the final semantic meaning of the unit, a fixed combination such as *-ar-stvo* that means 'person-abstract/collective' is accessed through the semantics of the SUFF2. Therefore, we maintain that the Slavic fixed and predictable suffix combinations are instances of semantically compositional morphological structure that is derived in terms of fixed combinations of cognitive concepts: lexical (such as nouns, adjectives, and verbs) and semantic (such as person, object, etc.), and does not have anything to do with syntax. Thus, we see such combinations as evidence for independency of morphology from syntax; that is, not for distributed but for autonomous morphology.

8. Conclusions

In this paper, the structure of the Slavic word was discussed in terms of affix ordering. The Slavic word was divided into domains, each domain with affix-order peculiarities of its own. The order of the Slavic suffixes was described with the help of binary combinations of suffixes of the type SUFF1-SUFF2, where SUFF1 and SUFF2 are any two neighboring suffixes in a domain. In most instances the SUFF1-SUFF2 combinations are fixed; that is, a SUFF1 selects only one particular SUFF2, this is especially true of the suffixes in the inflectional and evaluative word domains. In the nonevaluative derivational domain, SUFF1-SUFF2 combinations depend on the lexical- and semantic-category specification of a suffix in the sense that SUFF1 either combines with a single suffix of a major lexical category (N, ADJ, V) or in cases in which more than one SUFF2 of the same lexical category attaches to a SUFF1, (i) one of the SUFF2 suffixes applies by default (i.e., the majority of types are derived by that SUFF2); (ii) the available SUFF2 suffixes derive completely different meanings (concepts), e.g., an abstract noun and a place noun; or (iii) exactly the same semantics. Type (iii) appears to be a typical feature of the Russian derivational suffixes. Lexical and semantic categories are seen as cognitive by nature. Moreover, the cognitive grammar principle of entrenchment seems to play an important role in affix ordering in Slavic and the fixed and predictable suffix combinations are most probably rote-learned and accessed as wholes. Thus, this study speaks for the existence of word-like affixes and of affix combinations with a status of their own; that is, affix combinations that may be produced without reference to lexical bases, and that are pieces of semantically compositional but purely morphological structure.

References

Allen, Margaret. 1978. "Morphological Investigations." PhD diss., University of Connecticut.

Andrejčin, Ljubomir, ed. 1975. *Obraten rečnik na săvremennija bălgarski ezik*. Sofija: Bălgarska akademija na naukite [Reverse dictionary of Modern Bulgarian.]

Aronoff, Mark. 1976. *Word Formation in Generative Grammar*. Cambridge, MA: MIT Press.

Aronoff, Mark, and Nanna Fuhrhop 2002. "Restricting Suffix Combinations in German and English: Closing Suffixes and the Monosuffix Constraint." *Natural Language and Linguistic Theory* 20: 451–490.

Babko-Malaya, Olga. 1999. "Zero Morphology: A Study of Aspect, Argument Structure and Case." PhD diss., Rutgers University.

Bertinetto PM, Burani C, Laudanna A, Marconi L, Ratti D, Rolando C, Thornton AM (2005). *Corpus e Lessico di Frequenza dell'Italiano Scritto (CoLFIS)*. http://linguistica.sns.it/CoLFIS/Home.htm.

Bickel, Balthasar, Goma Banjade, Martin Gaenszle, Elena Lieven, Netra Prasad Paudyal, Ichchha Purna Rai, Manoj Rai, Novel Kishore Rai, and Sabine Stoll. 2007. "Free Prefix Ordering in Chintang." *Language* 83: 1–31.

Bybee, Joan L. 1985. *Morphology: A Study of the Relation Between Meaning and Form*. Amsterdam: John Benjamins.

Bulgarian National Corpus/Bălgarski Nacionalen Korpus. http://ibl.bas.bg/BGNC_bg.htm.

Croft, William. 2001. *Radical Construction Grammar: Syntactic Theory in Typological Perspective*. New York: Oxford University Press.

Doroszewski, Witold, ed. 1997. *Słownik Języka Polskiego*. Electronic version. Warszawa: Polskie Wydawnictwo Naukowe.

Dressler, Wolfgang. U., Willi Mayerthaler, Oswald Panagl, and Wolfgang U. Wurzel. 1987. *Leitmotifs in Natural Morphology*. Amsterdam: Benjamins.

Dressler, Wolfgang U. 1990. "The Cognitive Perspective of 'Naturalist' Linguistic Models." *Cognitive Linguistics* 1: 75–98.

Dressler, Wolfgang U. 2005. "Word-Formation in Natural Morphology." In *Handbook of Word-Formation*, edited by Pavol Štekauer and Rochelle Lieber, 267–284. Dordrecht: Springer.

Dubisz, Stanisław, ed. 2008. *Uniwersalny Słownik Języka Polskiego*. Warszawa: Wydawnictwo Naukowe PWN. Electronic version.

Fabb, Nigel. 1988. "English Suffixation is Constrained only by Selectional Restrictions." *Natural Language and Linguistic Theory* 6: 527–539.

Fillmore, Charles. 1982. "Frame Semantics." In *Linguistics in the Morning Calm*, edited by I.-S. Yang, 111–137. Seoul: Hanshin Publishing.

Geeraerts, Dirk. 2006. *Cognitive Linguistics: Basic Readings*. Berlin: Mouton de Gruyter.

Giegerich, Heinz. 1999. *Lexical Strata in English: Morphological Causes, Phonological Effects*. Cambridge: Cambridge University Press.

Hay, Jennifer. 2001. "Lexical Frequency in Morphology: Is Everything Relative?" *Linguistics* 39: 1041–1070.

Hay, Jennifer. 2002. "From Speech Perception to Morphology: Affix-ordering Revisited." *Language* 78: 527–555.

Hay, Jennifer. 2003. *Causes and Consequences of Word Structure*. London: Routledge.

Hay, Jennifer, and Ingo Plag. 2004. "What Constrains Possible Suffix Combinations? On the Interaction of Grammatical and Processing Restrictions in Derivational Morphology." *Natural Language and Linguistic Theory* 22: 565–596.

Halle, Morris, and Alec Marantz. 1993. "Distributed Morphology and the Pieces of Inflection." In *The View from Building 20*, edited by K. Hale and S. J. Keyser, 111–176. Cambridge, MA: MIT Press.

Jadacka, Hanna, ed.2001. *Słownik gniazd słowotwórczych współczesnego języka ogólnopolskieg. Tom 2: Gniazda odrzeczownikowe*. Kraków: Universitas.

Jackendoff, Ray. 1990. *Semantic Structures*. Cambridge, MA: MIT Press.

Kiparsky, Paul. 1982. "Lexical Morphology and Phonology." In *Linguistics in the Morning Calm: Selected Papers from SICOL-1981 (vol. 1)*, edited by I.-S. Yang, 3–91. Seoul: Hanshin.

Kuznecova, Ariadna I., and Tatjana F. Efremova. 1986. *Slovar' morfem russkogo jazyka*. Moskva: Russkij jazyk. [A Morpheme Dictionary of Russian]

Langacker, Ronald. 1987. *Foundations of Cognitive Grammar, Volume I, Theoretical Prerequisites*. Redwood City, CA: Stanford University Press.

Langacker, Ronald. 1991. *Foundations of Cognitive Grammar, Volume II, Descriptive Applications*. Redwood City, CA: Stanford University Press.

"*La Repubblica*" corpus. http://dev.sslmit.unibo.it/corpora/corpus.php?path=&name=Repubblica.

Manova, Stela. 2005. "Derivation Versus Inflection in Three Inflecting Languages." In *Morphology and its Demarcations*, edited by Wolfgang U. Dressler, Dieter Kastovsky, Oskar Pfeiffer, and Franz Rainer, 233–252. Amsterdam: Benjamins.

Manova, Stela. 2008. "On Some Recent Changes in Bulgarian Conjugation." In István, Ferincz (ed.). *Bulgarian Language and Literature at the Crossroads of Cultures*, Vol. 1, 22–29. Szeged: Szegedi Egyetemi Kiadó.

Manova, Stela. 2010. "Suffix Combinations in Bulgarian: Parsability and Hierarchy-based Ordering." *Morphology* 20: 267–296.

Manova, Stela, and Mark Aronoff. 2010. "Modeling Affix Order." *Morphology* 20: 109–131.

Manova, Stela. 2011a. *Understanding Morphological Rules: With Special Emphasis on Conversion and Subtraction in Bulgarian, Russian and Serbo-Croatian*. Dordrecht: Springer.

Manova, Stela. 2011b. "A Cognitive Approach to SUFF1-SUFF2 Combinations: A Tribute to Carl Friedrich Gauss." *Word Structure* 4: 272–300.

Manova, Stela, and Kimberley Winternitz. 2011. "Suffix Order in Double and Multiple Diminutives: With Data from Polish and Bulgarian." *Studies in Polish Linguistics* 6: 115–138.

Manova, Stela, in press. "Closing Suffixes." In *Word Formation in the European Languages*., edited by Peter Müller, Ingeborg Ohnheiser, Susan Olsen, and Franz Rainer. Handbooks of Linguistics and Communication Science. Berlin: De Gruyter Mouton.

Manova, Stela. 2014. "Affixation." *Oxford Bibliographies in Linguistics*, edited by Mark Aronoff. New York: Oxford University Press. DOI: 10.1093/OBO/9780199772810-0183.

Manova, Stela, and Luigi Talamo. Accepted. "On the Significance of the Corpus Size in Affix-Order Research." SKASE Journal of Theoretical Linguistics.

Markova, Angelina. 2011. "On the Nature of Bulgarian Prefixes: Ordering and Modification in Multiple Prefixation." *Word Structure* 4: 244–271.

Mestres-Missé, A.; A. Rodriguez-Fornells, and T. F. Münte. 2010. "Neural Differences in the Mapping of Verb and Noun Concepts onto Novel Words." *Neuromage* 49: 2826–2835.

Mohanan, K. P. 1986. *The Theory of Lexical Phonology*. Dordrecht: Reidel.

Muysken, Pieter. 1986. "Approaches to Affix Order." *Linguistics* 24: 629–643.

National Corpus of Polish/Narodowy Korpus Języka Polskiego. http://nkjp.pl/.

Plag, Ingo. 1996. "Selectional Restrictions in English Suffixation Revisited. A Reply to Fabb (1988)." *Linguistics* 34: 769–798.

Plag, Ingo. 1999. *Morphological Productivity: Structural Constraints in English Derivation*. Berlin and New York: Mouton De Gruyter.

Plag, Ingo. 2002. "The Role of Selectional Restrictions, Phonotactics and Parsing in Constraining Suffix Ordering in English." In *Yearbook of Morphology 2001*, edited by Geert Booij and Jaap van Marle, 285–314. Dordrecht: Kluwer.

Plag, Ingo, and Harald Baayen. 2009. "Suffix Ordering and Morphological Processing." *Language* 85: 109–152.

Rainer, Franz. 1988. "Towards a Theory of Blocking: The Case of Italian and German Quality Nouns." In *Yearbook of Morphology 1988*, edited by Geert Booij, and Jaap van Marle, 155–185. Dordrecht-Boston-London: Kluwer.

Reverse Dictionary of Modern Bulgarian. Andrejčin (ed.) 1975.

Rice, Keren. 2000. *Morpheme Order and Semantic Scope*. Cambridge: Cambridge University Press.

Rice, Keren. 2011. "Principles of Affix Ordering: An Overview." *Word Structure* 4: 169–200.

Russian National Corpus/Nacional'nyj korpus russkogo jazyka. www.ruscorpora.ru/.

Saloni, Zygmunt, ed. 2007. *Słownik Gramatyczny Języka Polskiego*. Warszawa: Wiedza Powszechna. Electronic version.

Selkirk, Elisabeth. 1982. *The Syntax of Words*. Cambridge, MA: MIT Press.

Siegel, Dorothy. 1974. *Topics in English Morphology*. Cambridge, MA: MIT Press.

Skalička, Vladimir. 1979. *Typologische Studien*, edited by Peter Hartmann. Wiesbaden: Braunschweig.

Stankov, Valentin, ed. 2002. *Nov pravopisen rečnik na bălgarskija ezik*. Sofija: Xejzăl [New Spelling Dictionary of Bulgarian].

Svenonius, Peter, ed. 2004. *Special Issue on Slavic Prefixes. Nordlyd* 32.2. Tromsø: University of Tromsø.

Stump, Gregory T. 2001. *Inflectional Morphology*. Cambridge: Cambridge University Press.

Szymanek, Bogdan, and Tetyana Derkach. 2005. "Constraints on the Derivation of Double Diminutives in Polish and Ukrainian." *Studies in Polish Linguistics* 2: 93–112.

Talamo, Luigi. 2015. "Suffix Combinations in Italian: Selectional Restrictions and Processing Constraints." In *Affix Ordering Across Languages and Frameworld*, edited by Stela Manova, 205–229. New York: Oxford University Press.

Tatevosov, Sergei. 2008. "Intermediate Prefixes in Russian." In *Formal Approaches to Slavic Linguistics 16: The Stony Brook Meeting 2007*, edited by Andrei Antonenko, John Bailyn, and Christina Bethin, 423–488. Ann Arbor: Michigan Slavic Publications.

Taylor, John R. 2002. *Cognitive Grammar*. New York: Oxford University Press.

Yandex dictionaries/Jandeks slovari. http://slovari.yandex.ru/.

Zaliznjak, Andrej A. 1977. *Grammatičeskij slovar' russkogo jazyka: slovoizmenenie*. Moskva: Russkij jazyk. [A grammatical dictionary of Russian.]

Zirkel, Linda. 2010. "Prefix Combinations in English." *Morphology* 20: 239–266.

Zirkel-Hilkenbach, L. 2011. "Can Complexity-Based Ordering Be Extended from English to German?" Paper presented at the 3rd Vienna Workshop on Affix Order. Vienna, January 15–16.

PART IV:

Description and Acquisition of Affix Order

10

Suffix Sets in Polish De-Nominal Derivatives

Iwona Burkacka

1. Introduction

Morphotactic issues in Polish are rarely analyzed, and morphotactics itself has a relatively short history in Polish linguistics.[1] This was noted by Kreja in the introduction to Z *morfologii i morfotaktyki współczesnej polszczyzny*, where he states that "in Polish linguistics morphotactics is not only new as a term, but also as a research question" (Kreja 1989: 7). Although this statement was made over twenty years ago it is still true, since there has been little attempt to produce an overview of the combinability of formatives and its constraints. A typical approach limits the scope of research to an analysis of particular prefixes or suffixes (e.g., Kreja 1989, 1996, and 2002). Sometimes these issues form a background for a discussion of Polish word formation (e.g., Szymanek 2010: 255–65), of productivity of word-formation types (Satkiewicz 1969), or constitute justification of a normative evaluation of a derived word or a neologism (Satkiewicz 1972, Jadacka 2005), or for contrastive studies (e.g., Szymanek & Derkach 2005, Manova & Winternitz 2011). The only exceptions are works by Michalewski (1984), Kreja (1989), Kowalik (1977, 1998) and an extensive article by Kowalik (1979).

Meanwhile morphotactics has become very popular worldwide, and a substantial body of research has been devoted to syntagmatic properties of formatives in a number of languages (see the overviews in Manova & Aronoff 2010 and Rice 2011). Thus the aim of this paper is to contribute to the current research on affix order with data from Polish. In what follows, I present the results of my research on suffix combinations in Polish denominal derivations and the factors governing the combinability of suffixes in such derivations (Burkacka 2006, 2010,

[1] Translated into English by Marta Usiekniewicz. Thank you to Stela Manova, Jadwiga Linde-Usiekniewicz, and two anonymous reviewers for stimulating discussion and helpful comments on this work.

2012).[2] The analysis is with a focus on semantic aspects of morphotactics. The research has been made possible as the beginning of the twenty-first century saw the publication of *Słownik gniazd słowotwórczych współczesnego języka ogólno polskiego* (*SGS*; Vogelgesang 2001, Jadacka et al. 2001, Skarżyński et al. 2004) that contains preliminarily analyzed and ordered word formation material (cf. section 3). The data selected for the analysis include derivatives formed from non-derived nouns. The selection was motivated by the following factors:

a. suffixation is an important word formation technique and the most common method of forming denominal derivatives;
b. denominal derivatives are numerous and diverse (they include nouns, adjectives, verbs, and adverbs);
c. the largest set of suffixes can be found in denominal derivatives;
d. nouns serve a very important function in describing and naming the world (cf. Markowski 1990);
e. the comparison of derivational nests of different parts of speech shows that denominal derivatives contain the longest and most diverse combinations of suffixes (long word formation sequences); in nests of lexemes belonging to other parts of speech the suffix sequences are rather short;
f. there is also an observable shift toward denominal derivation in contemporary Polish (especially in comparison with deverbal and deadjectival derivations, cf. Kleszczowa 2010), which results from contacts between languages (loanwords are mainly nouns: names of things, places, and phenomena; adaptation of nous is also easier), and partially from a reinterpretation of deadjectival structures (univerbisms) and decrease of derivational power of native verbs, since their word formation potential has been exhausted by earlier derivations (Kleszczowa 2010: 64).

The aim of this article is first to show the array of word formation suffixes presented in Polish derivatives, second to illustrate these with examples, and third to identify typical sequences of suffixes in Polish (i.e., those amply attested in the analyzed data). This will be accompanied by an overview of the source of the data, the *SGS*, and discussions of solutions adopted there (rules of analysis), which have obviously influenced the outcomes presented in this work.

This paper consists of the introduction, six sections, and a list of symbols and abbreviations. In section 2 the history of research into suffix combinability in

---

[2] Concerning derivational suffixes, such as *-arz*, *-erz*. Inflectional suffixes (such as *-a* in *lampa* 'lamp' or *lampka* 'little lamp'), which may serve as formatives (in paradigmatic derivation, e,g. *markiz* 'marquis'– *markiz-(a)* 'marchioness,' *lis* 'fox'– *liś-(i)* 'fox (modifier), i.e. a denominal adjective derived from the word fox,' cf. Waszakowa 1996, Szymanek 2010:233–236). Inflectional suffixes and paradigmatic formatives do not fall within the scope of my research. The term *suffix* refers here solely to derivational suffixes.

Poland is presented, section 3 introduces the source of data, section 4 is devoted to the suffix sets in Polish, section 5 discusses the frequency of those suffix sets, and in section 6 conclusions are drawn.

2. History of Research

A few book-length studies are entirely devoted to issues of suffix combinability in Polish: Kowalik (1977), Michalewski (1984), and Kreja (1989).

Michalewski (1984) focused on the phonetic conditions of nominal suffix combinability. He described the constraints on suffixation depending on the part of speech the base belongs to (e.g., the affix *-ość* combines only with adjectival bases) and the final sound sequence of the base; for example the suffix *-ek* appears in deverbal and denominal derivatives,[3] but in the former the bases need to end in a hard consonant (without /j/), and in the latter the ultimate consonant may also be soft (Michalewski 1984: 183). He enumerated the necessary alternations and noted that this phenomenon might decrease the transparency of the formation. In some cases the insertion of a connective (*interfix* in Michalewski's terminology) prevents allomorphy.[4] Another means of adjusting the stem to the suffix is the truncation of the base.[5] The researcher devotes a lot of space to these considerations. Yet some of the presented analysis is faulty; for example, deriving *łaciak* from the adjective *łaciaty* 'covered with patches' is an obvious mistake, since *łaciak* is 'a nail used to fix battens.'[6] Thus its base is the word *łata* 'batten,'[7] and not, as Michalewski suggests, the adjective *łaciaty* 'covered with patches' (deletion of *-at-*). His approach is asemantic because it takes into account only the abstract formal connections, which leads to a description of fictional relationships, and postulating sets of suffixes that do not exist in reality. In the current literature, his analysis of nouns denoting female humans in terms of deletion of *-nik* before *-nic(a)* is also considered questionable, since such formations can be accounted for in a simpler way by postulating a paradigmatic derivation.[8]

---

[3] The distribution of suffixes *-ek* and *-ik* has been amply studied (for example in Malicka-Kleparska 1985).

[4] A connective is an element added to expand the basic variants of formatives (Grzegorczykowa and Puzynina 1979: 16), it is also considered a type of individual quantitative thematic alternation (Grzegorczykowa and Puzynina 1979: 17), cf. Szymanek 1985.

[5] Similar comments about suffix combinability with derivational stems are present in works by Grzegorczykowa and Puzynina (1979) and Waszakowa (1994).

[6] This is the way other names of nails are being formed *gonciak* 'nail used to fix shingles,' *podkowiak* 'nail used to fix hooves,' *szyniak* 'nail used to fix tracks' (cf. Grzegorczykowa and Puzynina 1979: 157).

[7] A word formation paraphrase reveals that there is no immediate relationship between the words *łaciaty* 'covered with patches' and *łaciak* 'nail.' The derivative *łaciaty* is motivated by the lexeme ŁATA 'an irregular spot or patch in a color different than that of the background,' and *łaciak* is motivated by a homonym ŁATA 'a wooden slat used in carpentry.'

[8] For example *pracownik* 'workers' → *pracownica* 'women workers' [deletion of *-nik* and adding *-nic(a)*], in newest analysis we have only paradigmatic formative *-(a)* [paradigmatic derivation].

Kowalik's studies (1977, 1997) focus on the structure of simple adjectival suffixes and their combinability and contain the most comprehensive discussions of morphotactic issues. Another aspect of her research concerns the segmentation of nouns into morphemes and identification of simple suffixes. Kowalik (1998: 605) distinguishes the simplest, indivisible segments identical to word formation suffixes in synchronic analysis. Thus in her account, *nacjonalizacja* 'nationalization,' *racjonalizator* 'lit. rationalizator; improver' and *pensjonariuszka* 'female resident in some kind of facility (care, penitentiary, etc.)' contain seven elements (root and six elements on the right-side) identified on purely formal (i.e., asemantic) grounds: *nac-j-on-al-iz-ac-ja*, *rac-j-on-al-iz-at-or*, and *pens-j-on-ar-i-usz-ka* (Kowalik 1998: 605). Although these elements can be found in other Polish derivatives, an analysis in terms of base-formative relation shows that the actual number of derivational morphemes is much smaller. Hence:

- the noun *nacjonalizacja* contains four segments: root *nacj-* (originating in the noun *nacja* 'nation/nationality') and three suffixes: *-onaln(y)* (shortened by deleting *-n(y)* due to the addition of the subsequent suffixes), *-iz(ować)* and *-acj(a)*;[9]
- the word *racjonalizator* is even less complex: the base is the adjective *racjonalny* 'rational' (stem *racjonal-*, with deletion of *-n(y)*), and two suffixes: *-iz(ować)* and *-ator* (*racjonalny* 'rational' → *racjonalizować* 'to rationalize' → *racjonalizator* 'lit. rationalizator');
- similarly, *pensjonariuszka* 'female resident in some kind of facility' is a denominal derivative of *pensjonariusz* 'male resident of some kind of facility,' derived from the word *pensjonat* 'boarding house,' thus we arrive at the stem *pensjon-* (*-at* deleted) and two suffixes *-ariusz* and *-k(a)* (as a feminine suffix).

Moreover, many of the deletions mentioned by Kowalik may be dispensed with if the analysis includes the concept of the infix,[10] or assumes a different motivation.

---

[9] According to *SGS* it is the suffix *-ować* that has an inflectional character; *nacja* → *nacjonalny* 'national' → *nacjonalizować* 'to nationalize' → *nacjonalizacja* 'nationalization.'

[10] Provided that following Nagórko we posit infixes in adjectival diminutives and intensifications, e.g., *matka* 'mother' → *mat-eń-ka* 'dim. mother;' *lekk(i)* 'light' → *leci-ut-k(i)* 'very light' (Nagórko 2002: 179). The analysis of the derivative *mateńka* 'dim. mother' suggested by Kowalik postulates the suffix *-eńk(a)*, which in turn causes the truncation of the final element *-k(a)*, while the sequence *-utk(i)* is divided into intermorph *-ut-* and a simple suffix *-k(a)*. In her approach intermorphs are segments that cannot be classified as affixes. They are either optional expansions of a simple suffix or segments located between the base (root) and a simple suffix (Kowalik 1979: 24). They are semantically empty and are not morphologically conditioned (Kowalik 1977: 24); they can be either archaic or new (Kowalik 1977: 23–24, cf. Szymanek 2010: 252–24). Simple suffixes are suffixes whose "further morphological analysis is impossible" (Laskowski 1973: 136). Traditional morphological analyses of these examples would distinguish suffixes *-eńk(a)* and *-utk(i)* (deletion of *-k*-from the stem) (cf. Grzegorczykowa and Puzynina 1979 and Grzegorczykowa et al. (eds.) 1998).

The above-mentioned works examine derivatives in order to identify the simplest elements, hence they ignore the semantic aspect, while I am interested in a different approach to word formation. I take the existence of actual meaning relations between the base and the derivative, which are visible in the derivational paraphrase, as the base for identifying suffixes in derivatives. I also consider derivatives motivated by association ("onomasiological" or "associative" derivatives),[11] the essence of which are socially fixed metaphoric relationships that are a vital component of the naming processes. This is especially visible in colloquial language (e.g., *mrowiskowiec* 'a bloc of flats inhabited by a large number of people, like ants in an anthill;' derived from *mrowisko* 'anthill'); names of plants and animals (e.g., *słonecznik* 'sunflower, a plant which resembles a sun and follows the daily movement of the sun'; *jamnik* 'dachshund, a breed that was used during fox hunting because it could get into foxholes,' derived from *jama* 'foxhole'); and terminology (especially in botany and zoology, e.g., *bawełnica* 'plant louse (the female of this species is covered by white secretion that resembles raw cotton),' derived from *bawełna* 'cotton'; *bleskotka* 'chalcid wasp' (a wasp with a metallic sheen), derived from *blask* 'sheen'; *boczniak* 'edible tree mushroom that grows on the sides of tree trunks'; from *bok* 'side').

3. Sources of Data

This study is based on material from the second volume of *Słownik gniazd słowotwórczych współczesnego języka ogólnopolskiego*, *SGS* (Jadacka et al. 2001), which is a dictionary of word formation nests. (Word formation nests are exemplified later in this section.) The source provides a rich and diversified material, since it contains derivatives from both native and loanword bases, all of which are nouns. The derivatives exhibit both native (e.g., *-n(y)*,[12] *-eni(e)*) and foreign suffixes (e.g., *-i/yst(a)*,[13] *-i/yzm*), which had been incorporated into Polish affix sets. In denominal derivation suffixation is the most important technique. The nests vary in size and in semantics of lexemes that constitute their bases, which enables the analysis to cover a broad range of arrays. The second volume contains over 35,000 derivatives (36,240) gathered around 6,105 nonderived nouns with overall almost 8,000 senses (7,865) constituting a base for at least one derivative. After the preliminary research, I have excluded derivatives that could be considered as formed from more than one base. I focused on examining simple derivatives to maintain homogeneity of the material analyzed. Nevertheless, the volume of material remained substantial.

---

[11] Cf., e.g., Szymanek 2010: 29.

[12] Inflectional endings (grammatical morphemes) are given in parentheses.

[13] The spelling *-i/yst(a)* reflects the two variants of this suffix: *-ist(a)* occurs after stems ending with a soft consonant, and *-yst(a)*, appears after stems with a hard final consonant.

The above-mentioned second volume is a part of a four-volume dictionary, *SGS*. The two initial parts (volume 1, containing deadjectival nests, and volume 2, containing denominal nests) were published in 2001 and resulted from a research project conducted between 1997 and 1999 (funded by a grant from the State Committee for Scientific Research, KBN), though preparatory works had started earlier (between 1986 and 1990). Hanna Jadacka was the editor of both volumes. The author of the first was Teresa Vogelgesang (Vogelgesang 2001), and of the second Hanna Jadacka, Magdalena Bondkowska, Iwona Burkacka, Elżbieta Grabska-Moyle, and Tomasz Karpowicz (Jadacka et al. 2001).

The two subsequent volumes were published three years later (in 2004) and are a result of the work done by an interacademic research group (funded by a KBN grant, 2001–2004). Mirosław Skarżyński was the editor of these volumes. The third volume, published in two parts due to its size, contains nests motivated by nonderived verbs (authors: Mirosław Skarżyński, Małgorzata Berend, Magdalena Bondkowska, Iwona Burkacka, Hanna Jadacka, Monika Olejniczak, and Teresa Vogelgesang; Skarżyński et al. 2004). The fourth volume contains nests motivated by numerals, adverbs, pronouns, prepositions, "modulants," onomatopoeias, and exclamations (author: Mirosław Skarżyński) (Skarżyński 2004b). The dictionary of derivational nests was modeled upon the derivational nest dictionary by Tixonov (*Slovoobrazovatel'nyj slovar' russkogo jazyka* 'Word-formation dictionary of Russian') published in 1985.[14] In the introduction to this dictionary, Tixonov characterized the basic concepts of nest word formation: *word formation nest*, *word formation chain*, and *word formation paradigm*. The Polish nest dictionary applies findings of applicative grammar of Šaumjan and Soboleva,[15] especially their notation of word formation operations (cf. Jadacka 2001a: 7-21).[16] The nest methodology found its application in numerous nest dictionaries of Slavonic languages and in confrontative works, as it enables a multidimensional comparison of lexicons and word formation techniques (cf. the numerous works comparing Russian and Polish).[17]

The first volume of *SGS* also contains an extensive introduction outlining a brief history of word formation nest research in Poland, sources used for

---

[14] This was preceded by a tentative issue published 22 years earlier, *Škol'nyj slovoobrazovatel 'nyj slovar' russkogo jazyka. Pasobie dlja učaščixsja* (Moskva: Prosveščenie). The two earlier dictionaries of Russian, *Škol'nyj slovoobrazovatel'nyj slovar'* by Poticha, Moskva: Prosveščenie, 1964 and *Russian Derivational Dictionary* by Worth, Kozak, and Johnson, New York: Elservier Publishing Co., 1970, follow different assumptions. The former does not contain nests and the words are listed alphabetically with morphemes separated, and the latter organizes words around root morphemes (not lexemes) identified in a diachronic analysis.

[15] Presented in many works, e.g., Šaumjan 1965, and Soboleva 1970.

[16] Other works devoted to word formation use structural schemes, called graphs.

[17] For example Jochym-Kuszlikowa 1982, Czernecka 1985, and Stawnicka 1991, or concerning material from three languages, Fras 1985.

compiling the dictionary,[18] the symbols and abbreviations it uses,[19] the features of a nest description as a lexicographic method, and explanation of basic concepts (derivational nest, model, word formation chain). It also outlines the principles governing the description of various types of derivatives (simple and complex): an explanation of symbols and abbreviations used to indicate added bases and classification of parts of speech. One challenge to nest dictionaries is the impossibility of presenting analyses entailing mutual (bilateral) motivation and multiple motivations, because the derivative may be located only in one place of the nest. If a semantic analysis was consistent with two possibilities of establishing the motivation, the authors chose the one that was formally simpler as the direct one. Thus *słomianka* 'straw doormat' was paraphrased as '*wycieraczka słomiana*' (*wycieraczka* 'doormat' and *słomiana* 'adj. made of straw'), thus identifying the derivative as a deadjectival noun with the suffix *-k(a)*, and not paraphrased as '*wycieraczka ze słomy*' (lit. 'doormat from straw'), which would make the noun a denominal derivative with the suffix *-ank(a)*, although both paraphrases are equally valid on semantic grounds. However, the authors discarded the simplest possible formal variant if it was not supported by "completely correct paraphrase"; for example, *stoczniowiec* 'shipyard worker' was paraphrased as '*pracownik stoczni*' (N + N) ('an employee of a shipyard'), thus allowing to identify the suffix *-owiec*, and not as '*pracownik stoczniowy*' (N + Ad) ('shipyard employee') that would warrant identifying the (suffix *-ec*, Jadacka 2001: 27). By the same token, the authors assumed the rule of system domination only in cases of both possible paraphrases being semantically equally valid. The introduction also contains a list of morphophonological alternations occurring in native words (vowel and consonant changes) and in foreign words (consonant changes), and examples of the dictionary's uses.

The third volume of the *SGS* also features an introduction, mostly due to the fact that according to its authors, "a verb as a derivational base requires a modification of some of the solutions, and sometimes even a complete reworking of the method of description." Hence the third volume contains some changes and supplements. First, new entries were included among the dictionary's sources, such as the updated version of the *Nowy słownik języka polskiego PWN* (Sobol 2002), a new dictionary, *Słownik współczesnego języka polskiego* (Dunaj 1998) and

---

[18] Written sources include: single-volume *Mały słownik języka polskiego*, Skorupka, Auderska, and Łempicka (eds.), (Warszawa: Państwowe Wydawnictwo Naukowe 1968), ten-volume *Słownik języka polskiego*, Doroszewski (ed.), (Warszawa: Państwowe Wydawnictwo Naukowe 1958–1969), *Supplement* to this dictionary, three-volume *Słownik języka polskiego*, Szymczak (ed.), (Warszawa: Państwowe Wydawnictwo Naukowe 1978–1981), *Komputerowy słownik języka polskiego PWN*, (Warszawa: Państwowe Wydawnictwo Naukowe 1996); the dictionary also contains words of standard Polish not featuring in written sources but known to the editors.

[19] The list of symbols and abbreviations is on the last page. Each derivational step has been marked graphically as well (right-side indents); see an example of a derivational nest of the lexeme MYSZ 'mouse.'

a selective sample from other dictionaries.[20] Besides the specific nature of the material included (the way of presenting aspectual oppositions and intransitive reflexive verbs, as well as the description of derivatives with the postfix *się*), the crucial difference is the introduction of alternate prefix derivation to describe the four groups of verbs with the following general meanings:[21]

a. 'to redo something or to do something from the beginning,' such as *przerobić, przerabiać 1.* 'zrobić na nowo (to remake, to make anew)'; *przerobić, przerabiać* derives from a prefixed verb *zrobić* 'perf. to do,' and not from *robić* 'imperf. to do,' which is clearly supported by the above paraphrase; *przeważyć, przeważać* 'zważyć na nowo (to weigh one more time)' derives from *zważyć* 'perf. to weigh,' and not *ważyć* 'imperf. to weigh');
b. 'to do something completely,' such as *wygnić, wygniwać 1.* 'zgnić całkowicie (to rot completely)' derives from the perfective *zgnić* and not from the imperfective *gnić*;
c. 'to do something partially, to a certain degree,' such as *nadorać* 'zaorać kawałek (to plough a section)' or *podtruć, podtruwać* 'otruć częściowo (to partially poison)' derives from the respective perfective verbs: *zaorać* and *otruć*';
d. 'to do more than the expected norm (too much),' such as *objeść się, objadać się* 'zjeść za dużo (to eat too much)' derives from the perfective *zjeść* and not from the imperfective *jeść*) (Skarżyński 2004a: XII–XIII).

Also a new part of speech was introduced, namely the modulant, which lead to a reinterpretation of structures featuring the element *nie* 'negation, no.' In this volume they are classified as compositions, such as *nieciągły* 'not continuous' (and not as simple derivatives: prefix structures as in volumes 1 and 2). Additionally, a distinction was established between the numeral *pół* meaning 'half' (e.g., *półobrót* 'half-turn' and *pół*) derived from the adverb meaning 'incompletely/partially' (e.g., *półpłynny* 'semi-liquid') (Skarżyński 2004a: XIII–XIV).

In the analysis of the derivatives in volumes 1 and 2 of *SGS*, the authors followed a traditional version of word formation analysis (Grzegorczykowa & Puzynina 1979, Grzegorczykowa et al. (eds.) 1998, Waszakowa 1996, 1994), which was slightly changed in volumes 3 and 4 to include Śmiech's findings (1986) (the above mentioned alternate prefix derivation). In accordance with the methodology, formatives such as *-(Ø), -(a), -(i), -(ować)* are described as paradigmatic and not as zero suffixes

---

[20] *Uniwersalny słownik języka polskiego*, edited by Dubisz, (Warszawa: Wydawnictwo Naukowe PWN, 2003); *Praktyczny słownik języka polskiego*, edited by Zgółkowa, (Poznań: Wydawnictwo "Kurpisz," 1994–2005); *Nowy słownik ortograficzny PWN*, edited by Polański, (Warszawa: Wydawnictwo Naukowe PWN, 1997) as sources confirming the derivatives supplied by the authors.

[21] This is a significant change in relation to the traditional descriptions (e.g., Grzegorczykowa et al. 1998), in which prefixed verbs were derived from unprefixed bases, highlighting the formal relations and ignoring semantic factors.

(cf. Nagórko 2002: 180–81, Waszakowa 1996, Szymanek 2010: 233–36). Reanalyses of verbal derivation presented in volume 3 of *SGS* and discussed above obviously affect the way nominal derivation is presented, since derivational chains with nominal roots may contain several verbal derivates; thus the novel analysis of verbs has its impact on the way denominal chains are structured.

I follow the **formatives' identification principles** applied in the source of my material, *SGS*, volume 2, though I do introduce changes regarding alternate prefix derivation, the status of the *się* element and the connective (analogically to the solutions introduced in volumes 3 and 4 of the *SGS*, as presented in the introduction to volume 3).

The segments of formatives are established on the basis of appropriate entries in SGS. Let us follow its steps on the basis of the nest MYSZ 'mouse':[22]

MYSZ 'mouse'	
mysz-ka 1. 'small mouse'	S,S
myszecz-ka 'dim. small mouse'	S,S,S
myszk-(ować) 'to snoop around'*	S,S,V
po-myszkować 'perf. to snoop around, lit. to mouse around'	S,S,V,V
myszk-(ować) 2.* *łow.* 'to search for food or prey'	S,S,V
po-myszkować 'perf. to search for food or prey'	S,S,V,V
mysz-ka 2.* 'birthmark.,' 3.* 'a smell of old wine'	S,S
mysz-o-łów-(Ø) *zool.*'hawk, (lit. mouse hunter)'	(V+)S,S
mysz-o-płoch-(Ø) *zool.* 'ruscus (lit. mouse-scare)'	(V+)S,S
myś-(i) 'mousey/mouse-like'	S,Ad
mysi-królik* 'goldcrest, lit. mouse rabbit'	S,Ad(+S),S
myś-urek* *bot.* 'mousetail'	S,Ad,S
mysz-aty* 'mousy'	S,Ad
myszat-ość 'quality of being mousy'	S,Ad,S
mysz-owaty 1.* 'belonging to the mouse family'	S,Ad
myszowat-(e) 2. *zool uż. rzecz. pl.* 'belonging to the mouse family'	S,Ad,S

The above nest contains 11 derivational sequences, including:

¤ three two-step sequences (consisting of two steps, the lexeme and its derivative that is not a base for another derivative):
mysz – myszka 2., 3. 'mouse – little mouse,'
mysz – myszołów 'mouse – hawk/lit. mouse huner,'
mysz – myszopłoch 'mouse – ruscus;'

[22] The nest features in *SGS*, vol. 2: 916–17 (I have introduced simplifications regarding the notation of sources and supplied meanings of words [glossed in English; translator's note].) The list of symbols and abbreviations is on last page.

- six three-step sequences (consisting of a lexeme, immediate derivative motivated by that word and the 2nd order derivative):
  mysz – myszka 1. – myszeczka 'mouse – little mouse – dim. little mouse,'
  mysz – myszka 1. – myszkować 3. 'mouse – little mouse – to snoop around,'
  mysz – mysi – mysikrólik 'mouse – mousy – goldcrest/lit. mouse rabbit,'
  mysz – mysi – mysiurek 'mouse – mousy – moustail,'
  mysz – myszaty – myszatość 'mouse – mousy – quality of being mousy,'
  mysz – myszowaty – myszowatość 'mouse – mousy – belonging to mouse family;'
- two four-step sequences (consisting of four stages related by motivation):
  mysz – myszka 1. – myszkować – pomyszkować 'mouse – little mouse – to snoop around (perf. imperf.),'
  mysz – myszka 1. – myszkować 2 .– pomyszkować 'mouse – little mouse – to search for food (perf. imperf).'

After rejecting the compositions and single-suffix sets (two- or multistage sequences, but containing only one suffix), the list of sequences in this nest was limited to two derivational chains,[23] which became sources of the following suffix sets: *k(a)+k(a)* and *at(y)+ość*.[24]

I focus on sets with minimum two suffixes; that is, first-degree suffix derivatives (motivated by an unmotivated noun) and second-degree suffix derivatives, such as *EGZAMIN* 'exam' → *egzaminator* 'examiner' → *egzaminatorka* 'female examiner,' which results in a two-suffix set *ator+k(a)*. These include also derivatives of further steps, such as *NIT* 'bolt/rivet' → *nitować* 'to rivet' → *niter* 'lit. riveter' → *niterski* 'related to a riveter'; that is, a set *er+sk(i)*, or those sets in which suffixation is interwoven with other techniques, such as *KOŚĆ* 'bone' → *kostny* 'ossal' → *kostnieć* 'to ossify; fig. to become numb with cold' → *skostnieć* 'perfective' → *skostniały* 'ossified; fig. numb with cold' → *skostniałość* 'numbness with cold' (three-suffix set: *n(y)+x+x+ał(y)+ość*, in which the first symbol *x* indicates paradigmatic derivation, and the second – prefixation.

I am not interested in the combinability of suffixes and bases, as these issues for Polish have been analyzed in great detail (cf. Grzegorczykowa & Puzynina 1979, Waszakowa 1994).

Sometimes an addition of a suffix causes an obligatory alternation within the suffix attached earlier. This happens, for example, when suffix *-ek* is attached to the suffix *-ek*, which is a part of a derivational stem of the word *kwiatek* 'little flower' derived from an unmotivated noun *KWIAT* 'flower'; the final *k* /k/is replaced with

[23] This term is used interchangeably with *derivational sequence* (Jadacka 2001: 19).

[24] Derivatives, such as *mysi* 'mousy,' *myszkować* 'to snoop around,' *myszowate* 2 'belonging to mouse family' are labeled paradigmatic derivatives.

cz /č/: *KWIAT* 'flower' → *kwiatek* 'little flower' → *kwiateczek* 'dim. little flower,' so *ek+ek* = *eczek*. An alternation between *k* and *cz* /č/ before suffixes starting with a vowel (and before some suffixes starting with a consonant) is typical for derivation (sound *cz* is historically a palatalized variant of *k*), similarly to other alternations between palatalized and nonpalatalized consonants (e.g., *n : ń, n(y)+ak, kostny* 'ossal' → *kostniak* 'osteoma, benign tumor of the bones').[25]

Alternations accompanying derivation vary as to their regularity and depend on numerous factors related to the native or foreign origins of the derivational base (e.g., *t : ć* in native words, *but* 'shoe' → *bucik* 'little shoe'; *z : d* in loanwords, *korozja* 'corrosion' → *korodować* 'to corrode'),[26] colloquial and expressive character of derivatives (e.g., *g : ź, noga* 'leg' → *nózia* 'dim. leg'), type of the base's or previously added suffix's final consonant (e.g., alternations between stops and affricates are obligatory: *t, d, s, z: ć, dź, ś, ź* before nominal suffixes: *-ik, -iw(o), -idł(o), -in(a), -eń, -eż, -an, -ar(a), -or, -ur(a)*). These phenomena have been discussed in many works concerning morphophonology and word formation (e.g., Rubach 1981, Gussmann 1985, Grzegorczykowa et al. (eds.) 1998: 91–121, Kreja 1989), though most of the examples provided therein deal with the juncture of stem of the unmotivated word with the suffix, rather than with motivated words containing sets of suffixes. Some examples of alternations resulting from joining suffixes will be provided later, when suffix sets are discussed.

4. Suffix Sets in Polish

The analyzed material contains around 300 suffixes, two-thirds of which are native suffixes, while the remaining ones are foreign. There are some isolated suffixes; that is, appearing in a single formation (and thus not entering into any proportional relationship), such as *-aw* in the derivative *rękaw* 'sleeve' from *ręka* 'hand/arm.' Other suffixes appear very often, such as *-k(a), -ow(y), -ość*.

The suffixes form almost 2,000 generalized suffix sets (simple and complex),[27] of minimum two elements, starting at various steps.

The majority of suffix sets contain two suffixes (cf. Table 10.1), three-suffix sets are much rarer, and more numerous sets are very sporadic. The longest set

[25] We have the letters *ni* (for the signification of soft consonant in the word *kostniak*, not the letter *ń*) in the position before vowels. The orthographic rule allows soft consonants to be signified by the letter *i* or diacritics signs, e.g., *ń, ć, dź, ś*.

[26] Some works on derivation point to constrains of combinability or to phenomena related to joining suffixes. Waszakowa (1994), for example, enumerates typical alternations accompanying the combination of foreign suffixes or regular truncation accompanying suffixation, e.g., *-izm/yzm* truncated before suffixes *-ist(a)/yst(a), -ik(a)/yk(a)*, or *-ik/yk*; elements *-j(a)* and *-usz* and their expanded versions—without specifying before which suffixes this phenomenon takes place. Waszakowa points also to the change of character of some of the truncations, e.g., truncation *-n(y)* in native derivatives is irregular, while in derivatives with foreign suffixes it is regular (Waszakowa 1994).

[27] Phonetic variants were unified into a general format (e.g., *-izm* and *-yzm* are represented by *-i/yzm*)

TABLE 10.1
Suffix Distribution

Sequence	Example	Number	Percentage
Two suffixes	s_1+s_2	1402	77.9%
Three suffixes	$s_1+s_2+s_3$	375	20.8%
Four suffixes	$s_1+s_2+s_3+s_4$	23	1.3%
Five suffixes	$s_1+s_2+s_3+s_4+s_5$	1	0.1%

contains five suffixes distributed into five steps (with two truncations of suffix fragments): *al<n>(y)+iz(ować)+ac<j>(a)+or+sk(i)* (*CENTRUM* 2. 'center' → *centralny* 'central' → *centralizować* 'centralize' → *centralizacja* 'centralization' → *centralizator* 'an agent of centralization' → *centralizatorski* 'adj. related to the agent of centralization'). This material confirms the findings of Kowalik, who established that the language system eschews multisegmental structures (Kowalik 1997: 198).

Suffix sets may be continuous, with suffixes attaching directly one to another, without lacunae (i.e., some other formative interrupting the suffix sequence), as in the example of a five-element set quoted above, or may contain lacunae (i.e., other formatives appearing between suffixes), as suffixation is accompanied by other word formation techniques, such as *al<n>(y)+i/yz(ować)+x+x+acj(a)+n(y): CENTRUM* 'center' → *centralny* 'central' → *centralizować* 'to centralize' → *scentralizować* 'perf. to centralize' → *decentralizować* 'to decentralize' → *decentralizacja* 'decentralization' → *decentralizacyjny* 'adj. having the capacity to decentralize';[28] in this example *x* indicates the prefixes *s-* and *de-*, which are exponents of prefixation or of prefix-alternating derivation (as shown in Table 10.2).[29]

The majority of sequences are continuous (almost 90%), which means that suffixation is a homogenous technique. In the analyzed sets the lacunae (marked

TABLE 10.2
Suffix Set Characteristics

Suffix sequence type		Number	Percentage
Continuous suffix sequences		1591	88.3
Discontinuous suffix sequences	with 1 x	179	9.9
	with 2 x	30	1.7

[28] That the verb *decentralizować* is derived from *scentralizować* is clear from the derivational paraphrase "likwidować to, co zostało scentralizowane (to eliminate that what has become centralized)". This is a case of prefix-altering derivation; cf. section 3 concerning the changes introduced in *SGS-3*. Similarly in order to *rozminować coś* 'clear sth of mines' it is necessary for somebody to *zaminować* 'put in mines' first; thus the derivation sequence is. mine *military* 'mine' → *minować* 'to mine' → *zaminować* 'to mine (perf.)' → *rozminować* 'to clear of mines.'

[29] This example contains a modified sequence (in relation to *SGS-2*). I have introduced (according to the later findings of *SGS-3*) prefix-altering derivation. This was motivated by semantic considerations: one can *zdecentralizować* 'decentralize' something only if it has been *scentralizowane* 'centralized' previously (cf. comments in section 2, Skarżyński 2004a).

with *x*) usually appear once in a sequence (197 instances), and even less often twice in a sequence (only 30 instances). The discontinuous sequences amount to 11% of all sets.

It is worth noting that the calculations represent real number of suffixes in the sequence; that is, the suffixes that get truncated entirely during subsequent derivation, sometimes also with a fragment of a preceding suffix are not taken into account. For example, although in the sequence *al<n>(y)+<i/yzm>+i/yst(a)* (*sentyment* 'sentiment' → *sentymentalny* 'sentimental' → *sentymentalizm* 'sentimentalism' → *sentymentalista* 'sentimentalist') three suffixes can be identified, since one of them is deleted completely, I describe this sequence as containing two suffixes. The suffixation also reduces the length of the first affix, but the fragment -*al* remains, so it is included in the calculations.

Truncations of entire suffixes appear in 132 suffix sequences, most commonly in sets with two suffixes (in 96 sets),[30] less commonly in sets with three suffixes (33 times), and three times in sets with four suffixes. In some sets two truncations occur, but their number is marginal (only seven instances).

Usually, the sequences start at step I, where the first suffix combines with the stem of an unmotivated noun (83,2 % of sequences).

These numbers result from the degree of the nest complication (its depth). If the sequences are short, then the suffix sets need to appear at the beginning; that is, at step I. The longer the sequences in a nest, the more probable it is that a longer set would appear or that suffixation would take place at later steps. Obviously, the category of the base is also a factor.[31]

There are only 255 sequences beginning at step II (14% of sequences), and only 46 (2,5 %) at step III. The sequences identical to those starting at step I, but beginning at step II or III, have a smaller number of instantiations: they rarely attach to derived bases.

The configuration of suffixes in sets with two suffixes is presented in Table 10.3.

Data concerning sequences of sets with two suffixes generally follow the pattern described in the general overview: these sets are usually continuous (they constitute 86.8% of all two-element sequences and three-fourths of sequences starting at step I), and rarely contain other formatives. The most common sequence is s_1+s_2, for example at steps I and II *k(a)*$_d$+*k(a)*$_d$ [*CHMURA* 'cloud' → *chmurka* 'little cloud' → *chmureczka* 'dim. little cloud'] or at steps II and III *arz+k(a)*$_ż$ [*SZKŁO* 'glass' → *szklić* 'to glaze' → *szklarz* 'glazer' → *szklàrka* 'glazing'], less common s_1+x+s_2 [e.g., *ek+x+nic(a)*, *KRĄG* 'circle' → *krążek* 'little circle/roller' → *krążkować* 'to roller (metalworking)' → *krążkownica* 'rolling machine']. Only 21 sequences contain two

---

[30] This concerns sets containing two suffixes after the truncation. The sets with only one suffix after truncation were not included into the material, since they are considered to contain only one suffix.

[31] As well as its meaning, origin and history (cf. Burkacka 2012).

TABLE 10.3

Sequences of Sets with Two Suffixes

Beginning of the sequence	Distribution	Set	Number	Example
Step I and subsequent	I and II	s_1+s_2	957	ak+ow(y) *KRZYŻ* 'cross'→ *krzyżak* 'any cross-shaped device, e.g. trestle' → *krzyżakowy* 'adj. cross-shaped, transverse'
	I and III	s_1+x+s_2	120	i/ystyk(a)+truncation and der. parad.+k(a)$_{ż}$ *POLON<I>A* 'Latin word for Poland' → *polonistyka* 'Polish philology' → *polonista* 'male teacher of Polish'→ *polonistka* 'female teacher of Polish'
		s_1+<s>+s_2	32	al<n>(y)+<i/yzm>+i/yst(a) *CENTRUM*[a] 'center' → *centralny* 'central' → *centralizm* 'centralizm' → *centralista* 'centralist'
	I and IV	s_1+x+x+s_2	20	n(y)+der. parad.+pref.+ał(y), *KOŚĆ* 'bone' → *kostny* 'ossal' → *kostnieć* 'to ossify; fig. to numb with cold' → *skostnieć* 'perf. to ossify; fig. to numb with cold' → *skostniały* 'numb with cold'
		s_1+<s>+x+s_2	2	i/yczn(y)+<ość>+truncation and der. parad.+arz *ELEKTR<ON>* 'electron' → *elektryczny* 'electric' → *elektryczność* 'electricity' → *elektryka* 'wiring/electric installation'→ *elektrykarz* 'electrician'
		s_1+x+ <s>+s_2	3	ual<n>(y)+pref.+<i/yzm>+i/yst(a) *INTELEKT* 'intellect' → *intelektualny* 'intellectual' → *antyintelektualny* 'anti-intellectual'→ *antyintelektualizm* 'anti-intelectualism' → *antyintelektualista* 'anti-intellectualist'
	I and VI	s_1+<s>+x+x+ <s>+s_2	1	al<n>(y)+<i/yz>(ować)+pref.+pref. wym. +<acj>(a)+i/yst(a) *CENTRUM* 'center' → *centralny* 'central' → *centralizować* 'to centralize' → *scentralizować* 'perf. to centralize' → *decentralizować* 'to decentralize' → *decentralizacja* 'decentralization' → *decentralista* 'decentralist'
II and subsequent	II and III	s_1+s_2	166	acj(a)+n(y) *KORONA* 'crown' → *koronować* 'to crown' → *koronacja* 'crowning' → *koronacyjny* 'adj. crowning'
		<s>+s_1+s_2	49	<ow>(y)+i/yc(a)+owiec *NERW* 'nerve' → *nerwowy* 'nervous' → *nerwica* 'neurosis' → *nerwicowiec* 'neurotic'
	II and IV	<s>+s_1+x+s_2	7	<acj>(a)+aln(y)+pref.+ość *TON* 'tone'→ *tonacja* 'key' → *tonalny* 'tonal' → *atonalny* 'atonal'→ *atonalność* 'atonality'

Beginning of the sequence	Distribution	Set	Number	Example
III and subsequent	III and IV	s_1+s_2	36	ator+stw(o) *MASKA* 'mask'→ *maskować* 'to mask' → *demaskować* 'to unmask/expose' → *demaskator* 'unmasker/exposer' → *demaskatorstwo* 'unmasking/exposing'
		<s>+<s>+ s_1+s_2	4	<n>(y)+<i/yzm>+i/yz(ować)+acj(a) *ESTETYKA* 'aesthetics' → *estetyczny* 'adj. aesthetic'→ *estetyzm* 'aestheticism' → *estetyzować* 'to aesthecize'→ *estetyzacja* 'aesthetization'
		<s>+ s_1+s_2	1	der. parad.+<k>(i)+i/yz(ować)+acj(a) *LAIC<YZM>* 'secularity' → *laik* 'secular person' → *laicki* 'secular' → *laicyzować* 'secularize' → *laicyzacja* 'secularization'
		<s>+x+ s_1+s_2	1	<k>(i)+der. parad.+jar<ni>(a)+sk(i) *PERS<J>A* 'Persia' → *perski* 'Persian' → *pers* 'Persian cloth, chintz'→ *persjarnia* 'arch. chintz mill'[b] → *persjarski* 'refering to chintz mill'[c]
	III and V	s_1+x+s_2	2	acj(a)+pref.+n(y) *FORMA* 'form' → *formować* 'to form' → *reformować* 'to reform' → *reformacja* 'reformation' → *kontrreformacja* 'counterreformation' → *kontrreformacyjny* 'adj. counterreformation' [alternation *j: yj*]
IV and subsequent	IV and V	<s>+s_1+s_2	1	<i/yzm>+-i/yst(a)+i/yczn(y) *FORMA* 'form' → *formować* 'to form' → *reformować* 'to reform' → *reformizm* 'reformism' → *reformista* 'reformist' → *reformistyczny* 'adj. reformist'

[a] The string *-um* is considered inflectional; [b] archaic; [c] archaic.

different formants between suffixes $s_1+x+x+s_2$. Such sequences encompass four steps, so they require that the nest be deep. Sequences with only one change of the word formation technique are also rare; this phenomenon is present in 135 sequences of sets with two suffixes (9.6% of sequences with two suffixes).

Sequences with three suffixes have a very similar distribution. They mostly begin at step I (90.5%), less commonly at step II (only 32) or step III (only 2). The most diverse configurations appear in sets with the first suffix at the step I (cf. Table 10.4).

Among the presented chains the most numerous are those that contain only suffixes: $s_1+s_2+s_3$, e.g., *ak+n(y)+ość* (*CUD* 2. 'miracle, something unusual' → *cudak* 'freak' → *cudaczny* 'freaky' → *cudaczność* 'freakiness').[32] They constitute 92% of sets

[32] Alternation *k : cz*.

TABLE 10.4

Sequence of Sets with Three Suffixes

Beginning of sequence	Distribution	Set	Number	Example
Step I and subsequent	I, II and III	$s_1+s_2+s_3$	295	n(y)+ak+k(a)$_{ż}$ *KREW* 'blood' → *krewny* 'blood relative' → *krewniak* 'coll. blood relative' → *krewniaczka* 'coll. female blood relative'
	I, II and IV	$s_1+s_2+x+s_3$	13	al<n>(y)+i/ystyk(a)+ truncation and der. parad. +k(a)$_{ż}$ *ORIENT* 'Orient' → *orientalny* 'oriental' → *orientatalistyka* 'Oriental Studies' → *orientalista* 'Oriental Studies scholar' → *orientalistka* 'female Oriental Studies scholar'
		s_1+s_2+<s>$+s_3$	9	nik+arz+<sk>(i)+k(a)$_3$ *DZIEŃ* 'day' → *dziennik* 'diary; daily (newspaper)' → *dziennikarz* 'journalist' → *dziennikarski* 'adj. journalist' → *dziennikarka*[a] 'Faculty of Journalism'
	I, II and V	$s_1+s_2+x+x+s_3$	1	n(y)+i/yk+der. parad.+pref.+ówk(a) *LAS* 'forest' → *leśny* 'adj. forest' → *leśnik* 'forester' → *leśniczy* 'gamekeeper' → *nadleśniczy* 'district forester' → *nadleśniczówka* 'district forest manager's residence'
	I, III and IV	$s_1+x+s_2+s_3$	9	n(y)+pref-parad.+ąc(y)+o *SPOKÓJ* 'peace' → *spokojny* 'peaceful' → *uspokoić/uspokajać* 'to pacify/calm down perf./imperf' → *uspokajający* 'calming/soothing' → *uspokajająco* 'adv. calming/soothing'
		s_1+<s>$+s_2+s_3$	7	al<n>(y)+<i/yzm>+i/ystyk(a)+n(y) *REGION* 'region' → *regionalny* 'regional' → *regionalizm* 'regionalism' → *regionalistyka* 'Regional Studies' → *regionalistyczny* 'adj. Related to Regional Studies'
	I, IV and V	$s_1+x+x+s_2+s_3$	5	n(y)+der. parad.+pref.+ał(y)+ość *KOŚĆ* 'bone' → *kostny* 'ossal' → *kostnieć* 'ossify; fig. to become numb with cold' → *skostnieć* 'perf. to ossify; fig. to become numb with cold' → *skostniały* 'ossified; fig. numb with cold' → *skostniałość* 'numbness with cold'
	I, IV, VI	s_1+<s>+<s>$+s_2+x+s_3$	1	al<n>(y)+<i/yzm>+<i/ystyczn>(y)+z...i/yz(ować) się+truncation+ani(e) *MATERIA* 'matter' → *materialny* 'material' → *materializm* 'materialism' → *materialistyczny* 'materialist' → *zmaterializować się* 'reflexive perfective to materialize' → *zmaterializować* 'perfective materialize' → *zmaterializowanie* 'materialization'
	I, VI, VII	s_1+<s>$+x+x+$<s>$+s_2+s_3$	1	al<n>(y)+<i/yz>(ować)+x+x+<acj>(a)+i/ystyczn(y)+e *CENTRUM* 'center' → *centralny* 'central' → *centralizować* 'centralize' → *scentralizować* 'perfective centralize' → *decentralizować* 'decentralize' → *decentralizacja* 'decentralization' → *decentralistyczny* 'decentralist → *decentralistycznie* 'adv. decentralist'

Beginning of sequence	Distribution	Set	Number	Example
Step II and subsequent	II, III and IV	$s_1+s_2+s_3$	15	k(i)+i/yz(ować)+acj(a) *LAIC<YZM>* 'secularity' → *laik* 'secular person, non-professional person' [alternation *c:k*]→ *laicki* 'secular/non-professional' → *laicyzować* 'to secularize' → *laicyzacja* 'secularization'
		<s>+ s_1+s_2 $+s_3$	14	<sk>(i)+i/yfik(ować)+acj(a)+n(y) *ROSJA* 'Russia' → *rosyjski* 'Russian' [alternation *j:yj*] → *rusyfikować* 'to Russify' → *rusyfikacja* 'Russification' → *rusyfikacyjny* 'adj. related to Russification'
	II, IV, V	<s>+ s_1+x+s_2 $+s_3$	3	<n>(y)+i/yz(ować)+pref.+acj(a)+n(y) *MINERAŁ* 'mineral' → *mineralny* 'adj. mineral' → *mineralizować* 'to mineralize' → *demineralizować* 'to demineralize' → *demineralizacja* 'demineralization' → *demineralizacyjny* 'demineralizational'
Step III and subsequent	III, IV and V	$s_1+s_2+s_3$	2	n(y)+i/yc(a)+ow(y) *KRĄG* 'cicrcle' → *krążyć* 'to circle' → *okrążyć/okrążać* 'to cicrcle around perf./imperf.' → *okrężny* 'circular' → *okrężnica* 'colon' → *okrężnicowy* 'colonal'

[a] *Dziennikarka* 'female journalist' derives directly from *dziennikarz* 'journalist,' at the same step as *dziennikarski.*

with three suffixes (as shown in Table 10.5). Only 8% sets contains nonsuffix formatives, distributed at steps II, III, or IV (e.g., sequences: $s_1+x+s_2+s_3$ or $s_1+s_2+x+x+s_3$).

As many as 22 sets among the 23 sequences with four suffixes consist of suffixes only: $s_1+s_2+s_3+s_4$, such as *k(i)+k(i)+arz+ka*$_{\dot{z}}$*: SANIE* 'sleigh' → *sanki* 'sledge (for children)' → *saneczki* 'dim. sledge/sport sledge/toboggan' → *saneczkarz* 'tobogganist' → *saneczkarka* 'female tobogganist.' However, during suffixations numerous truncations occur. Combinations of complete suffixes are rare, as usually they are deprived of some noninflectional elements, most commonly *-n-* and *-i/yzm* is deleted from suffix *-aln(y)* and *-alizm, -janizm,* such as *al<n>(y)+i/yz(ować)+acj(a)+n(y): NORMA* 'norm' → *normalny* 'normal' → *normalizować* 'normalize' → *normalizacja* 'normalization' → *normalizacyjny* 'normalizing'; *al<izm>+i/yz(ować)+acj(a)+n(y): STRUKTURA* 'structure' → *strukturalizm* 'structuralism' → *strukturalizować* 'to structure' → *strukturalizacja* 'structuring' → *strukturalizacyjny* 'structural' [alternation *j:yj*]. The truncation of these elements occurs in five sequences with four one-syllable suffixes. The relatively short length of sequences containing four suffixes is quite common, for example *-eczkark-* [k(i)+k(i)+arz+k(a)$_{\dot{z}}$], *-eczkarsk-* [k(i)+k(i)+arz+sk(i)], *-nikowość-* [n(y)+i/yk+ow(y)+ość], *-nikarsko* [nik+arz+sk(i)+o].

It is also worth noting that word formation processes that occur at step V and VI of one of the sequences lengthen the word by only one syllable (prefix derivation and prefix-altering derivation); similarly, the addition of suffix *-iz(ować)* does

TABLE 10.5

Sequences of Sets with Four Suffixes

Beginning of sequence	Distribution	Set	Number	Example
Step I and sub-sequent	I, II, III and IV	$S_1+S_2+S_3+S_4$	19	al<izm>+iz(ować)+acj(a)+n(y) *STRUKTURA* 'structure' → *struktural<izm>* 'structuralism' → *strukturalizować* 'to structure' → *strukturalizacja* 'structuralization' → *strukturalizacyjny* 'structural'
	I, II, V and VI	$S_1+S_2+x+x+S_3+S_4$	1	al<n>(y)+i/yz(ować)+pref.+ pref. wym. +acj(a)+n(y) *CENTRUM* 'center' → *centralny* 'central' → *centralizować* 'to centralize' → *scentralizować* 'perfective to centralize' → *decentralizować* 'to decentralize' → *decentralizacja* 'decentralization' → *decentralizacyjny* 'decentralizational'
	I, III, IV and V	S_1+<s>+ $S_2+S_3+S_4$	2	i/yw(a)+<k>(a)$_3$+n(y)+i/yk +owat(y) *OL<EJ>* 'oil' → *oliwa*[a] 'olive oil' → *oliwka* 'olive tree' → *oliwny* 'referring to an olive tree' → *oliwnik* 'wild olive tree, Elaeagnus' → *oliwnikowaty* 'adj. wild olive tree-like'
Step II and sub-sequent	II, III, IV, and V	<s>+$S_{1>}+S_2+S_3+S_4$	1	<i/yzm>+i/yz(ować)+ac<j>(a)+ or+ski *MECHAN<IK>A* 'mechanics' → *mechanizm* 'mechanism' → *mechanizować* 'mechanize' → *mechanizacja* 'mechanization' → *mechanizator* 'mechanizer' → *mechanizatorski* 'adj. referring to mechanization'

[a] SGS considers *oliwa* 'olive oil' as a synchronic derivative of *olej* 'oil,' in spite of the fact that genetically *oliwa* is a loanword; the derivational relations is based on formal similarity and semantic relations (olive oil is a kind of vegetable oil).

not entail a significant enlargement of the structure. The derivational stem gains only one syllable *iz*. Thus the suffixes and other formatives do not form very large structures.

5. Discussion

The most noteworthy phenomenon observed are the differences in the frequencies of the sequences: ranging from several hundred of instances to a single one. Among sets with two suffixes, as many as 968 were found only once, though their structure seems simple enough, such as *usz+ow(y)*, *ur+ek*, *ul(a)+k(a)*$_d$, *k(a)*$_d$*+n(y)*.

It is also important to observe the position such sets occupy in the derivational history of a word. In most cases sets of two suffixes occur at the two first steps, which means that the first suffix of the set is attached directly to the nonderived word and the second directly to the first derivative. This can be seen when comparing the frequency of the sets at steps I and II with the sequence's total load (see Table 10.6). There are but few instances of suffixation occurring at further steps.

TABLE 10.6

The Frequency of Two-Suffix Sequences

Level	Sequence	Load		Example
		s_1+s_2 (step I and II)	total[a]	
Highest	*nie: n(y)+e*	333	354	CMENTARZ '*cemetery*' → cmentarny '*related to a cemetery*' → cmentarnie '*in a way related to cemetery*'
	owo: ow(y)+o	302	312	*CEL* 'purpose' → *celowy* 'purposeful' → *celowo* 'purposefully'
High	*ność: n(y)+ość*	225	235	*CEL* 'target' → *celny* 'accurate, well aimed' → *celność* 'accuracy'
	i/ycznie: i/yczn(y)+e	192	197	*CHOREOGRAF(I)A* 'choreography' → *choreograficzny* 'choreographic' → *choreograficznie* 'choreographically'
	owość: ow(y)+ ość	173	183	*KORONKA 1.* 'lace' → *koronkowy* 'lacy, lace-like' → *koronkowość* 'lace-like character'
	ówk(a): ow(y)+k(a)$_3$	159	162	*CYFRA* 'digit' → *cyfrowy* 'digital' → *cyfrówka* 'digital camera'
	eczk(a): k(a)$_d$+k(a)$_d$	133	133	*CHATA* 'cabin' → *chatka* 'small cabin' → *chateczka* 'dim. small cabin'
	kow(y): ek+ow(y)	125	133	*KOPER* 'dill' → *koperek* 'dim.dill' → *koperkowy* 'of dill'
	arsk(i): arz+sk(i)	117	125	*KRĘGLE* 'skittle, bowling pin' → *kręglarz* 'bowler' → *kręglarski* 'adj. bowling'
	owiec: ow(y)+ec	116	117	*KLATKA* 'stairwell' → *klatkowy* 'adj. of stairwells' → *klatkowiec* 'an appartment builidng with several stairwells'
	arstw(o): arz+stw(o)	115	121	*KAJAK* 'kayak' → *kajakarz* 'kayaker' → *kajakarstwo* 'kayaking'
	i/yczność: i/yczn(y)+ość	106	111	*CYKL* 'cycle' → *cykliczny* 'cyclic' → *cykliczność* 'cyclicity'
Medium	14	41 — 97		np. *k(a)$_d$+ow(y): KOLEJ* 'order' → *kolejka* 'queue' → *kolejkowy* 'refering to a queue'
	55	11 — 38		e.g. *nik+ow(y): KORA* 'bark' → *kornik* 'woodworm' → *kornikowy* 'related to a woodworm'
Low	411	2 — 10		e.g. *jer+sk(i): KASA* 'till' → *kasjer* 'teller' → *kasjerski* 'teller's'
Lowest	910	1		e.g. *us+k(a)$_3$:* *KRAK<ÓW>* 'Kraków' → *Krakus* 'coll. resident of Krakovian region' → *krakuska* 'traditional folk headwear worn by residents of Krakovian region'

[a] I.e., in sets starting at other steps than step I (e.g., II, III) or in sets with lacunae (e.g., s_1+x+s_2).

Not surprisingly, another regularity can be observed: the longer the sequence the less it is used (cf. Tables 10.6 and 10.7.). Sets of four or five suffixes are even rarer.

Among the sets of two suffixes, the set *n(y)+e* and *ow(y)+o* is the most frequent. When these suffixes are attached, the derivatives are either denominal adjectives or deadjectival adverbs. The sets *n(y)+ość*, *i/yczn(y)+e*, *ow(y)+ość*, *ow(y)+*$k(a)_3$, $k(a)_d+k(a)_d$ are less frequent in this group; the dominant derivatives are deadjectival nouns: nomina essendi, univerbalized structures, and second degree diminutives. The examples are given in Table 10.6. Usually the most frequent sequences (level highest and high) contain suffixes of different lexical categories and involve

TABLE 10.7

The Load of Sequences of Three Suffixes

Level	Sequence	Number of instances		Example
		$s_1+s_2+s_3$ (step I, II and III)	total	
highest	*i/yzacj(a): i/yz(ować)+acj(a)+n(y)*	16	21	*KLIMAT* 'climate' → *klimatyzować* 'climatize' → *klimatyzacja* 'climatization (i.e. air conditioning' → *klimatyzacyjny* 'refering to air conditioning'
Medium	*alizacj(a): al<n>(y)+ i/yz(ować)+ acj(a)*	9	11	*MATERIA* 'matter' → *material<n>y* 'material' → *materializować* 'materialize' → *materializacja* 'materialization'
	arsko: arz+sk(i)+o	6	6	*KRONIKA* 'chronicle' → *kronikarz* 'chronicler' → *kronikarski* 'referring to a chronicler' → *kronikarsko* 'like a chronicler'
	kowo: ek+ow(y)+o	6	8	*STÓG* 'cone' → *stożek* 'little cone' → *stożkowy* 'conical' → *stożkowo* 'conically'
	kowato: $k(a)_d$*+owat(y)+o*	6	6	*SIATKA* 'net' → *siateczka* 'little net' → *siateczkowaty* 'net-like' → *siateczkowato* 'in a net-like way'
	owcow(y): ow(y)+ec+ow(y)	6	7	*LÓD* 'ice' → *lodowy* 'glacial' → *lodowiec* 'glacier ' → *lodowcowy* 'glacial'
	i/yz(ować)+ator+sk(i)	6	6	*KOLONA<I> A* 'colony' → *kolonizować* 'colonize' → *kolonizator* 'colonizer' → *kolonizatorski* 'related to colonizers'
Rare	19	5–3		e.g. *n(y)+utk(i)+o:* *SPOKÓJ* 'calm (noun)' → *spokojny* 'calm (adjective)' → *spokojniutki* 'very calm' → *spokojniutko* 'very calmly'
Very rare	349	2–1		e.g. *arz+*$k(a)_ż+k(a)_d$ *KARCZMA* 'inn' → *karczmarz* 'inn-keeper' → *karczmarka* 'female inn-keeper' → *karczmareczka* 'dim. female inn-keeper'

either combination of adjectival and adverbial suffixes or of nominal and adjectival suffixes. Only two sequences contain suffixes of the same, that is nominal, category: $k(a)_d+k(a)_d$, *arz+stw(o)*.

Sets of three suffixes are less frequent: only sixteen instances have been found. This concerns the sequence *i/yz(owa*ć*)+acj(a)+n(y)*, which appears to be quite long (six syllables). In reality from the first suffix only the *-i/yz* remains, because the string *-owa*ć is inflectional; the second suffix is subject to alternation *j* : *yj*, so that the resulting sequence is only four syllables long (if the final vowel, representing the inflectional ending, is counted).

The sequence *al<n>(y)+i/yz(ować)+acj(a)* is quite frequent, and the *-n-* element is often deleted, as it generally happens in sets containing the suffix *-aln(y)*. Neither sequence is homogenic: they contain verbal, nominal, and adjective suffixes in different permutations. Other chains of three suffixes may contain other kinds of formatives, such as nominal, adjectival, and adverbial; the sets containing only nominal suffixes are rare (cf. the example from the line describing the very low level in Table 10.7), and so are sets containing two adjectival suffixes: *ow(y)+ec+ow(y)*.

The sequences of four or five suffixes have but singular instances in the data. They often contain the suffixes *-i/yz(ować), -acj(a), -n(y)*, and *-o* and *-e*. Only the string *-alistycznie: al<n>(y)+<i/yzm>+i/yst(a)+i/yczn(y)+e* was found twice. It should be noted that here also truncations are frequent, so the eventual length of the string is not just the result of juxtaposition of individual suffixes; for example in the sequence quoted above a part of the suffix *–i/yzm* and a part of the suffix *-aln(y)* are deleted (*FORMA* 'form' → *formal<n>y* 'formal' → *formal<izm>* 'formalism' → *formalista* 'formalist' → *formalistyczny* 'formalistic' → *formalistycznie* 'formalistically').

6. Conclusions

Almost 2,000 generalized suffix sets have been investigated. These were sets of minimum two elements, starting at various steps; about 300 different suffixes, native and foreign, were indentified within the sets.

It was demonstrated that the majority of suffix sets contain two suffixes; three-suffix sets are much rarer, and sets containing more than three suffixes are very sporadic. The longest set contained five suffixes distributed into five steps (with two truncations of suffix fragments).

There are differences in the frequencies of the sequences, ranging from several hundred of instances to a single one. Sets of four or five suffixes are rarer, they have but singular instances in the data. Two suffixes sets are the most frequent (e.g., *n(y)+e, ow(y)+o, n(y)+ość, i/yczn(y)+e, ow(y)+ość*, $ow(y)+k(a)_3$, $k(a)_d+k(a)_d$, *ek+ow(y), arz+sk(i)*; see: tab. 7), and among sets with three suffixes (- *i/yzacj(a)+acj(a)+n(y)*).

It was shown that suffix sets may be continuous, with suffixes attaching directly one to another, without lacunae (i.e., some other formative interrupting the suffix sequence) or may contain lacunae (i.e., other formatives appearing between suffixes), as suffixation is accompanied by other word formation techniques. The majority of sequences is continuous (almost 90%), which means that suffixation is a homogenous technique. The discontinuous sequences amount to 11% of all sets. Suffix sequences tend not to be interrupted by other formatives.

During suffixations numerous truncations occur. It has been shown that combinations of complete suffixes are rare, as usually they are deprived of some noninflectional elements; most commonly *-n-* and *-i/yzm* are deleted from the suffixes *-aln(y)* and *-alizm, -janizm*. Truncations of entire suffixes appear in 132 suffix sequences, most commonly in sets with two suffixes, less commonly in sets with three suffixes and in sets with four suffixes. In some sets two truncations occur, but the number of such sets is marginal. Truncation of suffixes or of their fragments plays an important role of limiting the length of the derived structures (cf. Burkacka 2007).

It has been shown that usually the sequences start at step I, where the first suffix combines with the stem of an unmotivated noun (83.2 % of sequences). Only 14% of sequences begin at step II and only 2.5 % at step III. This was to be expected in long sequences, but the similarity between sets with two and three suffixes is surprising. At first glance one would expect the sets with two suffixes to be structurally more diverse — for example to contain three instances of other word formation techniques — but this was not confirmed by the data. The reason may be that such procedure would have possibly entailed frequent shifts in part of speech between the base and the derivative, including some back and forth changes of categories. Such shifts are generally undesirable, as shown by Jadacka, who claimed that "word formation processes operate easier if each step does not involve a change of part of speech classification" (Jadacka 1995: 159).

Another regularity can be observed: the longer the sequence the less it is used.

The longer the sequence, the more likely it is to begin at step I. Thus to produce a long sequence, derivation should start with suffixation — though it must be added that single suffixes also appear most frequently at step I. Usually the most frequent sequences contain suffixes with different lexical-category specifications; they are usually combinations of adjectival and adverbial suffixes or of nominal and adjectival suffixes.

List of Symbols and Abbreviations Used

S – noun
Ad – adjective
Adv – adverb
V – verb

S,S	–	denominal noun
S,Ad	–	denominal adjective
S,V	–	denominal verb
(V+)	–	verbal added base, such as description (V+)S, S symbolizes a denominal noun, formed through composition, with the second, verbal base being attached to the underived noun
(+S)	–	nominal added bases, such as S,Ad(+S),S would corresponds to a derivative created from a denominal adjective and a noun
łow.	–	hunting terminology
zool.	–	zoological terminology
bot.	–	botanical terminology
uż. rzecz.	–	used as a noun
2., 3.	–	numbers of words' senses
*	–	asterisk – onomasiological derivative
→	–	direction of derivation
‘’	–	meaning of derivational paraphrase/English gloss
()	–	inflexional element, such as *(a)*, *(y)*
<>	–	truncation of noninflectional word fragment, such as -*k*- in the word *dzieć<>ak* ‘kid, derived from *dziecko* ‘child’
+	–	symbol indicating combination with another derivational formative
s_1	–	first suffix in a sequence
s_2	–	second suffix in a sequence
s_3	–	third suffix in a sequence
s_4	–	fourth suffix in a sequence
s_5	–	fifth suffix in a sequence
<s>	–	suffix deleted due to further suffixation
x	–	nonsuffixal formative
-k(a)$_{\dot{z}}$	–	suffix forming feminine nouns
-k(a)$_{d}$	–	suffix forming diminutive nouns
-k(a)$_{3}$	–	suffix forming neither feminine nor diminutive nouns (forming, e.g., names of professions or tools)
I	–	first step (derivatives motivated directly by an unmotivated noun)
II	–	second step (derivatives motivated directly by a derivative from the I step)
III	–	third step (derivatives motivated directly by a derivative from the II step)
IV	–	fourth step
V	–	fifth step
VI	–	sixth step, and so forth.
der. parad.	–	paradigmatic derivation
pref.	–	prefixal derivation

pref-parad. – prefixal-paradigmatic derivation
pref. wym. – prefix-altering derivation
capitals, for example FORMA – unmotivated nouns that are the center of derivational nests
SGS – *Słownik gniazd słowotwórczych współczesnego języka ogólnopolskiego*

References

Burkacka Iwona. 2006. "Ciągi sufiksalne w derywatach odrzeczownikowych." *Poradnik Językowy* 10:19–30.

Burkacka, Iwona. 2007. "Ucięcia w procesie sufiksacji." *LingVaria* 2(4):55–66.

Burkacka, Iwona. 2010. "Łączliwość rzeczownikowych sufiksów obcych." In *Słowa – kładki, na których spotykają się ludzie różnych światów*, edited by Iwona Burkacka, Radosław Pawelec, and Dorota Zdunkiewicz-Jedynak, 177–192. Warsaw: Nakładem Wydziału Polonistyki Uniwersytetu Warszawskiego.

Burkacka, Iwona. 2012. *Kombinatoryka sufiksalna w polskiej derywacji odrzeczownikowej*. Warsaw: Nakładem Wydziału Polonistyki Uniwersytetu Warszawskiego.

Czernecka, Jadwiga. 1985. "Gniazda słowotwórcze czasowników oznaczających przemieszczanie w przestrzeni (LETET'-LETAT,' LECIEĆ-LATAĆ) w języku rosyjskim i polskim." *Prace Językoznawcze UŚl.*, v. 11, *Problemy badań konfrontatywnych języka polskiego i rosyjskiego*, edited by Maja Szymoniuk, 91–111. Katowice.

Dunaj, Bogusław. (ed.). 1998. *Słownik współczesnego języka polskiego* (second edition). Warsaw: Reader's Digest Przegląd. Fras, Janina. 1985. "Gniazdo słowotwórcze ziemi w języku polskim, serbsko-chorwackim i rosyjskim." *Rocznik Slawistyczny*, v. XLV: 31–43.

Grzegorczykowa, Renata., Jadwiga Puzynina. 1979. *Słowotwórstwo współczesnego języka polskiego, rzeczowniki sufiksalne rodzime*. Warsaw: Państwowe Wydawnictwo Naukowe.

Grzegorczykowa, Renata., Roman Laskowski, and Henryk Wróbel (eds.). 1998. *Gramatyka współczesnego języka polskiego. Morfologia*, v. I i II. Warsaw: Wydawnictwo Naukowe PWN.

Gussmann, Edmund. (ed.). 1985. *Phono-Morphology. Studies in the Interaction of Phonology and Morphology*. Lublin: Redakcja Wydawnictw Katolickiego Uniwersytetu Lubelskiego.

Jadacka, Hanna. 1995. *Rzeczownik polski jako baza derywacyjna*. Warsaw: Wydawnictwo Naukowe PWN.

Jadacka, Hanna. 2001. 'Wstęp.' In *Słownik gniazd słowotwórczych współczesnego języka ogólnopolskiego*, edited by Hanna Jadacka et al. v. I i II. Kraków: Towarzystwo Autorów i Wydawców Prac Naukowych Uniwersitas.

Jadacka, Hanna, Magdalena Bondkowska, Iwona Burkacka, Elżbieta Grabska-Moyle, and Tomasz Karpowicz. 2001. *Słownik gniazd słowotwórczych współczesnego języka ogólnopolskiego*, Jadacka H. (ed.), v II *Gniazda odrzeczownikowe*. Kraków: Towarzystwo Autorów i Wydawców Prac Naukowych Uniwersitas.

Jadacka, Hanna. 2005. *Kultura języka polskiego. Fleksja, słowotwórstwo, składnia*. Warsaw: Wydawnictwo Naukowe PWN.

Jochym-Kuszlikowa, Ludwika. 1982. *Metoda analizy gniazdowej w konfrontatywnych badaniach słowotwórstwa współczesnego języka polskiego i rosyjskiego*. Kraków: Wydawnictwo Naukowe Wyższej Szkoły Pedagogicznej.

Kleszczowa, Krystyna. 2010. "Historycznie o podstawach słowotwórczych polskich rzeczowników," *LingVaria* 2(10):59–67.

Kowalik, Krystyna. 1977. *Budowa morfologiczna przymiotników polskich*, Wrocław: Ossolineum.

Kowalik, Krystyna. 1979. "Z morfotaktyki tematów rzeczownikowych współczesnego języka polskiego." *Polonica* V:25–57.

Kowalik, Krystyna. 1997. *Struktura morfonologiczna współczesnej polszczyzny*, Kraków: Instytut Języka Polskiego Polskiej Akademii Nauk.

Kowalik, Krystyna. 1998. *Morfotaktyka*. In *Gramatyka współczesnego języka polskiego. Morfologia*, edited by Renata Grzegorczykowa et al., 585–618. Warsaw: Wydawnictwo Naukowe PWN.

Kreja, Bogusław. 1989. *Z morfonologii i morfotaktyki współczesnej polszczyzny*, Wrocław: Ossolineum.

Kreja Bogusław. 1996. *Studia z polskiego słowotwórstwa*, Gdańsk: Wydawnictwo Uniwersytetu Gdańskiego.

Kreja, Bogusław. 2002. *Studia i szkice słowotwórcze*, Gdańsk: Wydawnictwo Uniwersytetu Gdańskiego.

Laskowski, Roman. 1973. "Budowa morfonologiczna polskich przyrostków rzeczownikowych." *Język Polski* LIII, Nr 2–3:132–141.

Malicka-Kleparska, Anna. 1985. "Parallel derivation and Lexicalist Morphology: the case of polish diminitivisation." In *Phono-Morphology. Studies in the Interaction of Phonology and Morphology*. edited by Edmund Gussmann, 95–112. Lublin: Redakcja Wydawnictw Katolickiego Uniwersytetu Lubelskiego.

Manova, Stela, and Mark Aronoff. 2010. "Modeling affix order." *Morphology* 20:109–131.

Manova, Stela, and Kimberley Winternitz. 2011. "Suffix order in double and multiple diminutives: with data from Polish and Bulgarian." *Studies in Polish Linguistics* 6:115–138.

Markowski, Andrzej. 1990. *Leksyka wspólna różnym odmianom polszczyzny*, v. I- II. Warsaw: Wydawnictwa Uniwersytetu Warszawskiego.

Michalewski, Kazimierz. 1984. *Dystrybucja polskich rzeczownikowych formantów przyrostkowych*. Łódź: Acta Universitatis Lodziensis. Folia Linguistica 9.

Nagórko, Alicja. 2002. *Zarys gramatyki polskiej*. Warsaw: Wydawnictwo Naukowe PWN.

Rice, Karen. 2011. "Principles of Affix Ordering: An Overview." *Word Structure* 4: 272–300.

Rubach, Jerzy. 1981. *Cyclic Phonology and Palatalization in Polish and English*. Warsaw: Wydawnictwa Uniwersytetu Warszawskiego.

Satkiewicz, Halina. 1969. *Produktywne typy słowotwórcze współczesnego języka ogólnopolskiego*. Warsaw: Wydawnictwa Uniwersytetu Warszawskiego.

Satkiewicz, Halina. 1972. "Zagadnienia poprawności słowotwórczo-semantycznej." In Halina Satkiewicz, *Kultura języka polskiego*, v. 2. 229–442. Warsaw: Państwowe Wydawnictwo Naukowe.

Skarżyński, Mirosław, Małgorzata Berend, Magdalena Bondkowska, Iwona Burkacka, Hanna Jadacka, Monika Olejniczak, Teresa Vogelgesang. 2004. *Słownik gniazd słowotwórczych współczesnego języka ogólnopolskiego*, v. III, edited by Mirosław Skarżyński. Kraków: Towarzystwo Wydawnicze „Historia Iagellonica."

Skarżyński, Mirosław. 2004a. "Wstęp." In Skarżyński Skarżyński, Mirosław, Małgorzata Berend, Magdalena Bondkowska, Iwona Burkacka, Hanna Jadacka, Monika Olejniczak, Teresa Vogelgesang. 2004. *Słownik gniazd słowotwórczych współczesnego języka ogólnopolskiego*, v. III, edited by Mirosław Skarżyński, VII–XIV. Kraków: Towarzystwo Wydawnicze „Historia Iagellonica."

Skarżyński, Mirosław. 2004b. *Słownik gniazd słowotwórczych współczesnego języka ogólnopolskiego*, v. IV. Kraków: Towarzystwo Wydawnicze „Historia Iagellonica."

Sobol, Elżbieta. (ed.). 2002. *Nowy słownik języka polskiego PWN*. Warsaw: Wydawnictwo Naukowe PWN.

Soboleva, Polina A. 1970. *Aplikativnaja grammatika i modelirowanie slovoobrazovanija*, Moskva.

Stawnicka, Jadwiga. 1991. "Wykorzystanie metody opisu gniazdowego w określeniu granic homonimii słowotwórczej (na materiale języka rosyjskiego i polskiego)." In *Z zagadnień językoznawstwa rosyjskiego i polskiego*, edited by Michał Blicharski, *Prace Językoznawcze* UŚl, v. 18, 50–64.

Szymanek, Bogdan. 2010. *A Panorama of Polish Word-Formation*. Lublin: Wydawnictwo Katolickiego Uniwersytetu Lubelskiego.

Szymanek, Bogdan, and Tatyana Derkach. 2005. "Constraints on derivation of double diminutives in Polish and Ukrainian," *Studies in Polish Linguistics* 2:91–110.

Śmiech, Witold. 1986. *Derywacja prefiksalna czasowników polskich*. Wrocław: Ossolineum.

Šaumjan, Sebastian K. 1965. *Strukturnaja lingvistika*. Moskva: Izdatel'stvo "Nauka."

Tixonov, Aleksand N. 1985. *Slovoobrazovatel'nyj slovar' russkogo jazyka*, v. 1 i 2. Moskva: Russkij jazyk.

Vogelgesang, Teresa. 2001. *Słownik gniazd słowotwórczych współczesnego języka ogólnopolskiego*, v. I *Gniazda odprzymiotnikowe*, edited by Hanna Jadacka. Kraków: Towarzystwo Autorów i Wydawców Prac Naukowych Uniwersitas.

Waszakowa, Krystyna. 1996. *Słowotwórstwo współczesnego języka polskiego. Rzeczowniki z formantami paradygmatycznymi*. Warsaw: Wydawnictwa Uniwersytetu Warszawskiego.

Waszakowa, Krystyna. 1994. *Słowotwórstwo współczesnego języka polskiego. Rzeczowniki sufiksalne obce*. Warsaw: Wydawnictwa Uniwersytetu Warszawskiego.

11

Reduplication, Repetition, Hypercharacterization, and Other Affix-Doubling in Child Language

Wolfgang U. Dressler, Katarzyna Dziubalska-Kołaczyk, Natalia Gagarina, and Marianne Kilani-Schoch

1. Introduction

The aim of this contribution is to focus on morphological aspects of reduplication, as well as of hypercharacterizing and other suffix-doubling types in child language; show how they differ from each other and from the respective adult languages; and to trace the interaction between these operations and emerging principles of affix ordering.

In the case of morphological reduplication, what takes place is a doubling of the base in total reduplication or a doubling of some part of the base in partial reduplication. Moreover, similar to syntactic reduplication, morphological reduplication typically involves some sort of iconic intensification of the meaning of the base, most so in intensification in the strict sense, as in It. *il fior fiore* 'the very cream of' (lit. 'the flower-flower of') or in the Sanskrit reduplicating verb category of intensives, or iconic repetition of action, as in Mokilese reduplication: *roar* 'to give a shudder,' *roarroar* 'to be shuddering.'

Strict strengthening of meaning occurs in recursive affix doubling or tripling, as in the recursive diminutive formation of Pol. *kot* 'cat' → *kot-ek* 'little cat' → *kot-ecz-ek* 'very small cat' (with morphonological palatalization of the first suffix-final velar), potential *kot-ecz-ecz-ek*, and so on.

Inflectional hypercharacterization involves a repetition of the (original) meaning of the more central affix by a more peripheral suffix, as in Breton pluralization of diminutives, for example *paotr* 'man,' diminutive *paotr-ig* 'lad,' plurals *paotr-ed* and *paotr-ed-ig-où* 'man-PL-DIM-PL.' As in this case, typically the hypercharacterized and the hypercharacterizing suffixes are not identical formally, although this occurs as well, such as in Sp. *café* 'coffee-house,' plurals *café-s* and substandard *café-s-es*.

Our topic is also relevant for the metatheoretical issue of recursivity: Hauser, Chomsky, and Fitch (2002) and Chomsky (2007) consider recursivity to be the most fundamental characteristic of the innate human language capacity. Therefore we should expect early emergence of repetitive morphological patterns in language acquisition. Strict recursivity in the sense of doubling or tripling of the same element may occur in reduplication in diminutive formation, but not in inflectional hypercharacterization. Recursivity in a looser sense of repetition -not by an identical affix but by a synonymous one- may occur in diminutive formation, but, again, not in inflectional hypercharacterization beyond one additional suffixation, especially in plural formation.

The question of what constrains affix order (cf. Manova and Aronoff 2010) is relevant for our topic only in the case of loose recursivity; that is, stacking of synonymous affixes in diminutive suffix accumulation and in inflectional hypercharacterization (cf. Dressler 2004b). We found two types of constraints: (1) suffix-specific constraints of the preceding suffix on which suffixes may follow, only very partially based on phonological properties of the suffixes; and (2) the constraint that no less productive suffix may follow a more productive suffix, which presupposes a model of graded morphological productivity, as proposed in Dressler (2004a) and Dressler and Ladányi (2000). This constraint can be referred to the issue of degree of parsability (Manova and Aronoff 2010: 122–123), because more productive suffixations are easier to parse than less productive ones.

The child data used for this paper are longitudinal and come from the international Cross-linguistic Project on Pre- and Protomorphology in Language Acquisition, organized and coordinated by the first author in behalf of the Austrian Academy of Sciences (cf. Dressler 1997, Dziubalska-Kołaczyk 1997, Bittner, Dressler and Kilani-Schoch 2003, Savickienė and Dressler 2007, Stephany and Voeikova 2009). This project is devoted to a comparative, longitudinal study of spontaneous speech corpora in early acquisition phases. Within the 18 project languages, our main data come from the authors' languages of French, German, Polish, and Russian. The child corpora come from:

GERMAN:[1]	Katharina (Kathi), single child, female, Vienna: 1;6–3;0
	Jan, second child, male, Vienna: 1,3–2;0
FRENCH:	Sophie, third child, female, Lausanne: 1;6–3;8
	Emma, single child, female, Lausanne: 1;4–2;11
POLISH:	Zosia, first child, female, Poznan: 1;7–3;2
	Filip, first child, male, Poznan: 1;7–3;7
	Magda, first child, female, Poznan, diary: 0;2–2;8
	Kasia, second child, female, Poznan, diary: 1;6–3;7

[1]We thank Sabine Laaha and Katharina Korecky-Kröll for the collection, transcription, coding, and raw analysis of the German data.

RUSSIAN: Varja, first child, female, Moscow: 0–2;6
Vanja, single child, male, St. Petersburg: 1;5–4;2
Filip, single child, male, St. Petersburg: 1;4–2;6
Liza, single child, female, St. Petersburg: 1;2–3;0
plus selected data from 15 further children

The data were transcribed and morphologically coded according to a version of CHILDES (cf. MacWhinney 2000) adapted for German by Sabine Laaha and Katharina Korecky-Kröll.

In addition, literature will be cited especially for German data from Elsen's (1991) monograph on her daughter Annalena, and for Czech examples from Pačesová (1968).

Morphological grammar emerges rather late in early child language. For the emergence of morphology, our project distinguishes three phases: premorphology, protomorphology, and morphology proper (or modularized morphology). We define the premorphological phase of language acquisition as the phase where morphological operations can already occur (both extragrammatical ones and rote-learned and isolated precursors of later grammatical rules). The protomorphological phase of language acquisition can be defined as the period when the child has detected morphology; that is, the principle of morphological (de)composition in form and meaning. Now the child is able to draw generalizations on morphological patterns and create analogies. The third phase of morphology proper (or modularized morphology, if one assumes ontogenetic modularization) is already qualitatively similar to adult morphology. This paper will focus, as our whole acquisition project, on protomorphology.

2. Reduplication

2.1. PHONOLOGICAL REDUPLICATION AND CONSONANT HARMONY

Since we deal with morphological reduplication only, we have to exclude from consideration phonological types of reduplication, which are restricted to premorphology (except when continued in playful rhymes): one common case are nonsense-word repetitions, as in Polish *dudubiduubulu* and *dudy budy*, stimulated by *dzwonek* 'bell' (Zosia 1;7) or Russian *papa ljapa* 'daddy X,' *ami mami* 'X mommy' (Voeykova 1997).

We also have to exclude consonant harmony and rarer vowel harmony, as in French *papouf* = *pantoufles* 'slippers' (Sophie 1;11), [piti] = *petit* (Emma 1;9). The phonological character is proven by the fact that harmony is often restricted to the anticipation or perseveration of single phonological features, especially in terms of consonant harmony, such as in German Jan's (1;3) *bam* = *Banane* 'banana' and (1;6) *dette* = *Kassette* 'cassette,' *kogo* = *Korb* 'basket' (cf. Dressler et al. 2005).

2.2. ONOMATOPOETIC REDUPLICATION

From early premorphology onward children take up traditional onomatopoetic words, such as the following selected examples (more in Dressler et al. 2005) referring to sounds or gestures of animals, for example:

CV.CV: German (Jan since 1;3) *wauwau* 'dog,' French, Russian
CVC.CVC: French (Emma 2;2) *boumboum* 'heart,' German, Russian
VC.VC: Russian (Vanja 2;1) *af-af* 'dog,' *am-am* 'eating.'
CVC: Polish (Zosia 1;8) *ciuch-ciuch* 'train.'

These reduplications share with morphological reduplications two prosodic properties that differentiate them from syntactic reduplications: they represent just one phonological word and have a unique main stress, which is word-final by default.

Such onomatopoetically reduplicated words are rarely further reduplicated, and if so, then presumably in word play. For instance in German (Kathi) we find *wawa* ~ *wauwau* → *wawawa*, *wawawawawa* 'dog' (1;10, 2;0), *wauwauwauwauwau* (2;0), *kwakkwakkwak* 'frog' (2;4). Russian examples are (Liza 1;11) *pi-pi-pi* and (Vanja 1;10) *bul′-bul′-bul′* ~ *bul′* 'cluck,' a Polish one is (Zosia 1;8) *tututu* (~*tutu*) 'train' and the main French example is (Sophie 1;8) *papam* (an elephant's song in a children book) → *papapam*. For Czech Pačesová (1968: 237) gives (1;5) *bububú* 'moo-moo.' At least triplication often seems to represent playful intensification.

Nontraditional onomatopoetic imitation of perceptual impressions used for denoting the referent in a sound-iconic way are from Polish: (Zosia 1;8.0) *dudi-dudi*, *dudi-didu*, (1;10) *titen*, *titon* 'clock,' *tato-taton taton-taton* 'sound of a clock,' and so on.

All these examples are instances of reduplications belonging to extragrammatical morphology (Dressler 2001) or to what Zwicky and Pullum (1987) called expressive morphology. Such operations start early in premorphology. If children use them much, then these operations decrease sharply in protomorphology and reach an adult-like level of frequency. Children engage in such extragrammatical patterns of word creation as a sort of formal exercise facilitating later acquired morphotactic operations of grammatical morphology in a sort of training phase (Protassova 1997, Voeykova 1997; but cf. Dressler, Kilani-Schoch, and Klampfer 2003).

Among these various types of extragrammatical reduplication there are early cases of recursive reduplication. Note that in all examples cited there is none that evidences a change of meaning due to recursive as opposed to simple reduplication.

2.3. PHONOLOGICAL SIMPLIFICATIONS

Reduplications may also be simplifying phonological modifications of nononomatopoetic adult target words. Reduplicative syllable copies replace target syllables that are difficult to produce or even to perceive for the child. This may result in radically different syllable shapes, first in disyllabic words, later especially in polysyllabic words. In this way the child simplifies word structures on the one hand, by avoiding word-final and syllable-final consonants and by producing a—at least partially—harmonic syllable. On the other hand, and what is more important, the child often tries to preserve the template, especially the number of syllables, of the target word.

These reduplications may be, in regard to syllable structure, either total or partial:

- Total CV: Jan (G.) 1;3: *nana* = *Nase* 'nose,' also *nane* (consonant harmony); Sophie (F.) 1;5 *pama* (= *pyjama*) and later reduplicated *papama* (1;7); Liza (R.) 1;8 diminutive *igigiki* = *ogurchiki* 'cucumbers-DIM'; Kasia (P.) *mokokoki/yla* ~ *lokomotywa* 'engine,' *kokika/kokaka* ~ *kochana* 'loved,' *fololoy* ~ *kolorowy* 'colored'; Czech (Pačesová 1968: 237) *kikikiki* (= *kykyryký*), *momonoc* (= *dobrou noc!* 'good night!').
- Partial CV: Jan (G.) 1;8,3 *bockdedecker* = *Doppeldecker* 'double-decker'; Sophie 1;9 [wawar] = *au revoir* 'goodbye'; CCV Liza (R.) 1;11: *ajasja* = *prineslá* 'she brought' (first /l/ lenited to /j/), and so on.

But also monosyllabic words can be reduplicated and more syllables can be produced than there are in polysyllabic targets (which may be hypercorrective in view of frequent syllable reductions), as in Nico's (G) *Pozizazei, Pozazazei* ~ *Polizei* 'police' (more in Dressler et al. 2005); Czech (Pačesová 1968: 237f) 1;5 *kukuk* ~ *kuk!* 'peep-bo!,' and a bisyllabic word is reduplicated in *hijaja* ~ *hijá!* 'gee up!.' Also such reduplications sharply decrease and even disappear in the transition to protomorphology, simultaneously with a sharp decrease in onomatopoetic reduplications and in the use other extragrammatical words (cf. Müller 1997, Klampfer, Korecky-Kröll, and Dressler 2001, Dressler, Kilani-Schoch, and Klampfer 2003).

2.4. MORPHOLOGICAL SPECIALIZATIONS

During the protomorphological phase such compensatory reduplications may be morphologically restricted, which may be interpreted as partial morphologization, as in the case of fillers (cf. Kilani-Schoch and Dressler 2000). Within our project, this is the case with two Lithuanian children and one Greek child:

The two Lithuanian girls Ruta and Egle (Wójcik and Smoczyńska 1997, Wójcik 2000: 137) replaced verb prefixes (particularly the reflexive semiclitic *si*, which turns up either as an infix *si* or as a suffix *-s*, thus having two clitic positions) with reduplicative CV or V, but never the negative prefix *ne-* nor the perfective prefix *pa-*: for example, *pa-si-puošti* 'to dress up' → *papapuoti*, *iš-si-maude* 'took a bath' → *amamaude*. Thus, there is a clear idiosyncratic morphological constraint on two single prefixes.

The Greek child Christos (Christophidou and Kappa 1998) replaced just the complementizer *na* before the subjunctive with reduplicative CV, such as *kóvi* 'cuts,' Subj. Aorist *na kópsi* → *kokópsi*, *pézume* 'we play,' Subj. *na péksume* → *pepétsiume*. Note that earlier on, Christos replaced *na* plus subjunctive with lengthening of the root vowel while dropping the clitic *na* altogether. This represents an iconic means of signaling the markedness of the subjunctive mood in contrast to the unmarked indicative.

2.5. PARADIGMATIC ICONICITY

So far we have only dealt with syntagmatic iconicity of reduplication. Now we come to paradigmatic iconicity; that is, cases when reduplication is used to express iconically the repetition or extension (in time) of events. Within the corpora of all Austrian children of our project there is just one example of the Viennese boy Bernd at 2;7 (Vollmann et al. 1997):

der Papa bau-bau-bau
'the daddy is building a lot' = 3.Sg.Pres. *bau-t*.

Here triplication of the uninflected root appears to express iconically continuity of the activity in extended time. This isolated example alone cannot, however, provide any useful evidence.

More examples have been found by Gagarina (1997, 2000) with several Russian children, where reduplication appears to express perceived iterativity or ongoingness, instead of using adult imperfective verbs or secondary imperfectivization devices that conventionally express iterativity and ongoingness. Examples where this iconic meaning is clearly proven by the context, including those where the child has an opposition between imperfective reduplicated and perfective simple forms, are given below. Interestingly, these examples of contrastive use of reduplication extend from early premorphology into protomorphology, which shows the stability of their use and might be explained by their high perceptual salience that facilitates their uptake from child-directed speech and use. This iconic relation between the form and repeated use supports the infrequent production of reduplications in nontarget contexts also during the protomorphological stage, when children learn canonical verb forms expressing iterativity versus ongoingness:

Premorphology:

- Anja 1;2.7: *bax-bax*, while knocking on various objects around her versus 1;2.21 *bax*, referring to an object that fell down.
- Filip 1;7.12: *bux* in reference to a toy train that fell down versus *bu-bu*, while knocking continuously with the same train.
- Filip 1;8 *sjapa n´am* '(to) eat the hat' versus 1;9 *n´am-n´am* 'I'm eating,' in response to the question 'What are you doing?.'
- Vanja 1;10: *bul´*, while the glass is in the water where it has sunk versus *bul´-bul´-bul´*, while the water is dripping out slowly from the glass.
- Maxim 1;4: *gom-gom* in reference to the boy Sasha's running (from adverb *begom* 'in running' with final stress).

- ¤ Liza 1;8 *kap-kap*, referring to digging (from *kapat´* [ka'pat´] 'to drop') with a shovel.
  Protomorphology:
- ¤ Liza 2;2: *fontanchiki kap-kap* 'fountains-DIM are dripping' (*kápat´*), cf. in premorphology Vanja (1;10) *kap-kap-kap* while holding his toys above the water in the bath-room and watching how the water is pouring out.
- ¤ Vanja 2;4: *pik-pik* 'jumping,' while he is jumping (iterative action), from *prygat´* 'to jump' opposed to the adult perfective *pryg-nut´*, which refers to a momentary action, cf. in premorphology Varja (1;6), while imitating how a rabbit is jumping.
- ¤ Anja 2;5: *Oj, kak bezhit, tak: bek-bek* 'Oh, how it (a puddle) is running, thus: running,' from *begat´* 'to run.'
- ¤ Anton 2;6: *Tuk-tuk*, then correct *stuchu* 'hammering, I'm hammering,' in response to the question 'what are you doing?,' which demands an answer in the imperfective aspect.

In sum, these examples of extragrammatical paradigmatic iconicity are rather rare. Their classification as expressing imperfective as opposed to perfective aspect does not imply that these children have already mastered the adult semantic and syntactic opposition, which would be a highly improbable assumption. Note that all examples show total reduplication, which fits to the cross-linguistic tendency (cf. Dressler 1968, Inkelas and Zoll 2005) that total reduplication is not only syntagmatically but also paradigmatically more iconic than partial reduplication, a case of second-order iconicity.

The examples of triplication as instances of recursive reduplication are just three, but two of them are relatively early (1;8, 1;10 vs. 2;7).

What we have not found in very early child language (premorphology) is any noniconic meaning of reduplication which is typical of adult grammatical reduplication; that is, meanings which do not signal greater extension, continuity, intensity, iterativity, or plurality (cf. Dressler 1968). This fits the generalization that extragrammatical morphology is characterized by more meaning–form iconicity than grammatical morphology. Moreover, young children eagerly reproduce extragrammatical adult onomatopoetic reduplications, but much less echo words of the type *zig-zag*, which are syntagmatically and paradigmatically less iconic than total, onomatopoetic reduplications (and this in periods where these children have no problem in producing different vowels in adjacent syllables). Note that none of the reduplicated verbs has the obligatory inflectional suffix, such as G. *bau-bau-bau* instead of 3.Sg. *bau-t*, and this at a time when these inflectional suffixes have been acquired and used at least for several months.

So far we can conclude that some of our data seem to be problematic for models of reduplication as affixation. Similarly, some of our data are problematic for Inkelas and Zoll's (2005) morphological doubling theory, especially the thesis of a complementary distribution of purely formal phonological copying and morphosemantic feature duplication, because children show both. Moreover, playful phonological reduplication exercises often seem to have an intensifying character, which may represent a transition to iconicity between meaning and form (cf. Inkelas and Zoll 2005: 13–16). Furthermore, their generalized phonology prediction "there is nothing unique about the phonology of reduplication constructions" (p. 69) does not hold for idiosyncratic phonological alterations in the premorphological phase. Thus their claim should be restricted from reduplication structures in general to "grammatical reduplication constructions." This would provide a further argument for considering many reduplicative operations in early childhood as extragrammatical.

3. Recursive Diminutive Formation With the Same or a Synonymous Suffix

Most of the examples we discuss come from various languages of the Crosslinguistic Project on Pre- and Protomorphology, as published in Savickienė and Dressler (2007). Recursive diminutive suffixation in the strict sense (with or without morphonological adaptations) is represented by suffix doubling, suffix tripling, and so on, as in the two Italian types stem-based *tant-in-ino* 'so much-DIM-DIM' and word-based *gufo-ino-ino-ino* 'owl-DIM-DIM-DIM,' both: *poch-ino-ino-ino* 'little-DIM-DIM-DIM' (cf. Dressler and Merlini Barbaresi 1994: 99, Noccetti et al. 2007: 127, 129).

In the data of four Italian children (Noccetti et al. 2007: 139) there is a single instance of *-in-ino* in protomorphology. There are three Spanish examples: *chiqu-in-ín* and *chiqu-it-ito* 'small-DIM-DIM,' *poqu-it-ito* 'little-DIM-DIM' in protomorphology (Marrero, Aguirre, and Albalá 2007: 167). This is a very meager result in terms of type frequency, for strict recursiveness.

Recursive diminutive formation in a looser sense is represented by the stacking of different but synonymous diminutive suffixes, as in It. *tanto* 'so much' → *tant-in-ello*. A type which has an intermediate status between both types is alternating suffixation, as in Columbian Spanish (due to preventive dissimilation): *corte* 'court-yard' → *cort-ico* → *cort-iqu-ito* → *cort-iqu-it-ico*, and so on.

According to the descriptions of longitudinal spontaneous corpora in Savickienė and Dressler (2007) recursive diminutive formation in the loose sense emerges in Russian already at the age of 1;7: **kam-us-it-ik* 'stone-DIM-DIM-DIM' from *kamen´*, **tjupan´tin´ki* 'tulips-DIM-DIM' from *tjul´pan*, **bant-it-ik* 'knot-DIM-DIM' (1;9) from *bant*, *kal-eč-ečki* 'rings-DIM-DIM' (2;10) from *kol´co* (Protassova and Voeikova 2007: 60, 61).

Among the Italian children we find in the later part of the protomorphological phase (i.e., in the second half of the third year of age) examples such as *fior-ell-ini* 'flowers-DIM-DIM,' *ors-ett-ino* 'bear-DIM-DIM,' *pezz-ett-ino* 'piece-DIM-DIM,' *scarp-ett-ine* 'shoes-DIM-DIM,' potential *penn-ell-ina* 'pen-DIM-DIM,' in all cases the most productive suffix *-ino* in second position, and in a rather surprising order *ov-in-ette* 'eggs-DIM-DIM' (because the most productive *-ino* may not be followed by the less productive *-etto*, unless in case of strong lexicalization and thus morphosemantic opacity of *X-ino*) in the phase of morphology proper (Noccetti et al. 2007: 134–6)

Spanish examples from protomorphology are *chiqu-it-ín*, *chiqu-it-ino* 'small-DIM-DIM' (Marrero, Aguirre, and Albalá 2007: 167)

Also, Greek examples are late and restricted to the overlapping category of hypocoristics (where the basic pragmatic meaning is still more evident than with diminutives of common nouns, cf. Savickienė and Dressler 2007): 2;8 *An-ul-itsa* 'Ann-DIM-DIM,' (plus in CDS *An-ul-áki*, *An-ul-ína*, *An-ul-íni*, *An-ul-in-íska/i*, *man-ul-ina* 'mum-DIM-DIM'), *pap-ul-ákis/-ínos* 'Grandfather-DIM-DIM' (Thomadaki and Stephany 2007: 108, 114, 116).

Although the Lithuanian girl Ruta's input contains many instances of diminutive accumulation (Savickienė 2007: 23), her output consists of just one late type (2 tokens):), *žmog-el-iuk-as* 'man-DIM-DIM-NOM.SG.' (2;4). But another girl, Monika, has 2 types, one of them earlier: *seli-ot-ytë* [širdutytė] 'heart:DIM:DIM' (1;10), *šaukšt-el-iukas* 'tea-spoon:DIM:DIM' (2.5).

In our French data, there is hypocoristic accumulation in Emma (2;8) *pap-oun-et* 'dad-DIM-DIM,' but it contrasts only with the simplex *papa*, not with the elsewhere attested hypocoristic *pap-oune* (only the simplex is attested in her recorded input).

The Hungarian diminutive and hypocoristic suffix combinations *-i-ke* or *-i-ka* (according to vowel harmony), *-i-ci*, *-u-ka* and *-u-ci* emerged already in the first investigated month (1;11): *cum-i-ka* 'pacifier-DIM-DIM,' *any-u-ci* 'mother-DIM-DIM' = *ap-u-ka* (several months later), *ap-u-ci* 'father-DIM-DIM' = *ap-u-ka* (several months later), hypocoristics *Mik-i-ci*, *Le(v)-i-ci*; the combination *-csi-ka* only later in *pul-csi-ka* 'pullover-DIM-DIM' (Bodor and Bardza 2007: 243, 249ff).

The two Viennese German diminutive suffix combinations *-i-li*, *-i-lein* do not occur in the output data of the two children investigated by Korecky-Kröll and Dressler (2007), although they are present in their inputs.

Child idiosyncrasies in multiple suffixations are attested early on in a Czech corpus (Pačesová 1968: 237ff): *aut-il-inko* 'car-DIM?-DIM,' *mam-uš-ka* 'mum-DIM-DIM' = *mam-ul-ka*, *mam-in-enka*, *mam-in-ečka*, *kluč-in-ek* 'boy-DIM-DIM,' *mal-in-kati* 'small-DIM-DIM' = *mal-il-in-enki*, *čaj-ič-ek* 'tea-DIM-DIM,' *čič-il-inka* 'pussy-DIM-DIM,' *tet-ul-inka* 'aunt-DIM-DIM,' and so on. For these examples, there is no evidence for the status of the elements *-il-*, *-in-*, *-uš-* as interfixes, suffixes or allomorphic amplifying elements.

We may conclude that strict recursivity of diminutive suffixation is extremely rare in early child language and that the looser type in the format of accumulation of different suffixes emerges relatively late in protomorphology with most children, if at all.

4. Hypercharacterization

4.1. GENERALITIES

In inflectional hypercharacterization of the suffixing type, a suffix signals a second time what is normally already signaled by another marker (either suffix of stem change, as in the case of umlaut). Thus either two synonymous suffixes follow each other in a nonadult-like way, as in children's G. *Erbs-i-ss-en* 'pea-DIM-PL-PL,' *Auto-s-n* 'car-PL-PL,' or a stem change is supplemented by suffixation as in the frequent English child production *feet-s* (generalized in Vernacular Black American English, cf. Hock 1986: 190, Plank 1981: 87).

Dressler (2004b) has claimed that the hypercharacterizing marker is always a productive marker and thus more productive than, or at least as productive as, the hypercharacterized marker, based on the establishment of degrees of productivity of adult morphological rules according to the application to new loan-words, abbreviations, and conversions (see Dressler 2004a, Dressler and Ladányi 2000). This claim holds both for diachronic change and for child language. This is only partially compatible with Hay's (2002) claim that "an affix may resist attaching to a complex word that is highly decomposable" (cf. Manova and Aronoff 2010: 122–123).

Our main questions will be whether the hypercharacterized marker is unproductive at the moment of hypercharacterization or simply less productive than the hypercharacterizing one, or rather equally productive. The answers to these related questions have implications for models of productivity (cf. Dressler 2004a, Dressler and Ladányi 2000, Bauer 2001) and psycholinguistic models of processing such as the dual-route model, which contrasts rule application and storage. Thus the plural *feet* would be stored as such; the addition of hypercharacterizing *-s* would be due to either application of a productive rule (e.g., in dual route models) or by analogy (e.g., in single route models, cf. below).

Another type of hypercharacterization is represented by Sp. *pie*, *café* 'foot; coffee (shop),' Pl. *pie-s*, *café-s*, colloquial Pl. *pie-s-es*, *café-s-es* (cf. Malkiel 1957: 79, 98), *mamá-s-es* 'mums' (Menn and MacWhinney 1984: 531).

The -s-pluralization is productive, but normally, even with new words this /s/ is preceded by an unstressed vowel (/e, o, a/ in *pec-es* 'fish,' *amigo-s*, *amiga-s* 'friends'). Thus the pattern of stressed word-final sequence /e/ + /s/ is exceptional.

If both markers are productive, then the hypercharacterizing one seems to represent the default, or at least the most amply used (the most normal one in the sense of Paul 1970: 163 and Plank 1981: 76), and the hypercharacterized one a nondefault, as in the colloquial Spanish cases.

Of these different constellations of competition between hypercharacterizing internal and hypercharacterized external morphological markers, those constellations where both markers have at least some productivity present problems to several psycholinguistic models of processing. Single-route models of the connectionist type (e.g., Plunkett and Juola 1999) cannot predict or explain any asymmetry in possible orders of such markers. The dual-route model of Clahsen (1999, cf. Clahsen et al. 2001) and Pinker (1999) fits only instances where the external marker represents the default, because only then the internal marker is part of a stored representation, whereas the external marker can be attached to it by rule (for similar dichotomies based on transparency, cf. Caramazza et al. 1988, Hay 2002). However, even in this case there exists the problem of the widely accepted Elsewhere Condition (Kiparsky 1982): it predicts that either the (nondefault) minor rule is applied or else just the major rule.

This problem does not exist in our model of degrees of productivity of morphological rules (Dressler 2004a, Dressler and Ladányi 2000): not only direct lexical access to stored forms (dominant for unproductive rules) competes with rule application, but rules may also compete among themselves. Productive inflectional rules have the syntactic function of supplying syntax with morphological symbols for syntactic categories, and this is achieved best in terms of indexicality if these markers are positioned on the outside boundary of a word form, especially in the case of a prototypical inflectional marker of contextual morphology (Dressler 1989, cf. Booij 1996). Since a fully productive marker fulfills this syntactic function better than a slightly productive one, we can predict that the more productive one should be external, and the less productive one internal. The same argument can be made for the external position of a default marker, because default markers are more reliable signs (in the sense of Morris 1971) than nondefault markers.

Most cases of hypercharacterization occur with nonprototypical markers of inherent inflection: plurals, past participles, and superlatives. Their syntactic functions involve less indexical reference to other syntactic constituents. Hence external position is functionally less relevant. In the case of pluralization of diminutives, there is competition between a nonprototypical inflectional and a nonprototypical derivational marker whose functional eligibility for external position is similar. This explains, first, why a diminutive suffix may follow a plural suffix (as in G. *Kind-er-chen* 'child-PL-DIM'), which, second, triggers the application of another plural marker in order to have inflection outsides derivation (as in Breton *paotr-ed-ig-où* 'man-PL-DIM-PL').

4.2. CHILD LANGUAGE DATA

Data from first language acquisition largely fit data from adult language without any further adjustment. Thus in a plural formation test with Viennese children of the age brackets between 3;1 to 6;0 (Klampfer, Korecky-Kröll, and Dressler 2001), the less productive or less default-like German plural marker (umlaut or suffix) is always internal, as in:

Vögel-n 'birds,' *Mänte(l)-n* 'coats,' *Äpfel-s* 'apples,' *Küh-n* 'cows' (umlaut plus suffix), *Hase-n-s* 'hare-s,' *Katze-n-s* 'cats.'

The first four examples have unproductive internal plural markers, which are in adult language incompatible with the plural suffixes *-n*, *-s* (cf. spontaneous productions in Vollmann et al. 1997: 70), comparable to E. *feet-s*, the other two have internal plurals which are fully productive both in adult and child language. Both types of hypercharacterization are also attested in data of equally not very young children from Germany (cf. Augst 1979: 229, Faust 1980: 390). For similar English examples see Menn and MacWhinney (1984: 531, with further references), instances of double-marked Finnish partitives are discussed by Laalo (2002: 99).

The majority of examples in the Viennese test data have a nontarget-like, internal plural suffix, as in: *Clown-e-s* 'clowns,' *Apfel-n-s* 'apples,' *Vogel-n-s* 'birds,' *Oma-n-s* 'grannies,' *Hammer-n-s* 'hammers,' *Bub-e-s* 'boys,' *Zug-e-s* 'trains,' *Wurm-e-s* 'worms,' *Maus-e-s* 'mice,' *Klo-n-s* 'WCs,' *Hemd-er-s* 'shirts,' *Mantl-er-n* 'coats,' *Vogl-er-n* 'birds.'

The first group of items belongs to inflection classes where the internal suffix is productive, thus they are potentially (but not actually) legal forms. Thus these internal plurals*?Clown-e,?Apfel-n*, and so on are formed via productive rules that compete with the productive –*s* marker, which the children at this time have acquired as universally applicable plural marker (a weak default). In the next two plurals the internal marker is productive, but not with the respective gender: -*e* is unproductive with feminine nouns (such as *Maus*) and -*(e)n* with neuters (like *Klo*). These two tokens were produced by children in the age-brackets of 2;6-3;0 and 3;1-3;6; that is, at times when children may not yet have acquired the precise conditions of productive rule application. Finally, the last three examples have the unproductive internal marker -*er*, used by older children (one 3;7-4;0, two 5;7-6;0): Presumably they had already acquired this rule, but may have mistaken it as productive, which in general explains why older children overgeneralize unproductive rules.

The only example of an -*s* marker in internal position within the test group is *Klo-s-e* 'WCs,' produced by a child of the 2;6-3;0 age-bracket. At this age, Viennese children often have not yet acquired the -*s* plural marker (cf. Vollmann et al. 1997, Sedlak et al. 1998, Korecky-Kröll and Dressler 2009) and produce spontaneously hypercharacteristic plurals, such as: *Auto-s-n* 'cars,' *Photo-s-n* 'photos,' *Erbs-i-ss-en* 'peas (diminutive)' and *Auto, Auto-s, Auto-n*, but not **Auto-n-s*. Therefore the CDS plurals *Auto-s, Photo-s, Erbs-i-s* behave like stored, rote-learnt forms to which they apply productively the already acquired productive -*(e)n* marker.

Our Viennese data lack examples of doubling of productive suffixes, as in:

E. *shoe-s-es, hope-d-ed*; Hung. Nom.Pl. *pingvin-k-ek* 'pinguins,' Hung. doubled sublative case with internal doubled possessive inflectional suffix: *rá-m-om-ra* 'to-my-my-to = to me' (MacWhinney 1975, 1985: 1102f).

Our Viennese data also lack instances of the sequence productive and unproductive suffix, as in MacWhinney's (1975, 1985: 1102) example: *ló-k-ak* 'horses' (adult: *lov-ak* with the second base *lov-* alternating nonproductively with the default base *ló*). Unless this is a blend between *lov-ak* and overregularized *ló-k*, this example shows free rule application not constrained by principles claimed for adult grammar, such as the Elsewhere Condition. It may also be the case that in the specific phase of language acquisition the contrast between productive -*ok* and unproductive -*ak* has not yet been acquired; that is, that at this point in the course of acquisition, the child may consider also -*ak* a productive plural suffixation.

5. Conclusion

In conclusion, we may state that our corpora do not contain data about the acquisition of grammatical reduplication. Extragrammatical reduplication of various kinds flourishes in the phase of premorphology, with the exception of reduplication with morphological meaning, which is rare but occurs in its majority in the phase of protomorphology.

For our data, questions of affix order are thus only directly relevant in cases of loose repetition; that is, in case of diminutive suffix accumulation and of hypercharacterization of one suffix by another one. But indirect relevance exists for extragrammatical reduplication as a preparatory exercise for the acquisition of affix (de)composition, and thus also affix order. For both adult and child language we found evidence for the claims that in hypercharacterization an unproductive suffixation may be followed by a productive suffixation (not by another unproductive one), and that a productive suffixation may be followed only by a more or equally productive suffixation. This is compatible with parsability accounts of affix ordering. The most interesting new evidence from child language data lies in the changing status of nonproductivity and productivity in the course of early language acquisition.

Language-specific arbitrary affix order constraints can be violated by children. Thus in adult Italian, the most productive suffix -*ino* may be followed by the less productive suffix -*ello*, as in *tant-in-ello*, but not by the equally less productive suffix –*etto*; however, one Italian child produced the violating form *ov-in-ette*, but no child a double diminutive *X-in-ello*. Obviously many more data are needed in order to draw safe conclusions. Also, on adult languages, detailed analyses about degrees of productivity of diminutive suffixations are often lacking. The state of research is on the one hand better for inflectional hypercharacterization in the sense of stacking of synonymous suffixes, but on the other hand relevant child data have been found only in few languages.

Our results have also implications for general questions of recursivity. First of all, loose repetition, as in the cases just discussed, have a different status according to whether one conceives of affixes as signs having a meaning and an expression part or whether one subscribes to the separation thesis of Beard (1995, cf. Stump 2001). Then the looser type of repetition via synonymous suffixes is normal recursivity on the level of morphosemantics, because this level is separated from the level of morphotactics.

The findings of our study shed some light on the question raised in the introduction on recursivity seen as a fundamental innate human property. In a previous study on the acquisition of compounding (Dressler, Lettner, and Korecky-Kröll 2010) we found late emergence of recursive compounding. In this study we found as well that morphological recursivity emerges late in the protomorphological phase. Now why does recursivity, as an allegedly innate capacity, manifest itself only late?

One explanation is the general preference for strictly binary constructions (cf. Kilani-Schoch and Dressler 2005). This holds also for the acquisition of inflection in agglutinating languages in which many inflectional suffixes may be concatenated. But in early protomorphology, both Turkish and Hungarian children produce only inflectional forms with one suffix; that is, strictly binary constructions.

A second answer may lie in the young child's difficulty in producing polysyllabic words. Consequently, young children produce truncated forms and amalgams (cf. Elsen 1991 and Rainer 2009). However, this is contradicted by the ample polysyllabic phonological reduplications produced even earlier in the premorphological phase of acquisition; that is, in an extragrammatical way not related to meaning. Therefore the problem of late appearance of morphological recursivity seems to be related to the problem of repeating meaningful elements, be it in compounding, in diminutive suffix accumulation, in inflectional hypercharacterization, or in iconically meaningful reduplications. The further question of whether recursivity is a specific grammatical or a general cognitive property is beyond the scope of this paper (cf. Dressler, Lettner, and Korecky-Kröll 2010).

References

Augst, Gerhard. 1979. "Neuere Forschungen zur Substantivflexion." *Zeitschrift für Germanistische Linguistik* 7:220–32.

Bauer, Laurie. 2001. *Morphological Productivity*. Cambridge: Cambridge University Press.

Beard, Robert. 1995. *Lexeme-Morpheme Base Morphology*. Albany: State University of New York Press.

Bittner, Dagmar, Wolfgang U. Dressler, and Marianne Kilani-Schoch, eds. 2003. *Development of Verb Inflection in First Language Acquisition*. Berlin: Mouton de Gruyter.

Bodor, Péter, and Virág Bardza. 2007. "Acquisition of diminutives in Hungarian." In *The Acquisition of Diminutives*, edited by Ineta Savickienė and Wolfgang U. Dressler, 231–62. Amsterdam: Benjamins.

Booij, Geert. 1996. "Inherent vs. contextual inflection and the split morphology hypothesis." *Yearbook of Morphology* 1995:1–16.

Caramazza, Alfonso, Alessandro Laudanna, and Cristina Burani. 1988. "Lexical access and inflectional morphology." *Cognition* 28:297–332.

Chomsky, Noam. 2007. "Approaching UG from below." In *Interfaces + Recursion = Language?* edited by U. Sauerland, and H.-M. Gärtner, 1–19. Berlin: Mouton de Gruyter.

Christophidou, Anastasia, and Ioanna Kappa. 1998. "Pre- and protomorphological fillers in Greek language acquisition." *Antwerp Papers in Linguistics* 95:193–214.

Clahsen, Harald. 1999. "Lexical entries and rules of language: A multidisciplinary study of German inflection." *Behavioral and Brain Sciences* 22:991–1060.

Clahsen, Harald, Sonja Eisenbeiss, Meike Hadler, and Ingrid Sonnenstuhl. 2001. "The mental representation of inflected words: an experimental study of adjectives and verbs in German." *Language* 77:510–43.

Dressler, Wolfgang U. 1968. *Studien zur verbalen Pluralität*. Wien: Böhlau.

Dressler, Wolfgang U. 1989. "Prototypical differences between inflection and derivation." *Zeitschrift für Phonetik, Sprachwissenschaft und Kommunikationsforschung* 42:3–10.

Dressler, Wolfgang U., ed. 1997. *Studies in Pre- and Protomorphology*. Wien: Österreichische Akademie der Wissenschaften.

Dressler, Wolfgang U. 2001. "Extragrammatical vs. marginal morphology." In *Extragrammatical and Marginal Morphology*, edited by Ursula Doleschal and Anna M. Thornton, 1–10. München: Lincom.

Dressler, Wolfgang U. 2004a. "Degrees of grammatical productivity in inflectional morphology." *Italian Journal of Linguistics* 15:31–62.

Dressler, Wolfgang U. 2004b. "Hypercharacterisation and productivity in inflectional morphology." In *Analecta Homini Universali Dicata, Fs. O. Panagl*, edited by Thomas Krisch et al., 515–24. Stuttgart: Verlag Hans-Dieter Heinz.

Dressler, Wolfgang U., Katarzyna Dziubalska-Kołaczyk, Natalia Gagarina, and Marianne Kilani-Schoch. 2005. "Reduplication in child language." In *Studies on Reduplication*, edited by Bernhard Hurch, 455–74. Berlin: Mouton de Gruyter.

Dressler, Wolfgang U., Laura E. Lettner, and Katharina Korecky-Kröll. 2010. "First language acquisition of compounds with special emphasis on early German child language." In *Cross-Disciplinary Issues in Compounding*, edited by Sergio Scalise and Irene Vogel, 323–44. Amsterdam: Benjamins.

Dressler, Wolfgang U., and Lavinia Merlini Barbaresi. 1994. *Morphopragmatics*. Berlin: Mouton de Gruyter.

Dressler, Wolfgang U., and Mária Ladányi. 2000. "Productivity in word formation (WF): A morphological approach." *Acta Linguistica Hungarica* 47: 103–44.

Dressler, Wolfgang U., Marianne Kilani-Schoch, and Sabine Klampfer. 2003. "How does a child detect morphology? Evidence from production." In *Morphological Structure in Language Processing*, edited by R. Harald Baayen, and Robert Schreuder, 391–425. Berlin: Mouton de Gruyter.

Dziubalska-Kołaczyk, Katarzyna, ed. 1997. "Pre- and protomorphology in language acquisition." *Papers and Studies in Contrastive Linguistics* 33.

Elsen, Hilke. 1991. *Erstspracherwerb. Der Erwerb des deutschen Lautsystems*. Wiesbaden: Deutscher Universitätsverlag.

Faust, Manfred. 1980. "Morphologische Regularisierung in Sprachwandel und Spracherwerb." *Folia Linguistica* 14:387–411.

Gagarina, Natalia. 1997. "Aspektual'naja semantika i funkcionirovanije vidov russkogo glagola v detskoj rechi." PhD diss., Hercen State Pedagogical University.

Gagarina, Natalia. 2000. "The acquisition of aspectuality by Russian children: The early stages." *ZAS Papers in Linguistics* 15:232–46.

Hauser, Marc D., Noam Chomsky, and William T. Fitch. 2002. "The faculty of language: What is it, who has it, and how did it evolve?" *Science* 298:1569–79.

Hay, Jennifer. 2002. "From speech perception to morphology: affix ordering revisited." *Language* 78:527–55.

Hock, Hans H. 1986. *Principles of Historical Linguistics*. Berlin: Mouton de Gruyter.

Inkelas, Sharon, and Cheryl Zoll. 2005. *Reduplication. Doubling in Morphology*. Cambridge: Cambridge University Press.

Kilani-Schoch, Marianne, and Wolfgang U. Dressler. 2000. "Are fillers as precursors of morphemes relevant for morphological theory? A case story from the acquisition of French." In *Morphological Analysis in Comparison*, edited by Wolfgang U. Dressler, Oskar Pfeiffer, and Markus Pöchtrager, 89–111. Amsterdam: Benjamins.

Kilani-Schoch, Marianne, and Wolfgang U. Dressler. 2005. *Morphologie naturelle et flexion du verbe français*. Tübingen: Narr.

Kiparsky, Paul. 1982. "Lexical phonology and morphology." In *Linguistics in the Morning Calm*, edited by I.-S. Yang, 3–91. Seoul: Hanshin.

Klampfer (= Laaha), Sabine, Katharina Korecky-Kröll, and Wolfgang U. Dressler. 2001. "Morphological potentiality in children's overgeneralization patterns: evidence from Austrian German noun plurals." *Wiener Linguistische Gazette* 67–69:25–43.

Korecky-Kröll, Katharina, and Wolfgang U. Dressler. 2007. "Diminutives and Hypocoristics in Austrian German (AG)." In *The Acquisition of Diminutives*, edited by Ineta Savickienė and Wolfgang U. Dressler, 207–30. Amsterdam: Benjamins.

Korecky-Kröll, Katharina, and Wolfgang U. Dressler. 2009. "The acquisition of number and case in Austrian German nouns." In *Development of Nominal Inflection in First Language Acquisition: A Cross-Linguistic Perspective*, edited by Ursula Stephany and Maria D. Voeikova, 265–302. Berlin: de Gruyter.

Laalo, Klaus. 2002. "Acquisition of case in Finnish: a preliminary overview." In *Pre- and Protomorphology*, edited by Maria Voeikova and Wolfgang U. Dressler, 83–103. München: Lincom.

Lettner, Laura, Katharina Korecky-Kröll, and Wolfgang U. Dressler. 2010. "Charakteristika von deutschen Nominalkomposita in der protomorphologischen Phase des Erstspracherwerbs." In *Wortbildung im Deutschen, zwischen Sprachsystem und Sprachgebrauch*, edited by Hilke Elsen and Sascha Michel, 191–208. Stuttgart: ibidem-Verlag.

MacWhinney, Brian. 1975. "Rule, rote, and analogy in morphological formations by Hungarian children." *Journal of Child Language* 2:65–77.

MacWhinney, Brian. 1985. "Hungarian Language Acquisition as an Exemplification of a General Model of Grammatical Development." In *The Crosslinguistic Study of Language Acquisition* II, edited by Dan Slobin, 1069–155. Hillsdale: Erlbaum.

MacWhinney, Brian. 2000. *The CHILDES Project: Tools for Analyzing Talk*. Vol. 1: *Transcription, Format and Programs*. Mahwah: Lawrence Erlbaum.

Malkiel, Yakov. 1957, 1958. "Diachronic hypercharacterization in Romance." *Archivium Linguisticum* 9: 79–113; 10:1–36.

Manova, Stela, and Mark Aronoff. 2010. "Modeling affix order." *Morphology* 20:109–31.

Marrero, Victoria, Carmen Aguirre, and María José Albalá. 2007. "The acquisition of diminutives in Spanish." In *The Acquisition of Diminutives*, edited by Ineta Savickienė and Wolfgang U. Dressler, 155–81. Amsterdam: Benjamins.

Menn, Lise, and Brian MacWhinney. 1984. "The repeated morph constraint." *Language* 60: 519–41.

Morris, Charles W. 1971. *Writings on the General Theory of Signs*. The Hague: Mouton.

Müller, Brigitta. 1997. "The acquisition of early German morphology: a case study." In *Studies in Pre- and Protomorphology*, edited by Wolfgang U. Dressler, 61–71. Wien: Österreichische Akademie der Wissenschaften.

Noccetti, Sabrina, Anna De Marco, Livia Tonelli, and Wolfgang U. Dressler. 2007. "The role of diminutives in the acquisition of Italian morphology." In *The Acquisition of Diminutives*, edited by Ineta Savickienė and Wolfgang U. Dressler, 125–53. Amsterdam: Benjamins.

Pačesová, Jaroslava. 1968. *The Development of Vocabulary in the Child*. Brno: Universita Purkyně.

Paul, Hermann. 1970 [1880]. *Prinzipien der Sprachgeschichte*. Tübingen: Niemeyer.

Pinker, Steven. 1999. *Words and Rules: The Ingredients of Language*. New York: Basic Books.

Plank, Frans. 1981. *Morphologische (Ir-)Regularitäten*. Tübingen: Narr.

Plunkett, Kim, and Patrick Juola. 1999. "A connectionist model of English past tense and plural morphology." *Cognitive Science* 23:463–90.

Protassova, Ekaterina. 1997. "Transition from babbling to word structure." *Papers and Studies in Contrastive Linguistics* 33:153–8.

Protassova, Ekaterina, and Maria Voeikova. 2007. "Diminutives in Russian at the early stages of acquisition." In *The Acquisition of Diminutives. A Cross-Linguistic Perspective*, edited by Ineta Savickienė and Wolfgang U. Dressler, 43–72. Amsterdam: Benjamins.

Rainer, Franz. 2009. *Carmens Erwerb der deutschen Wortbildung*. Wien: Verlag der Österreichischen Akademie der Wissenschaften.

Savickienė, Ineta. 2007. "Form and meaning of diminutives in Lithuanian child language." In *The Acquisition of Diminutives. A Cross-Linguistic Perspective*, edited by Ineta Savickienė and Wolfgang U. Dressler 13–41. Amsterdam, Philadelphia: John Benjamins.

Savickienė, Ineta and Wolfgang Dressler (eds.). 2007. *The acquisition of diminutives. A Cross-Linguistic Perspective*. Amsterdam, Philadelphia: John Benjamins.

Sedlak, Maria, Sabine Klampfer, Brigitta Müller, and Wolfgang U. Dressler. 1998. "The acquisition of number in Austrian German: a case study on the early ages." *Antwerp Papers in Linguistics* 95: 51–76.

Stephany, Ursula, and Maria D. Voeikova, eds. 2009. *Development of Nominal Inflection in First Language Acquisition: A Cross-Linguistic Perspective*. Berlin: de Gruyter.

Stump, Gregory T. 2001. *Inflectional Morphology: A Theory of Paradigm Structure*. Cambridge: Cambridge University Press.

Thomadaki, Evangelia, and Ursula Stephany. 2007. "Diminutives in Greek child language." In *The Acquisition of Diminutives*, edited by Ineta Savickienė and Wolfgang U. Dressler, 89–123. Amsterdam: Benjamins.

Voeykova, Maria D. 1997. "Early extragrammatical operations in the speech of a Russian child (from 1;4.11 to 2;0.3)." In *Studies in Pre- and Protomorphology*, edited by Wolfgang U. Dressler, 37–43. Wien: Österreichische Akademie der Wissenschaften.

Vollmann, Ralf, Maria Sedlak, Brigitta Müller, and Maria Vassilakou. 1997. "Early verb inflection and noun plural formation in four Austrian children: the demarcation of phases and interindividual variation." *Papers and Studies in Contrastive Linguistics* 33:59–78.

Wójcik, Paweł. 1997. *The Acquisition of Lithuanian Verb Morphology: A Case Study*. Warsaw: Universitas.

Wójcik, Paweł, and Magdalena Smoczyńska. 2000. "Acquisition of Lithuanian verb morphology: a preliminary report." In *Studies in Pre- and Protomorphology*, edited by Wolfgang U. Dressler, 83–100. Wien: Österreichische Akademie der Wissenschaften.

Zwicky, Arnold M., and Geoffrey K. Pullum. 1987. "Plain morphology and expressive morphology." *Berkeley Linguistic Society Papers* 13:339–342.

LANGUAGE INDEX

Note: The letters 'n' and '*t*' following locators refer to notes and tables respectively.

AUTHOR INDEX

Note: The letters 'n' and '*t*' following locators refer to notes and tables respectively.

SUBJECT INDEX

Note: The letters 'n' and '*t*' following locators refer to notes and tables respectively.